Advanced Science and Technology of Polymer Matrix Nanomaterials

Advanced Science and Technology of Polymer Matrix Nanomaterials

Editors

Peijiang Liu
Liguo Xu

Basel • Beijing • Wuhan • Barcelona • Belgrade • Novi Sad • Cluj • Manchester

Editors

Peijiang Liu
The Fifth Electronics Research
Institute of the Ministry of
Industry and Information
Technology
Guangzhou
China

Liguo Xu
College of Light Chemical
Industry and Materials
Engineering
Shunde Polytechnic
Foshan
China

Editorial Office
MDPI
St. Alban-Anlage 66
4052 Basel, Switzerland

This is a reprint of articles from the Special Issue published online in the open access journal *Materials* (ISSN 1996-1944) (available at: www.mdpi.com/journal/materials/special_issues/adv_sci_technol_polym_matrix_nanomater).

For citation purposes, cite each article independently as indicated on the article page online and as indicated below:

Lastname, A.A.; Lastname, B.B. Article Title. *Journal Name* **Year**, *Volume Number*, Page Range.

ISBN 978-3-7258-0272-2 (Hbk)
ISBN 978-3-7258-0271-5 (PDF)
doi.org/10.3390/books978-3-7258-0271-5

© 2024 by the authors. Articles in this book are Open Access and distributed under the Creative Commons Attribution (CC BY) license. The book as a whole is distributed by MDPI under the terms and conditions of the Creative Commons Attribution-NonCommercial-NoDerivs (CC BY-NC-ND) license.

Contents

About the Editors vii

Preface ix

Peijiang Liu, Liguo Xu, Jinlei Li, Jianping Peng and Zibao Jiao
Advanced Science and Technology of Polymer Matrix Nanomaterials
Reprinted from: *Materials* **2024**, *17*, 461, doi:10.3390/ma17020461 1

Radhakrishnan Sreena and Arputharaj Joseph Nathanael
Biodegradable Biopolymeric Nanoparticles for Biomedical Applications-Challenges and Future Outlook
Reprinted from: *Materials* **2023**, *16*, 2364, doi:10.3390/ma16062364 5

Ki Hoon Kim, Ji-Un Jang, Gyun Young Yoo, Seong Hun Kim, Myung Jun Oh and Seong Yun Kim
Enhanced Electrical and Thermal Conductivities of Polymer Composites with a Segregated Network of Graphene Nanoplatelets
Reprinted from: *Materials* **2023**, *16*, 5329, doi:10.3390/ma16155329 36

Zhuyun Xie, Dehai Xiao, Qin Yu, Yuefeng Wang, Hanyi Liao and Tianzhan Zhang et al.
Fabrication of Multifunctional Silylated GO/FeSiAl Epoxy Composites: A Heat Conducting Microwave Absorber for 5G Base Station Packaging
Reprinted from: *Materials* **2023**, *16*, 7511, doi:10.3390/ma16247511 48

Kuan-Ying Chen, Minsi Yan, Kun-Hao Luo, Yen Wei and Jui-Ming Yeh
Comparative Studies of the Dielectric Properties of Polyester Imide Composite Membranes Containing Hydrophilic and Hydrophobic Mesoporous Silica Particles
Reprinted from: *Materials* **2022**, *16*, 140, doi:10.3390/ma16010140 61

Anna V. Zhmurova, Galina F. Prozorova, Svetlana A. Korzhova, Alexander S. Pozdnyakov and Marina V. Zvereva
Synthesis and DC Electrical Conductivity of Nanocomposites Based on Poly(1-vinyl-1,2,4-triazole) and Thermoelectric Tellurium Nanoparticles
Reprinted from: *Materials* **2023**, *16*, 4676, doi:10.3390/ma16134676 80

Domenico Acierno, Lucia Graziosi and Antonella Patti
Puncture Resistance and UV aging of Nanoparticle-Loaded Waterborne Polyurethane-Coated Polyester Textiles
Reprinted from: *Materials* **2023**, *16*, 6844, doi:10.3390/ma16216844 95

Anna Daniela Dorsch, Walison Augusto da Silva Brito, Mihaela Delcea, Kristian Wende and Sander Bekeschus
Lipid Corona Formation on Micro- and Nanoplastic Particles Modulates Uptake and Toxicity in A549 Cells
Reprinted from: *Materials* **2023**, *16*, 5082, doi:10.3390/ma16145082 108

Xiongsheng Wang, Cuicui Yin, Juan Wang, Kaihong Zheng, Zhengrong Zhang and Zhuo Tian et al.
Suppressing Viscous Fingering in Porous Media with Wetting Gradient
Reprinted from: *Materials* **2023**, *16*, 2601, doi:10.3390/ma16072601 123

Cuicui Yin, Kaihong Zheng, Jiazhen He, Yongnan Xiong, Zhuo Tian and Yingfei Lin et al.
Turbulent CFD Simulation of Two Rotor-Stator Agitators for High Homogeneity and Liquid Level Stability in Stirred Tank
Reprinted from: *Materials* **2022**, *15*, 8563, doi:10.3390/ma15238563 . . . . **142**

Jianjian Song, Jianglin Zhu, Zhaoyong Wang and Gang Liu
Controlling Charge Transport in Molecular Wires through Transannular π–π Interaction
Reprinted from: *Materials* **2022**, *15*, 7801, doi:10.3390/ma15217801 . . . . **159**

Liguo Xu, Jintang Zhou, Zibao Jiao and Peijiang Liu
Special Issue: Advanced Science and Technology of Polymer Matrix Nanomaterials
Reprinted from: *Materials* **2022**, *15*, 4735, doi:10.3390/ma15144735 . . . . **168**

About the Editors

Peijiang Liu

Prof. Peijiang Liu studied materials at the Nanjing University of Aeronautics and Astronautics and wrote his PhD thesis (2018) on the electromagnetic wave absorption properties of advanced polymeric composites. From 2018 to 2021, he was employed at the South China Advanced Institute for Soft Matter Science and Technology at the South China University of Technology. Since 2021, he has been working at the Reliability Physics and Application Technology of Electronic Component Key Laboratory in the fifth Electronics Research Institute of the Ministry of Industry and Information Technology. Since 2022, he has been the Project Leader of that institute. His research activities are centered on advanced polymeric composites, reliability of materials, mechanical properties, electromagnetic wave absorption properties, polymer aging, and the templated synthesis method.

Liguo Xu

Liguo Xu, PhD, is an Associate Professor at the College of Light Chemical Industry and Materials Engineering at Shunde Polytechnic. He is mainly engaged in the field of functional polymer materials, including antifouling coatings, degradable materials, aggregation-induced luminescence materials, gel materials, flexible sensors, etc. He received his PhD from the South China University of Technology and was engaged in postdoctoral research. Thus far, he has published dozens of *SCI* papers in *Advanced Materials*, *Macromolecules*, *ACS Macro Lett.*, *Polymer Chemistry*, and other internationally renowned journals and has been authorized four invention patents.

Preface

It is with great pleasure and enthusiasm that we present the preface for this comprehensive scientific work, *"Advanced Science and Technology of Polymer Matrix Nanomaterials"*. This remarkable reprint encompasses a wide range of subjects within the field of polymer science, focusing specifically on the fascinating realm of nanomaterials.

The subject matter covered in this reprint delves into the intricacies of polymer matrix nanomaterials, exploring their synthesis, characterization, properties, and applications. It provides a comprehensive overview of cutting-edge research and technological advancements in this rapidly evolving field.

The aims and purposes of this scientific work are manifold. Firstly, it aims to provide a comprehensive and up-to-date resource for researchers, academicians, and industrial practitioners who are actively engaged in the study and development of polymer matrix nanomaterials. By offering a diverse collection of chapters authored by leading experts, this book seeks to contribute to the bridging of knowledge gaps and the fostering of collaboration within the field.

Furthermore, this book strives to enhance our understanding of the underlying principles governing the behavior and performance of polymer matrix nanomaterials. By exploring the unique properties and potential applications of these materials, it aims to inspire further innovation and advancements in various industrial sectors, including electronics, energy, healthcare, and environmental sustainability.

In composing this scientific work, our primary motivation was to consolidate the wealth of research and expertise available in the field of polymer matrix nanomaterials. By meticulously selecting and organizing contributions from renowned researchers, we aimed to create a comprehensive reference that integrates both fundamental concepts and state-of-the-art advancements.

We extend our gratitude and heartfelt acknowledgment to all the contributors who have generously shared their valuable insights and expertise. Their dedication and collaborative spirit have been instrumental in the realization of this remarkable publication.

Finally, it is our sincere hope that this reprint will serve as an invaluable resource for scientists, engineers, students, and professionals alike. We believe that its comprehensive coverage, detailed analyses, and meticulous presentation will facilitate further exploration, debate, and progress in the exciting field of polymer matrix nanomaterials.

Once again, we express our deepest appreciation to all those who have contributed to this reprint, and we humbly present it to the esteemed readers with the utmost enthusiasm and optimism.

Peijiang Liu and Liguo Xu
Editors

Editorial

Advanced Science and Technology of Polymer Matrix Nanomaterials

Peijiang Liu [1], Liguo Xu [2,*], Jinlei Li [3,*], Jianping Peng [4,*] and Zibao Jiao [5,*]

1. Reliability Physics and Application Technology of Electronic Component Key Laboratory, The Fifth Electronics Research Institute of the Ministry of Industry and Information Technology, Guangzhou 510610, China; cz2343222@163.com
2. College of Light Chemical Industry and Materials Engineering, Shunde Polytechnic, Foshan 528333, China
3. Science and Technology on Space Physics Laboratory, Beijing 100076, China
4. School of Materials Science and Hydrogen Energy, Foshan University, Foshan 528000, China
5. Jiangsu Urban and Rural Construction Vocational College, Changzhou 213147, China
* Correspondence: 21099@sdpt.edu.cn (L.X.); lei_manager@buaa.edu.cn (J.L.); pjp7@outlook.com (J.P.); jiaozibao@126.com (Z.J.); Tel.: +86-150-1302-3687 (L.X.); +86-152-5181-6557 (J.L.); +86-156-2508-1263 (J.P.); +86-139-7570-1188 (Z.J.)

1. Introduction

The advanced science and technology of polymer matrix nanomaterials are rapidly developing fields that focus on the synthesis, characterization, and application of nanomaterials in polymer matrices [1–5]. Combined together as an interdisciplinary area, they integrate principles from materials science, chemistry, physics, and engineering to create novel materials with enhanced properties.

In recent years, researchers have achieved significant advancements in the design and fabrication of polymer matrix nanocomposites. These materials consist of polymer matrices that are reinforced or modified with nanoscale fillers such as nanoparticles, nanofibers, and nanotubes [6]. The incorporation of these nanofillers into the polymer matrix leads to improved mechanical, electrical, thermal, and optical properties.

A key challenge in this field is achieving uniform dispersion and strong interfacial interactions between the polymer matrix and the nanofillers. Various techniques, including melt mixing, solution blending, templated synthesis, and in situ polymerization, have been employed to overcome this challenge [7–12]. These techniques enable precise control over the distribution of nanofillers within the polymer matrix, resulting in materials with tailored properties.

Polymer matrix nanomaterials find applications in a wide range of industries, including aerospace, electronics, energy, automotive, and biomedical [13–19]. For example, in the biomedical field, polymer matrix nanomaterials can be used for drug delivery systems, tissue engineering scaffolds, and biosensors [20]. In aerospace applications, nanocomposites offer lightweight and high-strength alternatives to conventional materials, leading to improved fuel efficiency and reduced emissions [21].

Furthermore, advances in characterization techniques such as transmission electron microscopy (TEM), X-ray diffraction (XRD), atomic force microscopy (AFM), and thermal analysis methods have allowed researchers to study the structure–property relationships of polymer matrix nanomaterials on the nanoscale [22–24]. These techniques provide valuable insights into the orientation, dispersion, and interfacial interactions of nanofillers within the polymer matrix.

In summary, the advanced science and technology of polymer matrix nanomaterials are rapidly evolving fields that offer exciting prospects for the development of innovative materials with enhanced properties. The precise control of nanofiller dispersion within polymer matrices, along with the use of advanced characterization techniques, enables researchers to tailor the properties of these materials for various applications [25,26].

In dealing with these challenging aspects of polymer matrix nanomaterials, the goal of the present Special Issue is to introduce the current knowledge on the designs, synthesis processes, characterizations, properties, and applications of polymer matrix nanomaterials.

2. Contributions

Kim et al. [27] prepared polypropylene composites filled with randomly dispersed graphene nanoplatelets (GNPs) and a segregated GNP network. Theoretical and experimental investigations were conducted to explore the enhancements in the thermal and electrical conductivities of the composites achieved through the selective localization of GNP fillers using a segregated structure and the formation of a conductive network.

Zhu et al. [28] successfully synthesized a series of molecular wires based on [2.2]paracyclophane-1,9-dienes and then elucidated the influence of transannular π–π interaction on carrier transport in these wires using the STM break junction technique. Both the current–voltage characteristics and single-molecule conductance could be systematically adjusted through the transannular π–π interaction.

Most of the current research on agitator design primarily focuses on enhancing solid–liquid mixing efficiency and homogeneity, while neglecting the stability of the liquid level. He et al. [29] utilized computational fluid dynamics modeling to compare the performance of two types of rotor–stator agitators in solid–liquid mixing operations. The evaluation included aspects such as power consumption, homogeneity, and liquid-level stability. The results indicated that the cross structure rotor–stator agitator achieved a significantly lower standard deviation of particle concentration σ of 0.15 compared to the A200 agitators, with a 42% reduction.

Yeh et al. [30] studied the impact of hydrophilic and hydrophobic mesoporous silica particles (MSPs) on the dielectric properties of composite membranes derived from polyester imide (PEI). The study revealed a clear trend in the dielectric constant of the membranes: PEI containing hydrophilic MSPs > PEI > PEI containing hydrophobic MSPs.

Yin et al. [31] conducted a numerical investigation on the displacement of immiscible fluid in porous media using the lattice Boltzmann method. The results demonstrated that the wetting gradient can control the displacement pattern and efficiency. By introducing a wetting gradient in porous media, the stability of the flow front can be enhanced. This finding was confirmed across a wide range of parameters, including different wetting gradients, capillary numbers, viscosity ratios, and porosities.

Arputharaj et al. [32] provided a comprehensive review of biopolymeric nanoparticles developed for biomedical applications, such as drug delivery, imaging, and tissue engineering. The authors also discussed important fabrication techniques, along with the challenges and future perspectives in this field. It is crucial to address the interaction between nanoparticles and the immune system, as well as their elimination from the human body, in future studies.

Pozdnyakov et al. [33] conducted an analysis of the structural characteristics and direct current (DC) electrical conductivity of organic–inorganic nanocomposites composed of thermoelectric Te0 nanoparticles and poly(1-vinyl-1,2,4-triazole). The findings revealed that the DC electrical conductivity of nanocomposites containing 2.8 and 4.3 wt% Tellurium at 80 °C exceeded the conventional boundary of 10^{-10} S/cm, separating dielectrics and semiconductors.

Bekeschus et al. [34] generated unilamellar vesicles using 1-palmitoyl-2-oleoyl-glycero-3-phosphocholine (POPC) and 1-palmitoyl-2-oleoyl-sn-glycero-3-phospho-L-serine (POPS). These vesicles were then incubated with pristine, carboxylated, or aminated polystyrene spheres to form lipid coronas around the particles. This study, for the first time, demonstrated the influence of different lipid types on differently charged micro- and nanoplastic particles and the resulting biological implications.

Acierno et al. [35] conducted a study to examine the impact of different types of nanoparticles on the UV weathering resistance of polyurethane (PU) treatment in polyester-based fabrics. The findings revealed that incorporating nanoparticles into impregnated fabrics did not significantly hinder polymer degradation following UV exposure. However, the nanoparticles appeared to enhance the reinforcement of PU polymers within the textile structure, thereby improving the overall mechanical strength, particularly after UV exposure.

Xie et al. [36] employed a simple solvent-handling method to fabricate silylated GO/FeSiAl epoxy composites. They subsequently explored the microwave absorption properties and thermal conductivity. Remarkably, it was observed that these composites achieved a reflection loss of up to −48.28 dB and an effective range of 3.6 GHz when operating at frequencies between 2.575 and 2.645 GHz, with a modest thickness of just 2 mm. These results underscored the high absorption performance of the composites, making them suitable for packaging 5G base stations.

The Guest Editors would like to extend their congratulations to all of the authors whose remarkable results have been published in this Special Issue. The papers presented here are expected to greatly contribute to the research community's understanding of the current status and trends in the advanced science and technology of polymer matrix nanomaterials. Moreover, the Guest Editors cordially invite all scientists working in this field to submit innovative articles for consideration in the second edition of the Special Issue on "Advanced Science and Technology of Polymer Matrix Nanomaterials (2nd Edition)".

Author Contributions: The manuscript was written through the contributions of all of the authors. All authors have read and agreed to the published version of the manuscript.

Funding: This work was financially supported by the 2022 Special Fund of Institute (22Z03), Featured Innovation Projects of General Colleges and Universities in Guangdong Province (2022KTSCX361), and Natural Science Foundation of Guangdong Province (no. 2022A1515110867).

Acknowledgments: The Guest Editors have acknowledged the authors for their vital contributions to this Special Issue and the Editorial staff of *Materials* for their extraordinary support.

Conflicts of Interest: The authors declare no competing interest.

References

1. Liu, P.; Yao, Z.; Zhou, J.; Yang, Z.; Kong, L.B. Small magnetic Co-doped NiZn ferrite/graphene nanocomposites and their dual-region microwave absorption performance. *J. Mater. Chem. C* **2016**, *4*, 9738–9749. [CrossRef]
2. Liu, P.; Yao, Z.; Ng, V.M.H.; Zhou, J.; Kong, L.B.; Yue, K. Facile synthesis of ultrasmall Fe_3O_4 nanoparticles on MXenes for high microwave absorption performance. *Compos. Part A* **2018**, *115*, 371–382. [CrossRef]
3. Jiao, Z.; Huyan, W.; Yang, F.; Yao, J.; Tan, R.; Chen, P.; Tao, X.; Yao, Z.; Zhou, J.; Liu, P. Achieving Ultra-Wideband and Elevated Temperature Electromagnetic Wave Absorption via Constructing Lightweight Porous Rigid Structure. *Nano-Micro Lett.* **2022**, *14*, 173.
4. Kango, S.; Kalia, S.; Celli, A.; Njuguna, J.; Habibi, Y.; Kumar, R. Surface modification of inorganic nanoparticles for development of organic–inorganic nanocomposites—A review. *Prog. Polym. Sci.* **2013**, *38*, 1232–1261.
5. Rozenberg, B.A.; Tenne, R. Polymer-assisted fabrication of nanoparticles and nanocomposites. *Prog. Polym. Sci.* **2008**, *33*, 40–112. [CrossRef]
6. Kumar, S.K.; Jouault, N.; Benicewicz, B.; Neely, T. Nanocomposites with Polymer Grafted Nanoparticles. *Macromolecules* **2013**, *46*, 3199–3214.
7. Liu, T.; Burger, C.; Chu, B. Nanofabrication in polymer matrices. *Prog. Polym. Sci.* **2003**, *28*, 5–26. [CrossRef]
8. Mallakpour, S.; Khadem, E. Recent development in the synthesis of polymer nanocomposites based on nano-alumina. *Prog. Polym. Sci.* **2015**, *51*, 74–93. [CrossRef]
9. Peng, J.; Liu, P.; Chen, Y.; Guo, Z.-H.; Liu, Y.; Yue, K. Templated synthesis of patterned gold nanoparticle assemblies for highly sensitive and reliable SERS substrates. *Nano Res.* **2023**, *16*, 5056–5064.
10. Liu, P.; Peng, J.; Chen, Y.; Liu, M.; Tang, W.; Guo, Z.-H.; Yue, K. A general and robust strategy for in-situ templated synthesis of patterned inorganic nanoparticle assemblies. *Giant* **2021**, *8*, 100076. [CrossRef]
11. Liu, P.; Ng, V.M.H.; Yao, Z.; Zhou, J.; Lei, Y.; Yang, Z.; Lv, H.; Kong, L.B. Facile Synthesis and Hierarchical Assembly of Flowerlike NiO Structures with Enhanced Dielectric and Microwave Absorption Properties. *ACS Appl. Mater. Interfaces* **2017**, *9*, 16404–16416.
12. Ahmad, R.; Griffete, N.; Lamouri, A.; Felidj, N.; Chehimi, M.M.; Mangeney, C. Nanocomposites of Gold Nanoparticles@Molecularly Imprinted Polymers: Chemistry, Processing, and Applications in Sensors. *Chem. Mater.* **2015**, *27*, 5464–5478. [CrossRef]
13. Naskar, A.K.; Keum, J.K.; Boeman, R.G. Polymer matrix nanocomposites for automotive structural components. *Nat. Nanotechnol.* **2016**, *11*, 1026–1030. [PubMed]

14. Bustamante-Torres, M.; Romero-Fierro, D.; Arcentales-Vera, B.; Pardo, S.; Bucio, E. Interaction between Filler and Polymeric Matrix in Nanocomposites: Magnetic Approach and Applications. *Polymers* **2021**, *13*, 2998. [PubMed]
15. Li, S.; Meng Lin, M.; Toprak, M.S.; Kim, D.K.; Muhammed, M. Nanocomposites of polymer and inorganic nanoparticles for optical and magnetic applications. *Nano Rev.* **2010**, *1*, 5214.
16. Shakiba, S.; Astete, C.E.; Paudel, S.; Sabliov, C.M.; Rodrigues, D.F.; Louie, S.M. Emerging investigator series: Polymeric nanocarriers for agricultural applications: Synthesis, characterization, and environmental and biological interactions. *Environ. Sci. Nano* **2020**, *7*, 37–67. [CrossRef]
17. Palza, H. Antimicrobial Polymers with Metal Nanoparticles. *Int. J. Mol. Sci.* **2015**, *16*, 2099–2116. [CrossRef] [PubMed]
18. Müller, K.; Bugnicourt, E.; Latorre, M.; Jorda, M.; Echegoyen Sanz, Y.; Lagaron, J.M.; Miesbauer, O.; Bianchin, A.; Hankin, S.; Bölz, U.; et al. Review on the Processing and Properties of Polymer Nanocomposites and Nanocoatings and Their Applications in the Packaging, Automotive and Solar Energy Fields. *Nanomaterials* **2017**, *7*, 74. [CrossRef]
19. Yang, F.; Yao, J.; Jin, L.; Huyan, W.; Zhou, J.; Yao, Z.; Liu, P.; Tao, X. Multifunctional Ti3C2TX MXene/Aramid nanofiber/Polyimide aerogels with efficient thermal insulation and tunable electromagnetic wave absorption performance under thermal environment. *Compos. Part B* **2022**, *243*, 110161.
20. Armentano, I.; Dottori, M.; Fortunati, E.; Mattioli, S.; Kenny, J.M. Biodegradable polymer matrix nanocomposites for tissue engineering: A review. *Polym. Degrad. Stab.* **2010**, *95*, 2126–2146. [CrossRef]
21. Joshi, M.; Chatterjee, U. 8—Polymer nanocomposite: An advanced material for aerospace applications. In *Advanced Composite Materials for Aerospace Engineering*; Rana, S., Fangueiro, R., Eds.; Woodhead Publishing: Sawston, UK, 2016; pp. 241–264.
22. Pandey, S.; Mishra, S.B. Sol–gel derived organic–inorganic hybrid materials: Synthesis, characterizations and applications. *J. Sol-Gel Sci. Technol.* **2011**, *59*, 73–94. [CrossRef]
23. Walters, G.; Parkin, I.P. The incorporation of noble metal nanoparticles into host matrix thin films: Synthesis, characterisation and applications. *J. Mater. Chem.* **2009**, *19*, 574–590.
24. Joshi, M.; Adak, B.; Butola, B.S. Polyurethane nanocomposite based gas barrier films, membranes and coatings: A review on synthesis, characterization and potential applications. *Prog. Mater Sci.* **2018**, *97*, 230–282. [CrossRef]
25. Liu, P.; Xu, L.; Li, J.; Peng, J.; Huang, Z.; Zhou, J. Special Issue: Advanced Science and Technology of Polymer Matrix Nanomaterials. *Materials* **2023**, *16*, 5551. [CrossRef]
26. Xu, L.; Zhou, J.; Jiao, Z.; Liu, P. Special Issue: Advanced Science and Technology of Polymer Matrix Nanomaterials. *Materials* **2022**, *15*, 4735.
27. Kim, K.H.; Jang, J.-U.; Yoo, G.Y.; Kim, S.H.; Oh, M.J.; Kim, S.Y. Enhanced Electrical and Thermal Conductivities of Polymer Composites with a Segregated Network of Graphene Nanoplatelets. *Materials* **2023**, *16*, 5329. [PubMed]
28. Song, J.; Zhu, J.; Wang, Z.; Liu, G. Controlling Charge Transport in Molecular Wires through Transannular π–π Interaction. *Materials* **2022**, *15*, 7801. [CrossRef] [PubMed]
29. Yin, C.; Zheng, K.; He, J.; Xiong, Y.; Tian, Z.; Lin, Y.; Long, D. Turbulent CFD Simulation of Two Rotor-Stator Agitators for High Homogeneity and Liquid Level Stability in Stirred Tank. *Materials* **2022**, *15*, 8563. [PubMed]
30. Chen, K.-Y.; Yan, M.; Luo, K.-H.; Wei, Y.; Yeh, J.-M. Comparative Studies of the Dielectric Properties of Polyester Imide Composite Membranes Containing Hydrophilic and Hydrophobic Mesoporous Silica Particles. *Materials* **2023**, *16*, 140. [CrossRef]
31. Wang, X.; Yin, C.; Wang, J.; Zheng, K.; Zhang, Z.; Tian, Z.; Xiong, Y. Suppressing Viscous Fingering in Porous Media with Wetting Gradient. *Materials* **2023**, *16*, 2601. [CrossRef] [PubMed]
32. Sreena, R.; Nathanael, A.J. Biodegradable Biopolymeric Nanoparticles for Biomedical Applications-Challenges and Future Outlook. *Materials* **2023**, *16*, 2364. [CrossRef] [PubMed]
33. Zhmurova, A.V.; Prozorova, G.F.; Korzhova, S.A.; Pozdnyakov, A.S.; Zvereva, M.V. Synthesis and DC Electrical Conductivity of Nanocomposites Based on Poly(1-vinyl-1,2,4-triazole) and Thermoelectric Tellurium Nanoparticles. *Materials* **2023**, *16*, 4676. [PubMed]
34. Dorsch, A.D.; da Silva Brito, W.A.; Delcea, M.; Wende, K.; Bekeschus, S. Lipid Corona Formation on Micro- and Nanoplastic Particles Modulates Uptake and Toxicity in A549 Cells. *Materials* **2023**, *16*, 5082. [CrossRef]
35. Acierno, D.; Graziosi, L.; Patti, A. Puncture Resistance and UV aging of Nanoparticle-Loaded Waterborne Polyurethane-Coated Polyester Textiles. *Materials* **2023**, *16*, 6844. [CrossRef] [PubMed]
36. Xie, Z.; Xiao, D.; Yu, Q.; Wang, Y.; Liao, H.; Zhang, T.; Liu, P.; Xu, L. Fabrication of Multifunctional Silylated GO/FeSiAl Epoxy Composites: A Heat Conducting Microwave Absorber for 5G Base Station Packaging. *Materials* **2023**, *16*, 7511. [CrossRef]

Disclaimer/Publisher's Note: The statements, opinions and data contained in all publications are solely those of the individual author(s) and contributor(s) and not of MDPI and/or the editor(s). MDPI and/or the editor(s) disclaim responsibility for any injury to people or property resulting from any ideas, methods, instructions or products referred to in the content.

Review

Biodegradable Biopolymeric Nanoparticles for Biomedical Applications-Challenges and Future Outlook

Radhakrishnan Sreena [1,2] and Arputharaj Joseph Nathanael [1,*]

[1] Centre for Biomaterials, Cellular and Molecular Theranostics (CBCMT), Vellore Institute of Technology (VIT), Vellore 632014, Tamil Nadu, India
[2] School of Biosciences & Technology (SBST), Vellore Institute of Technology (VIT), Vellore 632014, Tamil Nadu, India
* Correspondence: joseph.nathanael@vit.ac.in

Abstract: Biopolymers are polymers obtained from either renewable or non-renewable sources and are the most suitable candidate for tailor-made nanoparticles owing to their biocompatibility, biodegradability, low toxicity and immunogenicity. Biopolymeric nanoparticles (BPn) can be classified as natural (polysaccharide and protein based) and synthetic on the basis of their origin. They have been gaining wide interest in biomedical applications such as tissue engineering, drug delivery, imaging and cancer therapy. BPn can be synthesized by various fabrication strategies such as emulsification, ionic gelation, nanoprecipitation, electrospray drying and so on. The main aim of the review is to understand the use of nanoparticles obtained from biodegradable biopolymers for various biomedical applications. There are very few reviews highlighting biopolymeric nanoparticles employed for medical applications; this review is an attempt to explore the possibilities of using these materials for various biomedical applications. This review highlights protein based (albumin, gelatin, collagen, silk fibroin); polysaccharide based (chitosan, starch, alginate, dextran) and synthetic (Poly lactic acid, Poly vinyl alcohol, Poly caprolactone) BPn that has recently been used in many applications. The fabrication strategies of different BPn are also being highlighted. The future perspective and the challenges faced in employing biopolymeric nanoparticles are also reviewed.

Keywords: biopolymers; biopolymeric nanoparticles; biomedical; tissue engineering; drug delivery

Citation: Sreena, R.; Nathanael, A.J. Biodegradable Biopolymeric Nanoparticles for Biomedical Applications-Challenges and Future Outlook. *Materials* **2023**, *16*, 2364. https://doi.org/10.3390/ma16062364

Academic Editors: Peijiang Liu and Liguo Xu

Received: 8 February 2023
Revised: 9 March 2023
Accepted: 13 March 2023
Published: 15 March 2023

Copyright: © 2023 by the authors. Licensee MDPI, Basel, Switzerland. This article is an open access article distributed under the terms and conditions of the Creative Commons Attribution (CC BY) license (https://creativecommons.org/licenses/by/4.0/).

1. Introduction

Nanotechnology is the study that involves designing or fabricating materials and devices with at least one dimension of one billionth of a meter [1]. Multiple researchers have proved the advantages of the nano-dimension over the micrometer scale owing to the enhanced individual molecule interaction compared to the bulk [2]. Nanoparticles are zero-dimensional nanomaterials (0D) with a size range from 10 to 1000 nm. They are employed in many biomedical applications such as drug delivery [3], tissue engineering [4], biosensors [5], gene delivery [6], cell imaging and labeling [7,8] because of their enhanced surface-to-volume ratio and magnetic properties [9]. Nanoparticles have created an important role in the advancement of therapeutic applications since they exist in the same size range as that of proteins, and their small size and large surface help in the exposure of surface functional groups that can be tailored according to the requirement [10]. Nanoparticles obtained from biological sources are highly preferred because of their improved quality and stability compared to metal-based nanoparticles, where most are toxic to the human system [11]. Thus, nanoparticles can be obtained from biopolymers as a solution to the disadvantages posed by the counter-sources [2].

Biopolymers are the polymers obtained from living organisms such as plants, animals or microbes; they also include synthetic polymers obtained from renewable feedstock, bio-based monomers and also fossil fuels. Biopolymers can be classified into polysaccharides, polypeptides and polynucleotides based on the monomeric unit of the polymer, and are available in abundance and used extensively in the biomedical field, such as for wound healing drug/gene delivery, tissue engineering and cell imaging [12]. BPn appears to offer a solution to ameliorate the environmental effects and issues in biocompatibility and biodegradability caused by synthetic materials. The most important parameters that have a crucial impact on the fabrication of BPn are the surface charge, size, stability, compatibility with the cells and degradation [13]. Albumin was the first fabricated BPn [2]. Though polymeric nanoparticles have an issue of scaling-up and their capacity of drug loading is also comparatively low, researchers have widely employed them and tried ways to combat the disadvantages [10]. They have favorable properties such as biocompatibility, good anti-oxidant and anti-bacterial properties, and tailorable surface features [11]. BPn acquired from proteins and polysaccharides are superior when compared to synthetic materials, as the former can be easily metabolized naturally by the enzymes present in the digestive system, whereas the latter accumulates and leads to the formation of toxic by-products. Protein-based BPn can be surface modified, which can facilitate site-directed drug targeting [14,15]. One of the main limitations in employing biopolymeric nanoparticles from proteins or nucleic acids is that they are hydrophilic, whereas the polymers are mostly hydrophobic in nature and thus cause difficulties in drug encapsulation and degradation. Therefore, the preparation of biopolymeric nanoparticles is extremely critical [2]. Biodegradation of natural polymers occurs through biological processes, including enzymes such as collagenase in vivo and also via non-biological processes such as hydrolysis. It has been reported that the majority of natural polymers degrade with the help of enzymes. Polysaccharide-based biopolymers are degraded enzymatically within the human system with the help of enzymes such as lysozymes and amylases. Biodegradable synthetic biopolymeric nanoparticles degrade by hydrolysis of esters or urea linkages. It is also reported that polymers with polar groups degrade faster when compared to those with non-polar groups [16]. Table 1 shows a summary of the advantages and disadvantages of different sources of biopolymeric nanoparticles. Surface modification of the BPn is carried out to fine-tune the properties of the fabricated nanoparticles employed for biomedical applications. Some of the strategies employed include physical immobilization; modifications using chemicals such as grafting with amino, acrylate or acetyl group; and grafting induced by radiation such as ultrasonic waves. This type of modification enables improvement of the stability and the activity of the BPn and also aids in preventing aggregation, protecting them from any alteration [17]. BPn can be fabricated by employing different methods such as coacervation, desolvation and electro-spray techniques without employing the use of harsh organic solvents [13]. Figure 1 shows the schematic representation of biopolymeric nanoparticles employed for various applications. This review highlights the various biopolymeric nanoparticles employed for biomedical applications such as tissue engineering, drug delivery and images, and the various fabrication strategies are also discussed. The current status and the challenges in employing them are also highlighted.

Table 1. Summary of advantages and disadvantages of various biopolymeric nanoparticles.

Polymer	Advantages	Disadvantages	Reference
Albumin	Highly abundant, biodegradable, biocompatible, non-cytotoxic.	Immunogenic effects, very expensive, lack of efficacy.	[18,19]
Gelatin	Enhanced cell adhesion, proliferation and cell infiltration in the scaffolds, good stability and biodegradability, osteoconductive, non-immunogenic	Low stability in normal physiological conditions, poor bioactivity, brittle, fast degradation rate under physiological conditions	[16,20]
Silk fibroin	Biocompatible, osteoconductive, improves cell migration and angiogenesis, good elastic properties, moderate degradation rate.	Low mechanical strength, degradation of silk releases by-products that can cause immunogenic reactions, inability to induce osteogenesis.	[16,20,21]

Table 1. Cont.

Polymer	Advantages	Disadvantages	Reference
Collagen	Low immunogenicity, enhanced permeability properties, excellent cell adhesion, proliferation and differentiation properties, biodegradable, biocompatible.	Low mechanical strength, low structure stability, variability in different collagen sources.	[16,20,22]
Chitosan	Mucoadhesive nature, enhanced biocompatibility, osteoconductive, non-toxic, promotes cell adhesion, hemostatic potential, biodegradable, anti-bacterial activity.	In vivo degradation rate is very high, low mechanical strength, cross-linkers are required to maintain stability, solubility is less and viscosity is high at neutral pH, control of nanoparticle size is difficult.	[16,20,23]
Alginate	Biocompatible, biodegradable, cell compatible, gel-forming capability, low immunogenicity, mimics the extracellular matrix, low cost, ability of encapsulation.	Low mechanical properties, degradation is questionable sometimes, poor cell adhesion, sterilization is difficult.	[16,20,24]
Starch	Biodegradable, low cost, biocompatible, easily available, good cell adhesion.	Very high viscosity, low Mechanical properties, fragile, stability issues, water uptake is very high, modifying chemically can release toxic by-products.	[16,25]
Dextran	Biocompatible, anti-thrombotic property, good water solubility, functionalization can be carried out easily.	High cost, non-availability, very high permeability, encapsulated drugs are released very fast.	[26,27]
Poly-caprolactone	Compatible with cells, non-toxic, cell proliferation and angiogenesis can be controlled, good mechanical properties, improved cellular proliferation.	Bioactivity is less, poor cellular adhesion due to hydrophobic surface, use of toxic solvents.	[16]
Polyvinyl alcohol	Biocompatible, good elastic nature, water-soluble polymer, good tensile strength, improved flexibility, stability to various temperatures, low cost.	Lacks cell adhesion property, in growth of bone cells is significantly less, very high water uptake.	[16,28,29]
Polylactic acid	Biocompatible, cell compatible, degradation rate is good, by-products are non-toxic, properties can be easily tailored, eco-friendly.	Lack of cell adhesion and proliferation property, expensive, brittle (elongation at break is less than 10%), chemically inert.	[16,30]

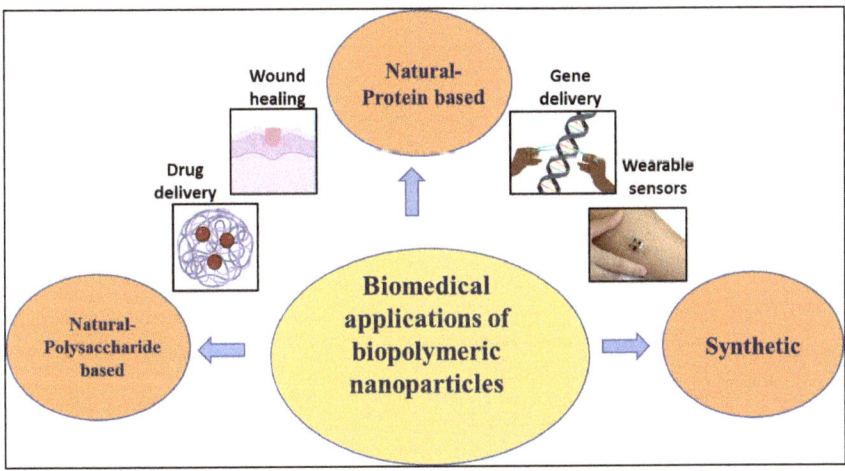

Figure 1. Schematic representation of the biomedical applications of biopolymeric nanoparticles.

2. Protein Based Biopolymeric Nanoparticles

Proteins are basically made of amino acids linked via peptide bonds, and their structure is stabilized by means of hydrophobic interactions and hydrogen and disulphide bonds [31]. These naturally derived polymers are highly preferred because of their excellent biocompatibility and good degradation characteristics. There are no harmful by-products since the degradation process is completely natural [32]; thus, the nanoparticles derived from the protein-based biopolymers are less toxic and easy to fabricate. The surface can also be easily tuned with respect to specific drug delivery applications [33]. A few other advantages of employing protein-derived biopolymeric nanoparticles for biomedical applications are that the fabrication is comparatively easier, and it has been reported to be more stable in vivo. The size distribution can be easily controlled and the process can be scaled up [2]. The defined primary structure in protein helps in the easy attachment of various drugs that play a key role in therapeutic applications [13]. The secondary structure of the protein determines the size of the proteins and also helps to fabricate nanoparticles precisely [10]. Some examples of nanoparticle-derived protein biopolymers employed for biomedical applications include silk fibroin, albumin, gelatin and collagen, which will be discussed in the following subsection.

2.1. Albumin

Albumin belongs to the family of globular proteins and acts as a carrier protein for endogenous or exogenous compounds. It is widely employed for treating a variety of diseases—especially cancer [34]. Researchers have tested the potential of albumin in various products and clinical trials. Albumin can be easily obtained from plants, animals and human beings. Ovalbumin, bovine serum albumin (BSA) and human serum albumin (HSA) are the three commonly used albumins for biomedical applications [35]. The main advantages of employing albumin are that it has good compatibility with human cells, it does not induce toxicity, and at the same time is also biodegradable and does not cause any adverse immune reactions. Thus, albumin is a very good candidate for fabricating nanoparticles. Various proteins that are expressed in a higher range in the tumor cells, such as secreted proteins acidic and rich in cysteine (SPARC), easily and very effectively bind to albumin [18]. A study has been performed where albumin nanoparticles were employed for the simultaneous delivery of two drugs, ibrutinib (IBR) and hydroxychloroquine (HCQ), for the treatment of glioma. Drug-loaded human serum albumin (HSA) nanoparticles were prepared by ultrasonication method. HCQ, as an inhibitor, blocks autophagosome degradation. IBR has a major role in glioma treatment by suppressing the malignant tumor growth but faces disadvantages such as poor bioavailability and drug exposure in the brain cells were found to be very limited. To overcome this, drug delivery using albumin nanoparticles was facilitated. The mean size of the drug-loaded HAS nanoparticles was found to be 160.1 ± 0.7 nm. The encapsulation efficiency (%) and the drug loading capacity (%) were found to be 97.2 ± 1.8 and 3.96 ± 0.06, respectively. The biodistribution analysis showed that the presence of HAS nanoparticles resulted in an increased accumulation (5.59 times higher than free drug) of IBR drug in the tumor. The fabricated drug-loaded nanoparticles showed high cytotoxicity against C6-luc cells in CCK-8 assay and apoptosis assay. In vivo analysis in mice showed that IBR-HCQ-HAS nanoparticles stayed for a prolonged time when compared to IBR-HAS nanoparticles. Thus, these results were found to be very promising for the treatment of glioma [36]. In another reported study, abaloparatide (ANPs) was encapsulated in bovine serum albumin nanoparticles by desolvation process, stabilized in chitosan by the self-assembly process, and then made into a nanofiber scaffold for bone tissue engineering applications. Electrospinning was carried out to fabricate polymeric nanofibers from a mixture of polycaprolactone (PCL), n-hydroxyapatite (n-HAp), aspirin (ASA) and abaloparatide. The schematic illustration of the synthesis of abaloparatide encapsulated in bovine serum albumin nanoparticles and the fabrication of electro spun nanofibers loaded with two drugs is depicted in Figure 2.

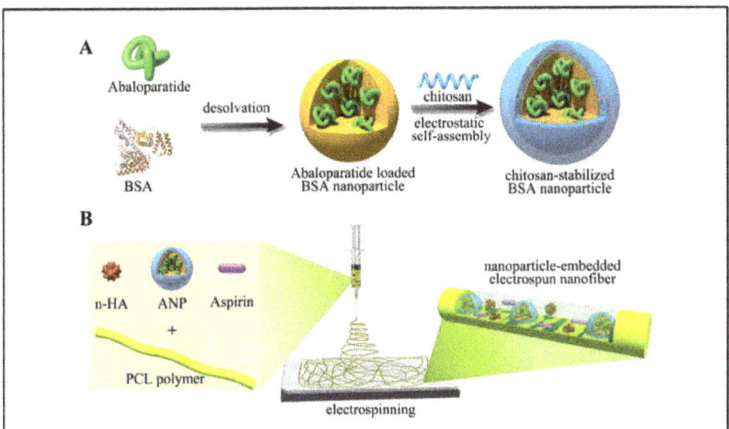

Figure 2. Schematic illustration of (**A**). Synthesis of chitosan stabilized BSA nanoparticle by desolvation approach; (**B**). Fabrication of PCL-based electro-spun nanofiber loaded with nano-hydroxyapatite, abaloparatide loaded BSA nanoparticle and aspirin. (Reproduced with permission from [37]; Copyright 2022, Elsevier).

The size range of the chitosan–abaloparatide nanoparticles was found to be 289 ± 34 nm. The scanning electron microscopy (SEM) images of the nanofiber matrix showed that they have irregular pore structures for the diffusion of oxygen and nutrients. In vitro release studies have shown that drug release was fast in the nanofibers with two drugs when compared to the one with a single drug. The ANPs/ASA/PCL/HA nanofiber scaffold showed that the release of the drug was slow because of the hydrophilicity and degradation characteristics. Cell adhesion was studied with the help of MC3T3-E1 and the morphology was observed by SEM, as reported in Figure 3.

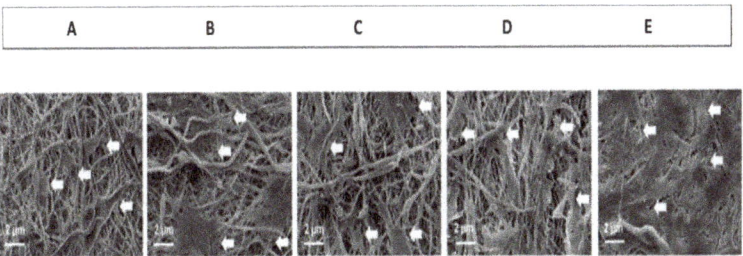

Figure 3. SEM images of MC3T3-E1 adhered on the nanofibrous scaffold after 2 days. (**A**) PCL/HA; (**B**) ASA/PCL/HA; (**C**) BSANps/PCL/HA; (**D**) ANPs/PCL/HA; (**E**) ANPs/ASA/PCL/HA The arrows indicate lamellipodia and cell outlines. (Reproduced with permission from [37]; Copyright 2022, Elsevier).

From Figure 3, we understand that the cells have properly spread on the nanofibrous scaffold and show fusiform morphology. Thus, the results of the dual drug-loaded nanofibers with chitosan-stabilized bovine serum albumin nanoparticles show excellent physical and chemical properties, good degradation rate, enhanced cell compatibility and osteogenic activity [37]. A study reported by Thangavel et al. used indocyanine green–paclitaxel encapsulated in human serum albumin nanoparticles that were functionalized with hyaluronic acid, as a ligand for drug delivery with image guiding capability directed to CD44 non-small cell lung cancer (NSCLC). The drug release analysis showed that paclitaxel was released more efficiently at pH 6.6 due to the acidic nature of the tumor micro-environment. Only 30% of the drug was released at pH 7.2 (blood circulation) after 46 h. The paclitaxel nanoparticles showed good anticancer activity against A549 and H299 cell lines; thus, image guided drug delivery was found to be very efficient without compromising the anticancer treatment efficiency [38]. Khella et al. studied the anti-tumor activity of MCF-7 and Caco-2 cell lines using carnosic acid encapsulated in bovine serum albumin nanoparticles. The results of the experiment showed excellent drug loading ability and the best release profile. Enhanced anti-tumor activity was found in both the cell lines; apoptosis results showed that the MCF-7 and Caco-2 cells were arrested at the G2/M phase (10.84% and 4.73%, respectively) [39]. One of the studies demonstrated the use of dexamethasone encapsulated in bovine serum albumin nanoparticles for enhanced anti-inflammatory activity in rats; a bimodal release of the drug and a significant anti-inflammatory activity was reported [40].

2.2. Gelatin

Gelatin is a natural biopolymer derived from animal collagen with favorable properties such as low cost, biocompatibility and biodegradability, that is derived from the hydrolysis of animal collagen [41,42]. Gelatin-based nanoparticles (GNPs) are very promising for a variety of biomedical applications such as tissue engineering and drug delivery because of their properties, such as easy availability, offering great stability and long-time storage in vivo [14]. GNPs have also been widely employed for treating brain disorders since they can cross the blood–brain barrier, and various properties such as mechanical properties, thermal and swelling behavior changes with respect to the amphoteric properties of gelatin [41]. Different cross linkers can be added to modify the physiochemical properties of gelatin nanoparticles [31]. A study reported the use of gelatin nanoparticles conjugated with polyethylene glycol for the simultaneous delivery of two drugs, doxorubicin and betanin, for cancer treatment. The particle size of the nanoparticles was found to be 162 nm. The encapsulation efficiency and the loading capacity was 82% and 20.5%, respectively. High cell cytotoxicity was observed after 48 h against the MCF-7 cancer cell line when two drugs were given rather than the individual drug, as depicted in Figure 4.

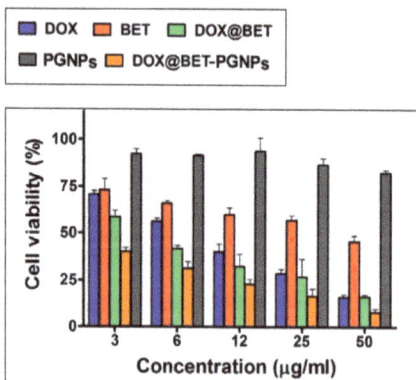

Figure 4. Cell toxicity results of MCF-7 cell line after treating them with individual and a combination of drugs encapsulated in gelatin nanoparticles. (DOX—doxorubicin; BET—betanin). (Reproduced with permission from [43]; Copyright 2019, Elsevier).

Cellular uptake results, as shown in Figure 5, revealed that high cellular uptake was witnessed with the nanoparticle-encapsulated drugs rather than the free form of the drug, proving that the nanoparticles have the ability to escape the endocytosis process.

The combination of the drugs encapsulated in gelatin nanoparticles showed excellent apoptotic activity. Thus, the multi-drug nanocarriers facilitate a new horizon to develop an enhanced treatment strategy for cancer [43]. A study reported by Yang et al. fabricated zoledronic acid (ZOL)-encapsulated gelatin nanoparticles integrated into a titanium scaffold for treating osteoporosis-based defects. The in vitro results showed enhanced osteoblast differentiation when ZOL concentration was 50 µmol L^{-1}. The in vivo studies in osteoporotic rabbits showed improved bone growth and osteogenesis [44]. One study reported the use of gold nanoparticles conjugated with gelatin nanoparticles for the purpose of bioimaging as well as a drug delivery system. The size of the nanoparticles was found to be 218 nm and showed no toxicity up to 600 µg mL^{-1}. The imaging of the nanoparticles in the skin tissue was carried out by using confocal laser scanning microscopy (CLSM), achieving a depth profile of 760 µm [45]. Gelatin nanoparticles were enteric coated to encapsulate 5-amino salicylic acid for oral drug delivery for the treatment of ulcerative colitis in one study. The nanoparticles' size ranged from 225 to 250 nm and were found to be spherical in nature. The administration of nanoparticles reduced mast cell infiltration and also maintained the colon tissue architecture. A significant reduction in the inflammatory markers such as TNFα, COX-2, IL1-β and nitrate levels was observed. The encapsulated drug showed enhanced therapeutic efficiency when compared to the free drug [46].

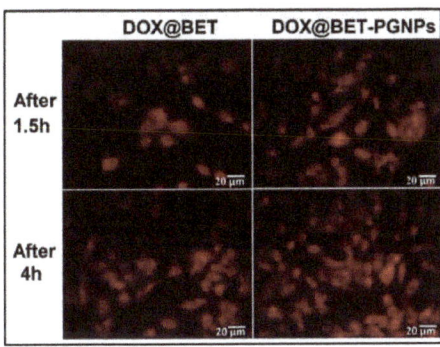

Figure 5. Cellular uptake studies of MCF-7 cell lines treated with the drug as well as nanoparticle encapsulated drugs. (DOX—doxorubicin; BET—betanin). (Reproduced with permission from [43]; Copyright 2019, Elsevier).

2.3. Silk Fibroin

Silk fibroin is a natural biopolymer obtained from the cocoons of Bombyx mori made of 5507 amino acid residues. The most important features that make silk fibroin an outstanding material for biomedical applications are its good biocompatibility with humans, very high mechanical strength and favorable biodegradable properties. It also helps to promote cell adhesion and proliferation. They can be used in various forms, such as films, hydrogels, fibers, spheres, mats, sponges and scaffolds, and are widely used in many applications such as wound healing; cancer therapy; drug delivery; and bone, skin and cartilage regeneration [47–49]. The controlled degradation rate and excellent biocompatibility of silk fibroin make it an excellent candidate for making nanoparticles [50]. In the nanoscale, silk fibroin shows improved physiochemical, mechanical and biological properties [51]. In one reported study, curcumin was encapsulated in silk fibroin nanoparticles for treating cancer; the therapeutic efficiency of the drug was enhanced by loading in a nanocarrier. The size of the nanoparticles ranged from 155 to 170 nm. The in vitro cytotoxicity assays revealed that the nanoparticles greatly reduced the viability of carcinogenic cells, and high cytotoxicity was seen more in neuroblastoma cells than hepatocarcinoma cells. The drug curcumin was found to be fluorescent when it was loaded into silk fibroin nanoparticles and not in the free state. The drug-loaded nanoparticles showed excellent anti-tumor and anti-oxidant activity [52]. Shen et al. developed a scaffold made of sodium alginate and silk fibroin loaded with silk fibroin nanoparticles for improving hemostasis and cell adhesion. The nanoparticles were obtained by the self-assembly process. The addition of nanoparticles to the scaffold system improved the compression strength, and reduced the degrading rate. The nanoparticles were found to be spherical and uniform in size. The cell adhesion and cell proliferation of L929 cells and HUVECs were studied by using a Live/Dead assay kit (Figure 6). It was found that at the end of 5 days, the cells showed proper adhesion, spreading, migration and proliferation. A greater number of cells were grown on the composite scaffold with silk fibroin nanoparticles (NP) when compared to the one without them (PM).

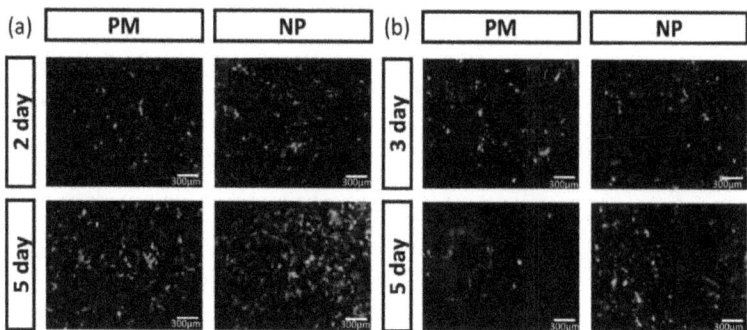

Figure 6. Viability and proliferation of the cells: (**a**) human umbilical vein endothelial cells (HU-VECs) (**b**) L929 cells in the scaffolds with and without nanoparticles. (Reproduced with permission from [53]; Copyright 2022, Elsevier).

Another study reported the use of simvastatin loaded into silk fibroin nanoparticles for the purpose of bone regeneration. The nanoparticles were found in the size range of 174 ± 4 nm and were spherical in morphology. A sustained drug release profile was seen for about 35 days. The in vitro cell studies revealed that the nanoparticles improved cell adhesion and proliferation, and also showed good alkaline phosphatase activity [54]. Rahmani et al. investigated the use of silk fibroin nanoparticles for the delivery of 5-fluoro uracil for the treatment of cancer. The size of the nanoparticles was found to be 286.7 nm and the loading efficiency was 52.32%. High loading efficiency and slow release of the drug were observed [55]. Doxorubicin and PX478 were co-loaded into silk fibroin nanoparticles that were functionalized with folic acid for the purpose of treating multi-drug-resistant tumors. The cellular uptake was increased and this nanoparticle combination significantly downregulated multiple genes to overcome multi-drug resistance. The lysosomal escape was achieved quickly, and doxorubicin could quickly enter the cells and kill the drug-resistant cells [56].

2.4. Collagen

Collagen is a structural biopolymer that is found abundant in the human body. It is the major part of the extracellular matrix and is found in tendons, ligaments, cartilage and skin [57,58]. Collagen has been widely employed in biomedical applications due to its properties such as biocompatibility, biodegradability, favorable gelling and surface behavior [59,60]. Nano collagen has an outstanding potential when compared to three-dimensional collagens in helping to withstand heavy loads with minimum tension due to the high surface-to-volume ratio [61]. The nanocollagen has notable properties, such as high contact area, reduced toxicity, easily sterilizable, increased retention of cells, and decreased effects of toxicity from the by-products as a result of degradation. They can be found in various forms, such as sheets, films, sponges, fibers, pellets, disks and nanoparticles [62,63]. One study reported the use of collagen nanoparticles from marine sponges fabricated by the process of alkaline hydrolysis. Estradiol–hemihydrate was loaded into the nanoparticle and the drug loading was found to be 13.1%. Prolonged drug release and improved drug absorption by the cells were observed. Thus, the presence of collagen nanoparticles facilitates exciting ways of drug delivery [64]. Appropriate cross-linking strategies have to be chosen to tailor the properties of collagen according to the intended application. The stability and degradation characteristics can be altered when the surface features are altered [31].

2.5. Elastin

Elastin is a natural biopolymer found in elastic fibers, especially in the extracellular matrix of skin, lungs, heart and blood vessels [65]. One of the main properties of elastin is that it can retain its original shape and insolubility even after stretching [66]. They are not always biocompatible and are very much difficult to alter. Thus, soluble elastin-like peptides are fabricated for a wide variety of biomedical applications [67]. Elastin nanoparticles have been employed as a nanocarrier for delivery drugs and genes and have proven to be very effective. The ability of elastin nanoparticles to self-assemble and respond to varying temperatures has allowed them to be employed for various therapeutic applications. The properties of the elastin nanoparticles can be tailored according to the intended application [14,67,68]. One study reported the use of elastin nanoparticles for the delivery of bone morphogenic proteins (BMPs). Poly (L-valine-L-proline-L-alanine-L-valine-L-glycine) pentapeptide is an elastin-like polymer where the central glycine molecule is replaced by alanine. A total of 94% of the BMP was successfully encapsulated into elastin-like polymer nanoparticles. The in vitro assays revealed that they are non-toxic and compatible with C2C12 cells [69]. Kim et al. reported a study where α-elastin nanoparticles were fabricated for protein delivery applications. The nanoparticles were grafted with polyethylene glycol to improve the colloidal stability; they were in the size range from 330 ± 33 nm. A sustained release of encapsulated insulin and bovine serum albumin (BSA) was observed for 72 h. The thermoresponsive nature enables the fabricated nanoparticles to be employed for a wide variety of drug delivery and tissue engineering applications [70]. The summary of protein based biopolymeric nanoparticles is given in Table 2.

Table 2. Summary of protein-based biopolymeric nanoparticles.

Protein	Overall Composition	Application	Key Findings of the Study	Reference
Albumin	Human serum albumin + ibrutinib and hydroxychloroquine (nanoparticles)	Co-drug delivery system for treatment of glioma	Improved bioavailability Prolonged survival time in in vivo treated mice High cytotoxicity against C6 cells	[36]
Albumin	Bovine serum albumin + abaloparatide + aspirin + polycaprolactone + hydroxyapatite (nanofibrous scaffold)	Bone regeneration	Improved degradation rate Slow drug release Enhanced compatibility Improved bone regeneration	[37]
Albumin	Human serum albumin (HSA) + indocyanine green (ICG) + paclitaxel (PTX) + hyaluronic acid (nanoparticles)	Image-guided drug delivery	Efficient drug release in the tumor environment Efficient anti-cancer activity	[38]
Albumin	Bovine serum albumin + carnosic acid	Anti-tumor activity of breast cancer and colon cancer.	Enhanced loading activity Improved release profile of the drug Enhanced anti-tumor activity Upregulation of *GCLC* gene and downregulation of *BCL-2* and *COX-2* gene.	[39]
Albumin	Bovine serum albumin + silymarin + curcumin + chitosan	Muco-inhalable drug delivery system	Significant reduction of interleukin-6 and c-reactive protein Efficient anti-viral activity in in vitro COVID-19 experiment	[71]

Table 2. Cont.

Protein	Overall Composition	Application	Key Findings of the Study	Reference
Albumin	Bovine serum albumin + poly-L-lysine + graphene oxide	Bone regeneration	Controlled release of BMP-2 (14 days) Improved matrix mineralization Enhanced Alkaline phosphatase (ALP) activity	[72]
Gelatin	Gelatin + concanavalin-A + cisplatin	Drug delivery for cancer therapy	Enhanced cellular uptake of nanoparticles Enhanced reactive oxygen species and apoptosis in cancer cells	[73]
Gelatin	Gelatin methacrylol nanoparticles + rhodamine	Cell imaging	Improved cell viability and cell proliferation in vitro Superior cell compatibility Enhanced cellular uptake Improved fluorescent properties	[74]
Gelatin	Amino cellulose + polycaprolactone + gelatin nanoparticles	Rheumatoid arthritis	Reduction in swelling and inflammation in rats. Maintaining cartilage and bone tissue architecture. Reduction of inflammatory markers	[75]
Gelatin	Gelatin + indocyanine + doxorubicin	Breast cancer treatment	Improved drug release Suppressed the tumor growth in vivo Enhanced degradation of matrix metalloproteinase-2	[76]
Gelatin	Polyethylene glycol grafted gelatin nanoparticles + doxorubicin + betanin	Cancer therapy	Enhanced cellular uptake Cell apoptosis induced in MCF cells; Controlled drug release observed	[43]
Silk fibroin	Curcumin + silk fibroin nanoparticles	Cancer therapy	Enhance anti-tumor activity Improved anti-oxidant activity Curcumin was found to be fluorescent when encapsulated	[52]
Silk fibroin	Silk fibroin + sodium alginate + silk fibroin nanoparticles (scaffold)	Wound healing	Improved cell adhesion Enhanced hemostasis Improved platelet adhesion Excellent biocompatibility and improved cell adhesion and proliferation	[53]
Silk fibroin	Silk fibroin + simvastatin (nanoparticles)	Bone regeneration	Sustained release profile Improved ALP production Enhanced production of osteoblast cells	[54]
Silk fibroin	Silk fibroin + 5 fluorouracil (nanoparticles)	Drug delivery	Improved loading efficiency Slower release of the drug	[55]
Silk fibroin	Silk fibroin nanoparticles + PX478 + doxorubicin	Reverse multi-drug resistance	Increased cellular uptake Downregulation of genes-MDR1, VEGF and GLUT-1	[56]
Silk fibroin	Silk fibroin + tamoxifen (nanoparticles)	Breast cancer	The particle size was found to be 186.1 nm Encapsulation efficiency was found to be 79.08% Biphasic release profile was observed	[77]

Table 2. Cont.

Protein	Overall Composition	Application	Key Findings of the Study	Reference
Collagen	Collagen + estradiol–hemihydrate	Transdermal drug delivery	Enhanced drug loading capacity Increased sustained drug release Improved drug absorption	[64]
Elastin	Elastin-like polymeric nanoparticles + bone morphogenic protein	Drug delivery system	Improved encapsulation efficiency Compatible with C2C12 cells	[69]
Elastin	α-elastin + methoxy polyethylene glycol + BSA/Insulin	Protein delivery	Encapsulation at low temperatures with simple mixing Sustained release for 72 h The nanoparticles are of normal size distribution	[70]
Elastin	Elastin-like recombinamers + docetaxel + RGD peptide	Drug delivery system	High yield of 70% Monodispersed nanoparticles-40 nm Very much effective against breast cancer cell line	[78]

3. Polysaccharide Based Polymeric Nanoparticles

Polysaccharides are long carbohydrate molecules made of monosaccharide units that keep repeating and are linked by glycosidic bonds. Some examples of polysaccharides include chitosan, alginate, dextran, starch, heparin and hyaluronic acid. These naturally derived biopolymers form the main constituent of the extracellular matrix. The main advantages of polysaccharides are that they are highly stable, compatible with human cells and have favorable degradable properties. Carbohydrate-based nanoparticles, along with immobilization techniques, help in improving biocompatibility. Due to their small size and high surface-to-volume ratio, nanoparticles have wide applications, such as delivering drugs, proteins and nucleic acids. Polysaccharide-based nanoparticles can be fabricated by various methods and the properties can be tailored by modifying the structure according to the intended application [2,10,13,15]

3.1. Chitosan

One of the most important cationic biopolymers employed for various biomedical applications is chitosan. This hetero polymer is made of N-acetyl-D-glucosamine, which is an acetylated unit, and D-glucosamine, which is a deacetylated unit linked by β-1,4 linkages. It is a hydrophilic biopolymer with the ability to open tight junctions of the cell membranes that are degraded by the presence of enzymes such as lysozymes, proteases and lipases [10,79]. The positive charge of the chitosan nanoparticles is due to the presence of amine groups that has the ability to adhere to the negatively charged mucosal membrane and aid in the release of the encapsulated drugs in a sustained manner. A complex formation is induced by the electrostatic interactions along with hydrogen bonding and hydrophobic interactions, and thus, the mucoadhesive property of the chitosan nanoparticles are highly exploited for oral drug delivery applications. The nanoparticles also have cell compatibility in both in vitro and in vivo models [80]. The bioavailability and stability issues are overcome with the surface modification of the chitosan nanoparticles. The chitosan nanoparticles show improved bioavailability, increased specificity and reduced toxicity, and the properties vary with size. Due to all these properties, they are employed in applications such as nanomedicine, biomedical and pharmaceutical industries [81]. They can be fabricated by a variety of methods such as emulsification, precipitation, ionic or covalent cross-linking, solvent diffusion method and solvent evaporation [82,83]. Dev et al. fabricated chitosan nanoparticles along with Poly lactic acid for encapsulation of the anti-HIV drug called Lamivudine; the nanoparticles were found to be around 300 nm and the drug encapsulation efficiency was 75.4%. They were found to be non-toxic to mouse fibroblast cells (L929 cells) [84]. Hydrophilic drugs such as 5-fluorouracil and leucovorin have been encapsulated in chitosan nanoparticles for the treatment of colon cancer. The drug-loaded nanoparticles were in a wide size range of 34–112 nm. The drugs loaded into the nanoparticles initially had a burst release followed by a continuous and constant release of the drugs. Encapsulation efficiency and the drug loading capacity of the drugs were found to be very efficient because of the strong interaction between the biopolymer and the drugs [85]. Chitosan nanoparticles were incorporated into the silk fibroin hydrogel scaffolds for the repair of cartilage defects. The incorporation of tumor growth factor (TGFβ) and bone morphogenic protein (BMP) was carried out to repair the articular defects. Enhanced cell viability, cytocompatibility and chondrogenesis was observed [86]. Curcumin was encapsulated into chitosan nanoparticles and finally incorporated into nanofiber mats containing polycaprolactone and gelatin. The nanoparticles were in the size range of 278 ± 60 nm. The encapsulation efficiency and the drug loading capacity were found to be 93 ± 5% and 4.2 ± 0.2%, respectively. Drug release of the nanocomposite was observed up to 240 h. The cell compatibility of the nanocomposite was studied with the help of human endometrial stem cells (EnSCs), as indicated in Figure 7.

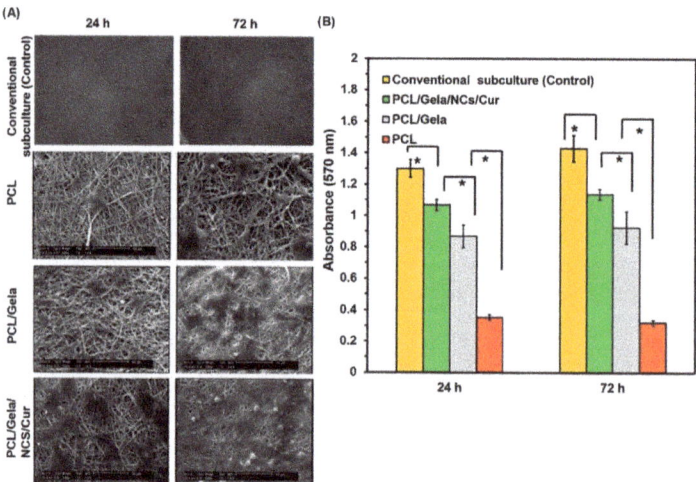

Figure 7. (**A**) Scanning electron microscopy images of human endometrial stem cells attached in PCL, PCL/gelatin and PCL/gelatin/chitosan nanoparticles/curcumin-loaded fibrous mats for 24 h and 72 h; (**B**) Results of cellular growth obtained through MTT assay. (Vertical bars: standard deviations; * p-value < 0.05) (Reproduced with permission from [87]; Copyright 2020, Elsevier).

Higher cellular growth was found in the PCL/gelatin/chitosan nanoparticles/curcumin nanofiber mats. An increase in cell adhesion and proliferation of the nanofiber mats was observed at the end of 72 h. The hybrid composite was found to be biocompatible, as observed through MTT assay.

3.2. Alginate

Alginate is one of the most important anionic biopolymers obtained from seaweeds such as brown algae. They are linear and are made of units of α-L-guluronic acid and β-D-mannuronic acid linked by 1,4 glycosidic linkages. The presence of carboxyl and hydroxyl groups in their structure facilitates easy modification according to the intended application. It can be transformed into any form, such as nanoparticles, hydrogels, microparticles and porous scaffolds. The ability of alginate to form gels without the addition of any toxic substance at normal conditions has enabled it to be used in a wide variety of therapeutic applications. It is also very easily available, not toxic, and has favorable cell compatibility and biodegradable properties. Alginate nanoparticles are fabricated by means of pre-gelation with calcium; they are widely being explored in the field of tissue engineering, regenerative medicine, wound healing, biosensors, genetic transfection and environmental applications. The nanoparticles show improved biocompatibility, degradation properties and also mucoadhesiveness properties; they are combined with other polymers to modulate their physiochemical, mechanical and biological properties [13,31,88–90]. A study reported the use of alginate nanoparticles along with an antibiotic called polymyxin B sulphate to be one of the layers for the biomembrane designed for wound healing. The biomembranes showed low toxicity and were found to be biocompatible with the fibroblast cells; the in vivo analysis showed promising outcomes [91]. Alginate nanoparticles, along with chitosan, were employed for the delivery of the drug called nifedipine. The nanoparticles had an average diameter of 20 to 50 nm. The drug release was found to be pH responsive, i.e., the percentage of the drug varies with respect to the pH. Initial burst release followed by continuous controlled release was observed. Fick's diffusion was found to be the reason for the drug release [92]. Curcumin diethyl disuccinate was encapsulated in chitosan/alginate nanoparticles for anti-cancer therapy. A sustained release profile of the drug and improved bioavailability was observed. The drug was found to be stable when exposed to digestive fluids. The main mechanism behind the release of the drug was found to be diffusion. It was found that the cellular uptake was enhanced and showed cytotoxicity against the HepG2 cell line [93]. Zohri et al. reported a formulation where chitosan and alginate nanoparticles were used as a non-viral vector for gene delivery applications and optimized using the D-optimal design. The nanoparticles were found to be compatible with cells and a transfection efficiency of 29.9% was observed [94]. One study reported the sustained release of the drug esculentoside from chitosan/alginate nanoparticles that were embedded in a collagen/chitosan scaffold for the treatment of burn wounds. The highest encapsulation efficiency of 78.20% was observed. The composite scaffold showed good anti-inflammatory activity. The in vitro assays showed that M2 macrophages were activated, which promoted quick healing of the burn wounds. The in vivo evaluation of the nanocomposite in the burn wounds also showed promising results (Figure 8).

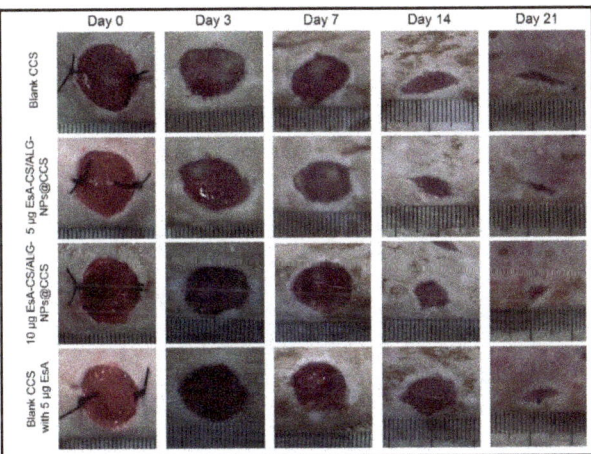

Figure 8. Images of the healing of burn wounds after transplantation with the blank Collagen/chitosan scaffold, 5 µg drug–chitosan/alginate nanoparticles @ collagen/chitosan scaffold, 10 µg drug–chitosan/alginate nanoparticles @collagen/chitosan scaffold, and blank collagen chitosan scaffold with 5 µg drug at days 0, 3, 7, 14, and 21 (Reproduced with permission from [95]; Copyright 2023, American Chemical Society).

Drug concentrations in the nanocomposite scaffold showed better healing properties than the blank scaffold. The wound was almost completely healed at the end of day 21 [95].

3.3. Starch

Starch is a natural, biodegradable biopolymer obtained from various plants such as potato, wheat, rice or corn, and it is made of amylose and amylopectin. It is widely employed for biomedical applications such as tissue engineering or wound healing. It is easily available since it is the second most abundant biomass present on the earth. The important favorable characteristics that make them a suitable candidate for various applications are that they have swelling characteristics, rheological properties, degradable properties, solubility and biocompatibility. Starch-based nanoparticles are used as fillers with other polymer matrices and help to improve the various physiochemical and mechanical properties. Studies also have reported that starch-based nanoparticles increase encapsulation efficiency. They can be fabricated by a variety of methods such as precipitation, micro fluidization and enzyme hydrolysis, homogenization and emulsification. They have enhanced absorptive capacity and biological penetration rate and are thus employed as carriers to deliver bioactive compounds [10,13,96–99]. One study reported that CG-1521 was encapsulated in starch nanoparticles for the treatment of breast cancer. Improved therapeutic index and bioavailability were reported due to the presence of nanoparticles. The release rate of the drug was reduced and the cytotoxicity was enhanced towards the MCF-7 cell line. Cell cycle arrest and apoptosis were witnessed in the MCF cell line in in vitro study. The drug delivery of the drug was found to be promising without interfering with the mechanism of drug action [100]. Curcumin was loaded onto starch nanoparticles derived from green bananas. The nanoparticles were found to be about 250 nm in size and the encapsulation efficiency was found to be 80%. More controlled release of curcumin was observed because of the strong hydrogen bond interaction [101]. Starch nanoparticles grafted with folate and biotin for the delivery of Doxorubicin and siRNA. A high amount of cytotoxicity was observed against the A549 cell line (human lung cancer cell line). The lowest amount of cell proliferation was observed and the mechanism behind cellular uptake was found to be either clathirin or caveolae-mediated [102]. A nano-based drug delivery system was designed by using starch nanoparticles conjugated with aptamer loaded with para coumaric acid for the treatment of breast cancer. The nanoparticles were found to be less agglomerated and the particle size was found to be 218.97 ± 3.07 nm. The encapsulation efficiency was found to be 80.30 ± 0.53%. Rapid and burst release of the drug was observed for the initial five hours. Higher cytotoxicity was observed towards MDA-MB-231 cells [103]. Triphala Churna, an ayurvedic drug, was encapsulated in starch nanoparticles for the purpose of releasing various drugs and bioactive compounds. The nanoparticles were in the size range of 282.9 nm. Improved fast drug release was observed at pH 7.4, and enhanced drug encapsulation was observed. The anti-oxidant and anti-bacterial results of the drug-loaded starch nanoparticles showed promising results. The drug showed improved activity and the mechanism of the drug was not altered though it was encapsulated in starch nanoparticles [104]. Methacrylated starch-based nanoparticles have been employed as hydrogels by photopolymerization. Dense and stiff hydrogels that are compatible with human cells were fabricated and reported in a study by Majcher et al. The shear modulus was found to be increased by at least five times [105].

3.4. Dextran

Dextran belongs to a family of microbial polysaccharides obtained from lactic acid bacteria (LAB) and their enzymes in the presence of sucrose. This exopolysaccharide is linked by D glucose units majorly by α-1,6 bonds. The physio–chemical properties vary with respect to the strain producing it. The favorable rheological, thermal properties, biocompatibility and biodegradability, enable dextran to be employed in a lot of applications. Dextran has been employed in biomedical applications such as wound healing, tissue engineering, imaging and as drug carriers. The ability of dextran nanoparticles to form a stable backbone has shown promising results to be employed as a nano drug carrier [106,107]. One study reported the use of the anticancer drug doxorubicin encapsulated in carboxymethyl dextran nanoparticles for cancer treatment. The nanoparticles were in the size of 242 nm and had an encapsulation efficiency of greater than 70%. Rapid release of the drug was observed initially. In vitro assays revealed that the fabricated nanoparticles showed higher cytotoxicity towards the SCC7 cancer cell line. A high anti-tumor effect was exhibited from the drug-loaded nanoparticles [108]. A dextran nanoparticle of about 13 nm was crosslinked with Zirconium (Zr-89) to be used as a positron emission tomography (PET) imaging agent for the purpose of imaging macrophages. The half-life was found to be 3.9 h, and they primarily imaged only the tissue macrophages and not the white blood cells. The in vivo imaging results showed that the tumoral uptake was very high and was able to surpass the reticuloendothelial system [109]. Acryloyl crosslinked dextran dialdehyde (ACDD) nanoparticles grafted with glucose oxidase for the fabrication of a pH-responsive insulin delivery system. A controlled release of insulin of 70% was observed in the artificial intestinal fluid conditions for 24 h. In the presence of glucose, the release was found to be 90% under artificial intestinal fluid conditions. The mechanism behind the release of the drug was found to be non-Fickian diffusion [110]. Butzbach et al. reported a study where photosensitizer was encapsulated in spermine and acetyl-modified dextran nanoparticles and grafted with folic acid on the surface that is specifically expressed in the tumor cells. Cellular uptake against He-La KB cells and cytotoxicity induced by light were observed [111]. Another study reported the use of dextran nanoparticles conjugated with acitretin for the treatment of psoriasis-like skin disease. A low dosage of the drug does not induce and side effects. In vitro results showed that keratinocyte proliferation was enhanced. The mechanism behind that was that the STAT-3 phosphorylation was efficiently inhibited [112]. Cerium oxide nanoparticles were coated with dextran for use as a contrast agent in the gastrointestinal tract and bowel diseases. Enhanced imaging in the inflammation sites. No toxicity was observed and was protective against oxidative damage. The oral dose (>97%) was cleared after 24 h [113]. In another study, dextran nanoparticles were cross-linked with colon-specific oligoester that responds to enzymes was fabricated. 5-Fluoro uracil was encapsulated in the dextran nanoparticles for the treatment of cancer. The nanoparticles were in the size range of 237 ± 25 nm. The encapsulation efficiency of the drug was found to be 76%. The drug was found to release only in the presence of the enzyme dextranase. 75% of the drug was released up to 12 h of incubation. The dextran nanoparticles were found to be compatible with the HCT116 colon cancer cell line and were found to be cytotoxic in the presence of the enzyme dextranase [114]. The summary of the polysaccharide based polymeric nanoparticles is given in Table 3.

Table 3. Summary of polysaccharide-based polymeric nanoparticles.

Polysaccharide	Overall Composition	Application	Key Findings of the Study	References
Chitosan	Chitosan + polylactic acid + lamivudine	Drug delivery	Drug release was found to be higher when higher percentage was loading. The nanoparticles were found to be non-toxic to the L929 cell line. The degradation rate increases with respect to pH	[84]
Chitosan	Chitosan + 5-fluorouracil and leucovorin	Drug delivery	Improved encapsulation efficiency and drug loading capacity. Release profile can be modulated by changing the parameters	[85]
Chitosan	Chitosan + ellagic acid	Oral cancer therapy	Particle size was found to be 176 nm; Encapsulation efficiency was found to be 94 ± 1.03%. Sustained release of the drug was observed. Cytotoxicity was observed in KB cell line	[115]

Table 3. Cont.

Polysaccharide	Overall Composition	Application	Key Findings of the Study	References
Chitosan	Chitosan + tetracycline+ gentamycin + ciproflaxin	Drug delivery	Superior antibacterial properties; improved physiochemical and mechanical properties; greater penetration of nanoparticles observed in the fiber	[116]
Chitosan	Chitosan + 5-fluorouracil	Drug delivery	Negative binding energy makes it energetically suitable; high drug loading capacity; reduced toxicity and increased reactivity	[117]
Chitosan	Chitosan + dexamethasone	Drug delivery	Particle size ranged from 277 to 289 nm; Drug release increased up to 8 h and was constant upto 48 h. Mild cytotoxicity was observed against L929, HCEC and RAW 264.7 cells. Effective anti-inflammatory activity against RAW macrophages	[118]
Chitosan	Chitosan + sodium alginate + polyvinyl alcohol + rosuvastatin	Drug delivery	Enhanced mechanical properties of the hydrogel film. The size of the nanoparticles ranged between 100–150 nm. Encapsulated drug was released within 24 h. High cell viability of fibroblast cells observed after 72 h of incubation	[119]
Alginate	Alginate + rifampicin/isoniazid/ pyrazinamide/ethambutol	Anti-tuberculosis drug carrier	High drug encapsulation ranging from 70 to 90%. Improved bioavailability of the drugs Promising in vivo results	[120]
Alginate	Chitosan + alginate nanoparticles + curcumin diethyl disuccinate	Drug delivery	Enhanced stability; good bioavailability; improved cellular uptake; cytotoxicity against Hep G2 cell line.	[93]
Alginate	Chitosan oligosaccharide + alginate nanoparticles + astaxanthin	Drug delivery	Encapsulation efficiency and drug loading capacity were found to be 71.3% and 6.9%. Exhibited stability in acidic, alkaline and ultraviolet light. Sustained drug release was observed. Improved bioavailability and anti-oxidant activity.	[121]
Alginate	Chitosan + alginate nanoparticles	Gene delivery	Particle size of 111 nm; no toxicity observed; transfection efficiency of 29.9%	[94]
Alginate	Chitosan + alginate nanoparticles + esculentoside	Wound healing	Enhanced healing rate; improved anti-inflammatory activity; Sustained drug release rate	[95]
Starch	Starch nanoparticles + citric acid (nanocomposite)	-	The size of the nanoparticles ranged from 50 to 100 nm. Improved storage modulus and glass transition temperature. Decrease in water vapor permeability	[122]

Table 3. Cont.

Polysaccharide	Overall Composition	Application	Key Findings of the Study	References
Starch	Starch + CG-1521	Breast cancer treatment	Slow release of the drug; Improved cytotoxicity towards MCF-7 cell line. Cell cycle arrest was induced and apoptosis was seen in MCF-7 cells	[100]
Starch	Starch nanoparticles + curcumin	Drug delivery	Enhanced encapsulation efficiency (80%) Controlled release observed	[101]
Starch	Starch nanoparticles + doxorubicin + siRNA	Cancer therapy	Low cell proliferation; enhanced cytotoxicity against A549 cell line; decreased expression of IGFR 1 protein	[102]
Starch	Starch nanoparticles + para coumaric acid	Breast cancer	Increased cytotoxicity towards MDA-MB-231 cells; burst release observed initially; enhanced encapsulation efficiency	[103]
Starch	Starch nanoparticles + triphala churna	Drug delivery system	Enhanced encapsulation efficiency Improved anti-bacterial and anti-oxidant activity; initial drug release was found to be very fast	[104]
Dextran	Dextran nanoparticles + doxorubicin	Cancer therapy	Enhanced anti-tumor effect; high cytotoxicity towards SCC7 cancer cell line; improved encapsulation efficiency	[108]
Dextran	Zirconium-89 labeled dextran nanoparticles	In vivo imaging	Enhanced tumor uptake; half-life of 3.9 h. Targets only tissue macrophages	[109]
Dextran	Dextran nanoparticles + glucose oxidase	Insulin delivery	Controlled release of insulin -90% under artificial intestinal fluid conditions; mechanism—Non-Fickian diffusion	[110]
Dextran	Dextran nanoparticles + acitretin	Treatment of psoriasis skin disease	Average size of 100 nm; sustained release of 80%. Enhanced proliferation level of keratinocytes; improved inhibition of STAT-3 phosphorylation	[112]
Dextran	Carboxymethyl dextran nanoparticles + Cy-5 labeling	Retinoblastoma	Enhanced ocular bioavailability; more affinity toward ocular tumor	[123]
Dextran	Dextran nanoparticles + Cerium oxide nanoparticles	CT contrast imaging agent	Oxidative stress protection; no toxicity observed; majority of the drug released in 24 h	[113]

4. Synthetic Biopolymeric Nanoparticles

This type of biopolymer is either obtained by modifying the natural polymers or by chemically synthesized from the monomers in such a way that they do no leave any toxic by product. It can be either obtained from renewable feedstock or from fossil fuels. They are more advantageous than natural polymers and are employed in a variety of applications because of their stability and flexibility. They also facilitate controlled release, non-immunogenic and can be easily cleared from the body. One of the disadvantages of synthetic biopolymers are that they lack cell adhesion sites and chemical modifications are required to improve their property. Some examples of synthetic biopolymers include polycaprolactone (PCL), Polylactic acid (PLA), Polyvinyl alcohol (PVA) and Polyethylene glycol (PEG), which are widely being studied for various biomedical applications. The nanoparticles synthesized out of them have improved properties such as biocompatibility, biodegradability and stability. The higher surface-to-volume ratio enables higher reactivity and a capability to easily modify the functional groups and, thereby, the governing properties [124].

4.1. Polycaprolactone Nanoparticles

Polycaprolactone (PCL) is a polymer that is biodegradable and belongs to the family of aliphatic polyesters, and is fabricated by using the polymerization technique using a monomer and an initiator. It is widely used in many biomedical applications such as tissue engineering, wound healing and drug delivery because of its favorable features such as biocompatibility, biodegradability, bioresorbability and rheological properties. PCL is also approved by the Food and Drug Administration (FDA). It is used to deliver multiple drugs and also further includes peptides, proteins and bioactive molecules for various therapeutic applications. The degradation of PCL takes about 2 to 3 years and the byproduct is also metabolized by the body [125–128]. The drugs have been encapsulated in PCL nanoparticles to improve the bioavailability, specificity and the therapeutic index [129]. One study reported the encapsulation of carboplatin in PCL nanoparticles for the purpose of intranasal delivery. The drug-loaded nanoparticles were fabricated by a double solvent evaporation method. They were in the size of 311 ± 4.7 nm. The encapsulation efficiency was found to be $27.95 \pm 4.21\%$. The drug release profile showed a biphasic pattern where there was an initial burst release followed by controlled continuous release. In vitro analysis exhibited an increased cytotoxicity activity against human glioblastoma cells—LN229 cell line. Nasal perfusion studies performed in situ in Wistar rats showed that the absorption capacity of the drug was higher in the case of an encapsulated drug rather than a free drug [130]. PCL, along with Tween 80, was fabricated into nanoparticles and used for loading the drug docetaxel for the purpose of cancer therapy. The nanoparticles were found to be spherical in shape and about 200 nm in diameter. 10% of the drug was encapsulated and nearly 35% got released in a period of 28 days. This combination showed high cellular uptake and exhibited enhanced cytotoxicity towards the C6 glioma cancer cell line [131]. Geranyl cinnamate was encapsulated in PCL nanoparticles to improve its stability and prevent it from thermal degradation. They were fabricated by solvent evaporation method and the particles were found to be spherical with a size of 177.6 nm. The drug-loaded nanoparticles showed stability for 60 days. The drug release occurs only in the presence of an external trigger, such as oil phase or an enzyme to degrade the polymer matrix [132]. Hybrid nanoparticles made of PCL and hydroxyapatite were fabricated to improve osteogenesis. Enhanced cell proliferation and differentiation was observed. A low amount of cell cytotoxicity was reported. Osteogenic markers such as Run x-2 and osteopontin were moderately expressed and sialoprotein was highly expressed after 10 days [133]. Hao et al., reported a study where PCL nanoparticles was grafted with polyethylene glycol and loaded with indocyanine green and 5-fluorouracil for the treatment of skin cancer. This system was integrated with a hyaluronic acid microneedle system. The cell proliferation of A431 and A375 was very well inhibited. The whole system showed an enhanced photothermal effect. Controlled release of the drug and its promising anti-tumor ability was reported [134]. Dorzolamide was encapsulated on to PCL nanoparticles coated with chitosan for ocular drug delivery. The size and the encapsulation efficiency of the nanoparticles were found to be 192.38 ± 6.42 nm and $72.48 \pm 5.62\%$. Drug release was found to be a biphasic patter with an initial burst release for 2 h followed by a sustained release for 12 h. Improved permeation rate and mucoadhesive behavior when compared to the control group. Histopathology analysis revealed that they were completely safe to use and did not induce any toxicity [135]. PCL nanoparticles were grafted with polyethylene glycol and were used to load the drug Cabazitaxel for the treatment of colorectal cancer. Improved bioavailability and biocompatibility were reported. Enhanced drug loading capacity, anti-tumor effect and stability were observed [136]. PCL nanoparticles were employed for the simultaneous delivery of two drugs such as Paclitaxel and IR780, for the treatment of ovarian cancer. The nanoparticles were found to have a high drug-loading capacity and the release of the drug was facilitated by the presence of light. They specifically target ovarian cancer cells and accumulated the drug in an in vivo mouse model [137].

4.2. Polylactic Acid Nanoparticles

Polylactic acid (PLA) is an FDA-approved biodegradable polymer derived from sources such as corn starch and sugarcane. It is linear and lipophilic in nature and can be obtained from the polycondensation of a monomer called lactic acid. The only degradation product, lactic acid, is either metabolized or eliminated via urine. It is widely used for biomedical applications such as tissue engineering, wound healing, implants and as drug delivery carriers. The disadvantage is that it has poor stability in heat and is very brittle. PLA nanoparticles are fabricated to encapsulate drugs or used as a filler in other polymer matrices. The nano form of PLA improves the stability and reactivity [124]. One study reported the use of PLA nanoparticles to encapsulate quercetin for therapeutic applications. The drug-loaded nanoparticles were prepared by the solvent evaporation method. The drug was loaded to improve the stability, permeation rate and solubility. The size of the particles was found to be 250 ± 68 nm and the encapsulation efficiency to be 40%. The drug release pattern was found to be initially burst followed by sustained release of the drug. The enhanced anti-oxidant activity was reported [138]. Rifampicin was loaded into PLA nanoparticles for the treatment of anti-bacterial actions. They were fabricated by nanoprecipitation method and a two phase drug release was observed. Enhanced antibiotic delivery was reported [139]. Enrique Niza et al., fabricated polyethylene imine coated PLA nanoparticles loaded with a bioactive compound called Carvacrol for enhanced anti-bacterial and anti-oxidant activity. The size and the encapsulation efficiency of the nanoparticles was found to be 100 nm and 30%. Burst release of 15% of the drug followed by sustained drug release at the end of 8 h. Enhanced anti-microbial activity and stability was reported [140]. Berberine is an anti-cancer drug that was loaded into PLA nanoparticles by using coaxial electrospray technique for sustained drug release. The size of the fabricated nanoparticles was found to be 265 nm and the encapsulation efficiency was found to be 81%. High cell cytotoxicity and cellular uptake was reported against HCT116 cell line [141]. PLA nanoparticles was used to encapsulate two drugs daunorubicin and glycyrrhizic acid for simultaneous delivery to treat leukemia. Enhanced encapsulation and loading capacity were observed. Improved drug uptake and further facilitated an increase in apoptosis rate [142]. A novel drug delivery system was designed for the treatment of cancer using PLA nanoparticles loaded with PLX4032 which is an anti-cancer drug. Enhanced loading efficiency and the cancer cells were destroyed. This theranostic device was used for the purpose of cancer treatment [143].

4.3. Poly Vinyl Alcohol Nanoparticles

Poly vinyl alcohol (PVA) is a water soluble polyhydroxy polymer that is semi-crystalline and can be obtained from polyvinyl acetate by hydrolysis reaction. They are widely employed for biomedical applications because of their properties such as low cost, compatibility with cells, highly elastic in nature and has tensile strength that matches with that of the articular cartilage. The disadvantages are that it has very less growth of osteoblast cells since it lacks self-adhesion sites [16,144,145]. PVA nanoparticles can be fabricated by techniques such as nanoprecipitation or by emulsion technique. The nanoparticles enable widely in cancer treatment by delivering the drug to the tumor site because of the leaky vessels. PVA nanoparticles aid in improving the bioavailability and the stability of the loaded drug [146]. Zinc oxide/PVA nanoparticles were fabricated by sol–gel method for the purpose of reducing the level of glucose. The nanoparticles were found to be spherical in shape and varying amounts of polyvinyl alcohol had an impact on the photocatalytic activity. The in vivo analysis also showed promising results of reduced glucose levels in rats affected with diabetes [147]. Bovine serum albumin was encapsulated in polyvinyl alcohol nanoparticles for the purpose of delivering peptides. The nanoparticles were fabricated by water in an oil emulsion technique and the diameter of the particles were found to be 675.56 nm. The encapsulation efficiency of the drug was 96.26%. The release of the protein, governed by the diffusion process, was held in a sustained manner that lasted up to 30 h. The stability of the drug was raised when it was loaded onto polymeric nanoparticles [148]. The summary of the synthetic biopolymeric nanoparticles is given in Table 4.

Table 4. Summary of synthetic bio polymeric nanoparticles.

Synthetic Biopolymer	Overall Composition	Application	Key Findings of the Study	References
Polycaprolactone	Polycaprolactone nanoparticles + carboplatin	Intra nasal delivery	Size- 311.6 ± 4.7 nm; Biphasic pattern of drug release-initial burst release followed by slow and controlled release. Cytotoxic towards human glioblastoma cell line. Better nasal absorption than free drug	[130]
Polycaprolactone	Polycaprolactone + Tween 80 + docetaxel	Cancer therapy	Enhanced cellular uptake; Improved cytotoxicity against C6 glioma cells; 35% of the drug released in 28 days.	[131]
Polycaprolactone	Polycaprolactone nanoparticles + paclitaxel	Cancer therapy	Enhanced encapsulation efficiency; the size was found to be 140 nm. Cell viability reduced against SKOV-3 cell line	[132]
Polycaprolactone	Polycaprolactone nanoparticles + α-tocopherol	-	Decrease in encapsulation efficiency, particle size when the ultrasonication time was increased.	[149]
Polycaprolactone	Polycaprolactone + hydroxyapatite	Bone tissue engineering	Enhanced cell proliferation and differentiation; Moderate expression of markers such as Runx-2 and osteopontin. High expression of sialoprotein at the end of 10 days.	[133]
Polycaprolactone	Polycaprolactone + chitosan + dorzolamide	Ocular drug delivery	Biphasic pattern of drug release; Enhanced drug permeation rate; Improved mucoadhesion; It was found to be non-cytotoxic and safe to use.	[135]
Polylactic acid	Polylactic acid + quercitrin	Therapeutic effect	Size- 250 ± 68 nm; encapsulation efficiency −40%; drug release -burst release followed by sustained release. Enhanced anti-oxidant activity.	[138]
Polylactic acid	Polylactic acid + rifampicin	Antibacterial activity	Biphasic drug release; Improved antibiotic efficiency	[139]
Polylactic acid	Polylactic acid + polyethylene imine coating + carvacrol	Anti-oxidant and Antibacterial activity	Enhanced anti-oxidant and antimicrobial activity. Improved stability rate	[140]

Table 4. *Cont.*

Synthetic Biopolymer	Overall Composition	Application	Key Findings of the Study	References
Polylactic acid	Polylactic acid + berberine	Drug delivery system	Technique: coaxial electrospray; high cellular uptake; cell cytotoxicity against HCT116 cell line; slow release profile of the drug was reported	[141]
Polylactic acid	Polylactic acid + daunorubicin + glycyrrhizic acid	Leukemia	Inhibited leukemia cells; enhanced drug uptake; improved apoptosis rate	[142]
Polyvinyl alcohol	ZnO + polyvinyl alcohol nanoparticles	Treatment of diabetes	Exhibited photocatalytic activity In vivo analysis reported lower glucose level	[147]
Polyvinyl alcohol	Bovine serum albumin + polyvinyl alcohol nanoparticles	Delivery of proteins	High drug loading capability; drug release up to 30 h controlled by diffusion process; Enhanced stability of the loaded drug	[148]

5. Fabrication of Biopolymeric Nanoparticles

The fabrication of the biopolymeric nanoparticles can be either by top-down or bottom-up approaches. The synthesis technique greatly influences the size and the poly-dispersity index of the nanoparticles. An appropriate fabrication process is chosen by considering the required features of the polymeric nanoparticles. Some of the fabrication techniques employed for biopolymeric nanoparticles, such as emulsification, precipitation, coacervation and spray deposition are discussed in the following section [150]. The fabrication strategies of biopolymeric nanoparticles are schematically represented in Figure 9.

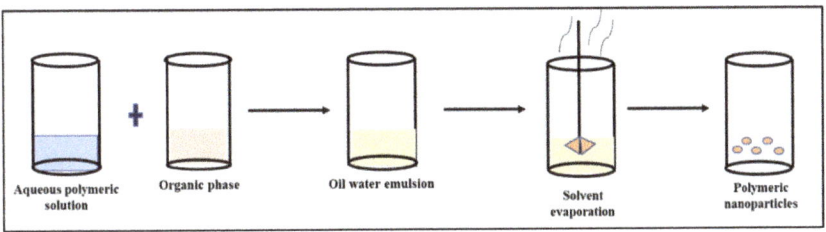

(A)

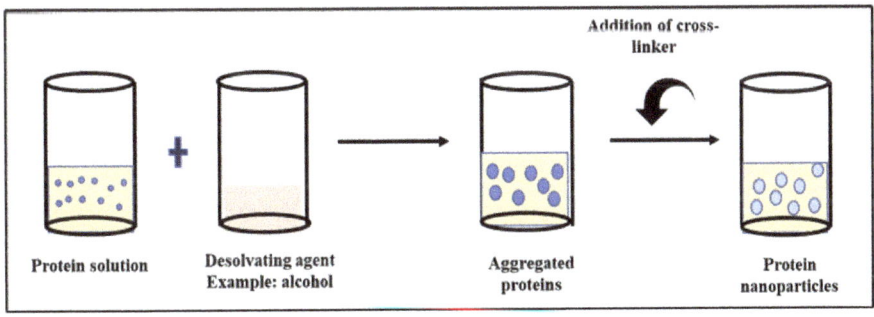

(B)

Figure 9. *Cont.*

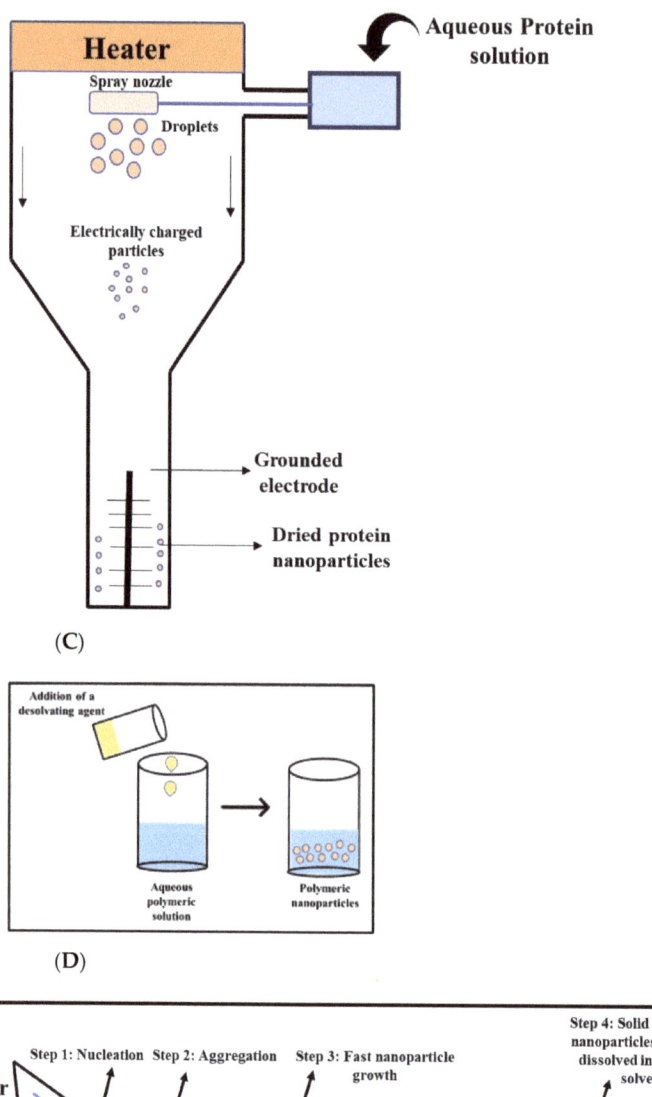

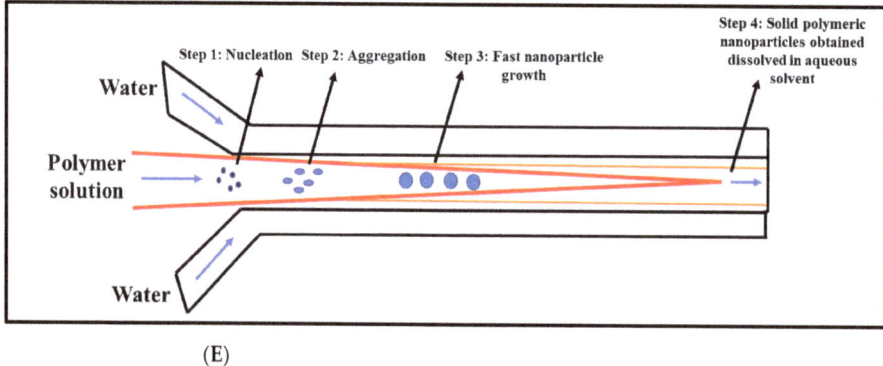

Figure 9. Schematic representation of various methods for biopolymeric nanoparticle fabrication. (**A**) Emulsification, (**B**) phase drying/coacervation, (**C**) spray drying, (**D**) desolvation, (**E**) microfluidics.

5.1. Emulsification

This method involves the formation of droplets in the nano range when the aqueous and the organic phase are mixed together in a ratio 2:1. The aqueous phase is usually made of water and a surfactant that is hydrophilic. The organic phase is made of a surfactant that is lipophilic, oils obtained from plants and a solvent that can dissolve in water. They can be either water in water (W/W) or water in oil emulsion (W/O) phase. The W/W phase is employed for fabricating hydrogel-based protein or polysaccharide nanoparticles, and an additional crosslinking, such as treatment using transglutaminase or acidification for the internal phase, can be employed. The W/O emulsion phase helps to fabricate nanoparticles that are stable with a high yield. The nanoparticles fabricated usually have high drug loading capacity and entrapment efficiency. The solvent in the organic phase can be removed by using the evaporation technique. One of the main disadvantages in this technique is employing and removing the organic solvent since the residues in the end can lead to toxicity [13,150,151].

5.2. Desolvation

Desolvation is also known as anti-solvent precipitation and is widely employed for the fabrication of biopolymeric nanoparticles from proteins as well as polysaccharides. Solute precipitation is facilitated when the quality of the solvent employed for dissolving the polymer is reduced. Factors such as pH, the concentration of the cross-linking agent (e.g., glutaraldehyde), and ionic strength can be optimized to control the size of the particles. The solvents include water, supercritical CO_2 or any organic solvent. The driving force behind the formation of nanoparticles is the imbalance in the interactions between the solute, solvent and anti-solvent. This method is highly preferred since this method does not use high-end equipment and is of low cost [2,13,150,152,153].

5.3. Coacervation

This method is similar to the phase separation technique, where there is a separation of the polymer-rich and low-polymer-content phases. The rich polymer phase, known as coacervates, is formed when oppositely charged biopolymers interact that can facilitate the encapsulation of the active ingredient. The solvents usually employed include acetone or ethanol. The fabricated nanoparticles are usually stabilized by adding cross-linking agents such as glutaraldehyde. The factors that have to be noted to control the particle size are the molecular weight and the quantity of the polymer. The main drawbacks of the method are that they have low stability and controlling the size of the biopolymeric nanoparticles is very critical [2,13,154,155].

5.4. Spray Deposition

The spray deposition method is also known as electrohydrodynamic atomization, which employs the generation of droplets that are charged as a result of the atomization process by the application of an electrical field. The nanoparticles are dried on the substrate and are strongly bonded. No particular surfactant or template is required for the process. The size of the nanoparticles is altered by the variation in the voltage supply, charge, flow rate, and the distance between the substrate and the needle. This method is highly preferred for the fabrication of biopolymeric nanoparticles, especially drug nanocrystals, since there is no alteration in the biological properties [2,11,13,156].

5.5. Microfluidics

Biopolymeric nanoparticles can be synthesized by using microfluidic technology with the aid of micro-reactors that have inner dimensions of less than 1 mm. These microreactors are similar to lab-on-chip devices and are usually made of polymers such as polydimethyl siloxane (PDMS) or glass. They can be either single-phase or multi-phase flow systems. The mechanism behind the formation of polymeric nanoparticles in the microfluidic channel is usually the self-assembly or nanoprecipitation method. The main advantages of this technique being employed for polymeric nanoparticle formation is that they have high reproducibility, low reagent requirement and enhanced control of experimental parameters. The disadvantages of the technique include the design of microfluidic channel being very complex, and there are chances that the nanoparticles can diffuse through the polymeric matrix and cause clogging in the channel [157–159].

6. Challenges and Future Perspective

Biopolymeric nanoparticles are widely employed for a wide variety of biomedical applications such as tissue engineering, drug delivery systems, imaging and sensor systems for theranostic kits. The properties such as degradability, cell compatibility, improved stiffness and strength makes them very much suitable for various applications [13]. This field is gaining high interest and is reflected in terms of publications by researchers. They are either being patented or in the process of being commercialized. For example, the product Ecosphere® from the company Ecosynthetix, in 2008, developed starch nanoparticles for adhesive purposes owing to its higher surface-to-volume ratio and improved reactivity [99]. Biopolymeric nanoparticles are used in the treatment of cancer owing to their selective tumor-targeting ability. The properties can be tuned appropriately and are supposedly the most suitable candidate for biomedical applications. The high surface-to-volume ratio enhances the molecules' association and facilitates a high drug encapsulation rate. Surface modification of the biopolymeric nanoparticles can be carried out to improve the circulation time and immunogenic properties. A more efficient drug delivery system can be designed with combined therapeutic and diagnostic for the treatment of various diseases [89]. Some polysaccharide and protein-based biopolymers, such as alginate and bovine serum albumin, have mucoadhesive nature and the small size makes penetration to the target size easier [160]. Focusing on this direction helps to bring in various technological advances in the biomedical sector. One of the main challenges to employing these for biomedical applications is nanoparticle toxicity. There are no standard assessment methods for nanoparticle toxicity. The nanoparticles can accumulate over time in the system and cause side effects. The toxicity differs with the dose and the time of exposure. Though multiple products exist in the market containing nanoparticles, a scientific gap exists since there are no strict regulations. Thus, proper regulatory measures are required when nanoparticles are being dealt with for medical applications. Upscaling the technology or commercialization also plays a key role and remains to be a challenge. Currently, researchers are highly focused on biopolymeric nanoparticles to be employed for biomedical applications with improved efficiency and reduced toxicity [4,13,161,162].

7. Conclusions

The use of biopolymeric nanoparticles has proven to be economical, environmentally friendly and promising in the technical aspect for a wide range of applications, especially in the medical domain. A lot of research work is going on employing protein, polysaccharide and synthetic-based biopolymer systems owing to their positive features such as biocompatibility and biodegradability. Nanotechnology is highly blooming in the 21st century and nanoparticles have the innate ability to be modified according to the required application. Biopolymeric nanoparticles are found to be highly stable and show improved biocompatibility, degradation rate and surface reactivity. It is very critical and important to produce biopolymeric nanoparticles of favorable size and properties to be employed in fabricating novel drug delivery systems for sustained drug release. The choice of the nanoparticle depends on the application and the properties can be tuned according to the intended application. Surface modification of the biopolymeric nanoparticles aids in the enhancement of the circulation time and prevents immunogenic reactions. This review focused on the various biopolymeric nanoparticles fabricated for biomedical applications such as drug delivery, imaging and tissue engineering. The important fabrication techniques, along with the challenges and the future perspective in this domain, were also discussed. The initial stage for the development of the biopolymeric nanoparticles requires expensive instruments and up-scaling the technology is also challenging. Thus, future researchers should focus on this and on ways to make sure that the nanoparticles do not induce bioaccumulation in the human system. It is also necessary to develop nanoparticles with enhanced efficacy. A deep and clear understanding of nanoparticle–immune system interaction and the elimination from the human system is an important concern and must be addressed in the future.

Author Contributions: Conceptualization: R.S. and A.J.N.; writing—original draft preparation: R.S.; writing—review and editing: R.S. and A.J.N.; supervision: A.J.N. All authors have read and agreed to the published version of the manuscript.

Funding: A.J.N. would like to acknowledge the financial support from the Department of Biotechnology, Government of India, through Ramalingaswami Re-entry fellowship (D.O. No. BT/HRD/35/02/2006) and VIT seed grant (2021–2022) (SG20210234).

Institutional Review Board Statement: Not applicable.

Informed Consent Statement: Not applicable.

Data Availability Statement: Data sharing is not available.

Acknowledgments: A.J.N. would like to acknowledge the financial support from the Department of Biotechnology, Government of India, through Ramalingaswami Re-entry fellowship (D.O. No. BT/HRD/35/02/2006) and VIT seed grant (2021–2022) (SG20210234).

Conflicts of Interest: The authors declare no conflict of interest.

References

1. Silva, G.A. Introduction to nanotechnology and its applications to medicine. *Surg. Neurol.* **2004**, *61*, 216–220. [CrossRef] [PubMed]
2. Sundar, S.; Kundu, J.; Kundu, S.C. Biopolymeric nanoparticles. *Sci. Technol. Adv. Mater.* **2010**, *11*, 014104. [CrossRef] [PubMed]
3. Couvreur, P. Nanoparticles in drug delivery: Past, present and future. *Adv. Drug Deliv. Rev.* **2013**, *65*, 21–23. [CrossRef]
4. Hasan, A.; Morshed, M.; Memic, A.; Hassan, S.; Webster, T.J.; Marei, H. Nanoparticles in tissue engineering: Applications, challenges and prospects. *Int. J. Nanomed.* **2018**, *13*, 5637–5655. [CrossRef]
5. Luo, X.; Morrin, A.; Killard, A.; Smyth, M.R. Application of Nanoparticles in Electrochemical Sensors and Biosensors. *Electroanalysis* **2006**, *18*, 319–326. [CrossRef]
6. Tian, H.; Chen, J.; Chen, X. Nanoparticles for Gene Delivery. *Small* **2013**, *9*, 2034–2044. [CrossRef]
7. Kolosnjaj-Tabi, J.; Wilhelm, C.; Clément, O.; Gazeau, F. Cell labeling with magnetic nanoparticles: Opportunity for magnetic cell imaging and cell manipulation. *J. Nanobiotechnology* **2013**, *11*, S7. [CrossRef]
8. Bhirde, A.; Xie, J.; Swierczewska, M.; Chen, X. Nanoparticles for cell labeling. *Nanoscale* **2011**, *3*, 142–153. [CrossRef]
9. Issa, B.; Obaidat, I.M.; Albiss, B.A.; Haik, Y. Magnetic Nanoparticles: Surface Effects and Properties Related to Biomedicine Applications. *Int. J. Mol. Sci.* **2013**, *14*, 21266–21305. [CrossRef]
10. Nitta, S.K.; Numata, K. Biopolymer-Based Nanoparticles for Drug/Gene Delivery and Tissue Engineering. *Int. J. Mol. Sci.* **2013**, *14*, 1629–1654. [CrossRef] [PubMed]
11. Vodyashkin, A.A.; Kezimana, P.; Vetcher, A.A.; Stanishevskiy, Y.M. Biopolymeric Nanoparticles–Multifunctional Materials of the Future. *Polymers* **2022**, *14*, 2287. [CrossRef]
12. Yadav, P.; Yadav, H.; Shah, V.G.; Shah, G.; Dhaka, G. Biomedical Biopolymers, their Origin and Evolution in Biomedical Sciences: A Systematic Review. *J. Clin. Diagn. Res.* **2015**, *9*, ZE21. [CrossRef]
13. Verma, M.L.; Dhanya, B.; Sukriti; Rani, V.; Thakur, M.; Jeslin, J.; Kushwaha, R. Carbohydrate and protein based biopolymeric nanoparticles: Current status and biotechnological applications. *Int. J. Biol. Macromol.* **2020**, *154*, 390–412. [CrossRef] [PubMed]
14. Elzoghby, A.O.; Samy, W.M.; Elgindy, N.A. Protein-based nanocarriers as promising drug and gene delivery systems. *J. Control. Release* **2012**, *161*, 38–49. [CrossRef]
15. Mizrahi, S.; Peer, D. Polysaccharides as building blocks for nanotherapeutics. *Chem. Soc. Rev.* **2012**, *41*, 2623–2640. [CrossRef] [PubMed]
16. Reddy, M.S.B.; Ponnamma, D.; Choudhary, R.; Sadasivuni, K.K. A Comparative Review of Natural and Synthetic Biopolymer Composite Scaffolds. *Polymers* **2021**, *13*, 1105. [CrossRef]
17. Yoha, K.S.; Priyadarshini, S.R.; Moses, J.A.; Anandharamakrishnan, C. Surface Modification of Bio-polymeric Nanoparticles and Its Applications. In *Advanced Structured Materials*; Springer: Berlin/Heidelberg, Germany, 2020; pp. 261–282. [CrossRef]
18. Meng, R.; Zhu, H.; Wang, Z.; Hao, S.; Wang, B. Preparation of Drug-Loaded Albumin Nanoparticles and Its Application in Cancer Therapy. *J. Nanomater.* **2022**, *2022*, 3052175. [CrossRef]
19. Pulimood, T.B.; Park, G.R. Debate: Albumin administration should be avoided in the critically ill. *Crit. Care* **2000**, *4*, 1–5. [CrossRef]

20. Sergi, R.; Bellucci, D.; Cannillo, V. A Review of Bioactive Glass/Natural Polymer Composites: State of the Art. *Materials* **2020**, *13*, 5560. [CrossRef]
21. Liu, J.; Ge, X.; Liu, L.; Xu, W.; Shao, R. Challenges and opportunities of silk protein hydrogels in biomedical applications. *Mater. Adv.* **2022**, *3*, 2291–2308. [CrossRef]
22. Dong, C.; Lv, Y. Application of Collagen Scaffold in Tissue Engineering: Recent Advances and New Perspectives. *Polymers* **2016**, *8*, 42. [CrossRef] [PubMed]
23. Garg, U.; Chauhan, S.; Nagaich, U.; Jain, N. Current Advances in Chitosan Nanoparticles Based Drug Delivery and Targeting. *Adv. Pharm. Bull.* **2019**, *9*, 195–204. [CrossRef] [PubMed]
24. Gheorghita Puscaselu, R.; Lobiuc, A.; Dimian, M.; Covasa, M. Alginate: From Food Industry to Biomedical Applications and Management of Metabolic Disorders. *Polymers* **2020**, *12*, 2417. [CrossRef]
25. Zarski, A.; Bajer, K.; Kapuśniak, J. Review of the Most Important Methods of Improving the Processing Properties of Starch toward Non-Food Applications. *Polymers* **2021**, *13*, 832. [CrossRef]
26. Das, M.; Shukla, F.; Thakore, S. Carbohydrate-derived functionalized nanomaterials for drug delivery and environment remediation. In *Handbook of Functionalized Nanomaterials*; Elsevier: Amsterdam, The Netherlands, 2021; pp. 339–364. [CrossRef]
27. Varghese, S.A.; Rangappa, S.M.; Siengchin, S.; Parameswaranpillai, J. Natural polymers and the hydrogels prepared from them. In *Hydrogels Based on Natural Polymers*; Elsevier: Amsterdam, The Netherlands, 2020; pp. 17–47. [CrossRef]
28. Gaaz, T.S.; Sulong, A.B.; Akhtar, M.N.; Kadhum, A.A.H.; Mohamad, A.B.; Al-Amiery, A.A. Properties and Applications of Polyvinyl Alcohol, Halloysite Nanotubes and Their Nanocomposites. *Molecules* **2015**, *20*, 22833–22847. [CrossRef] [PubMed]
29. Jain, N.; Singh, V.K.; Chauhan, S. A review on mechanical and water absorption properties of polyvinyl alcohol based composites/films. *J. Mech. Behav. Mater.* **2017**, *26*, 213–222. [CrossRef]
30. Casalini, T.; Rossi, F.; Castrovinci, A.; Perale, G. A Perspective on Polylactic Acid-Based Polymers Use for Nanoparticles Synthesis and Applications. *Front. Bioeng. Biotechnol.* **2019**, *7*, 259. [CrossRef] [PubMed]
31. Wong, K.H.; Lu, A.; Chen, X.; Yang, Z. Natural Ingredient-Based Polymeric Nanoparticles for Cancer Treatment. *Molecules* **2020**, *25*, 3620. [CrossRef]
32. DeFrates, K.; Markiewicz, T.; Gallo, P.; Rack, A.; Weyhmiller, A.; Jarmusik, B.; Hu, X. Protein Polymer-Based Nanoparticles: Fabrication and Medical Applications. *Int. J. Mol. Sci.* **2018**, *19*, 1717. [CrossRef]
33. Mahmoudi, M.; Lynch, I.; Ejtehadi, M.R.; Monopoli, M.P.; Bombelli, F.B.; Laurent, S. Protein−Nanoparticle Interactions: Opportunities and Challenges. *Chem. Rev.* **2011**, *111*, 5610–5637. [CrossRef]
34. Stein, N.C.; Mulac, D.; Fabian, J.; Herrmann, F.C.; Langer, K. Nanoparticle albumin-bound mTHPC for photodynamic therapy: Preparation and comprehensive characterization of a promising drug delivery system. *Int. J. Pharm.* **2020**, *582*, 119347. [CrossRef] [PubMed]
35. Hornok, V. Serum Albumin Nanoparticles: Problems and Prospects. *Polymers* **2021**, *13*, 3759. [CrossRef] [PubMed]
36. Yang, Z.; Du, Y.; Lei, L.; Xia, X.; Wang, X.; Tong, F.; Li, Y.; Gao, H. Co-delivery of ibrutinib and hydroxychloroquine by albumin nanoparticles for enhanced chemotherapy of glioma. *Int. J. Pharm.* **2023**, *630*, 122436. [CrossRef]
37. Lin, P.; Zhang, W.; Chen, D.; Yang, Y.; Sun, T.; Chen, H.; Zhang, J. Electrospun nanofibers containing chitosan-stabilized bovine serum albumin nanoparticles for bone regeneration. *Colloids Surf. B Biointerfaces* **2022**, *217*, 112680. [CrossRef] [PubMed]
38. Thangavel, K.; Lakshmikuttyamma, A.; Thangavel, C.; Shoyele, S.A. CD44-targeted, indocyanine green-paclitaxel-loaded human serum albumin nanoparticles for potential image-guided drug delivery. *Colloids Surf. B Biointerfaces* **2022**, *209*, 112162. [CrossRef]
39. Khella, K.F.; El Maksoud, A.I.A.; Hassan, A.; Abdel-Ghany, S.E.; Elsanhoty, R.M.; Aladhadh, M.A.; Abdel-Hakeem, M.A. Carnosic Acid Encapsulated in Albumin Nanoparticles Induces Apoptosis in Breast and Colorectal Cancer Cells. *Molecules* **2022**, *27*, 4102. [CrossRef] [PubMed]
40. El Tokhy, S.S.; Elgizawy, S.A.; Osman, M.A.; Goda, A.E.; Unsworth, L.D. Tailoring dexamethasone loaded albumin nanoparticles: A full factorial design with enhanced anti-inflammatory activity In vivo. *J. Drug Deliv. Sci. Technol.* **2022**, *72*, 103411. [CrossRef]
41. Yasmin, R.; Shah, M.; Khan, S.A.; Ali, R. Gelatin nanoparticles: A potential candidate for medical applications. *Nanotechnol. Rev.* **2017**, *6*, 191–207. [CrossRef]
42. Elzoghby, A.O. Gelatin-based nanoparticles as drug and gene delivery systems: Reviewing three decades of research. *J. Control. Release* **2013**, *172*, 1075–1091. [CrossRef]
43. Amjadi, S.; Hamishehkar, H.; Ghorbani, M. A novel smart PEGylated gelatin nanoparticle for co-delivery of doxorubicin and betanin: A strategy for enhancing the therapeutic efficacy of chemotherapy. *Mater. Sci. Eng. C Mater. Biol. Appl.* **2019**, *97*, 833–841. [CrossRef]
44. Yang, X.-J.; Wang, F.-Q.; Lu, C.-B.; Zou, J.-W.; Hu, J.-B.; Yang, Z.; Sang, H.-X.; Zhang, Y. Modulation of bone formation and resorption using a novel zoledronic acid loaded gelatin nanoparticles integrated porous titanium scaffold: An in vitro and in vivo study. *Biomed. Mater.* **2020**, *15*, 055013. [CrossRef]
45. El-Sayed, N.; Trouillet, V.; Clasen, A.; Jung, G.; Hollemeyer, K.; Schneider, M. NIR-Emitting Gold Nanoclusters–Modified Gelatin Nanoparticles as a Bioimaging Agent in Tissue. *Adv. Healthc. Mater.* **2019**, *8*, 1900993. [CrossRef] [PubMed]
46. Ahmad, A.; Ansari, M.; Mishra, R.K.; Kumar, A.; Vyawahare, A.; Verma, R.K.; Raza, S.S.; Khan, R. Enteric-coated gelatin nanoparticles mediated oral delivery of 5-aminosalicylic acid alleviates severity of DSS-induced ulcerative colitis. *Mater. Sci. Eng. C* **2021**, *119*, 111582. [CrossRef]

47. Zhao, Z.; Chen, A.; Li, Y.; Hu, J.; Liu, X.; Li, J.; Zhang, Y.; Li, G.; Zheng, Z. Fabrication of silk fibroin nanoparticles for controlled drug delivery. *J. Nanoparticle Res.* **2012**, *14*, 736. [CrossRef]
48. Lujerdean, C.; Baci, G.-M.; Cucu, A.-A.; Dezmirean, D.S. The Contribution of Silk Fibroin in Biomedical Engineering. *Insects* **2022**, *13*, 286. [CrossRef] [PubMed]
49. Kundu, J.; Chung, Y.-I.; Kim, Y.H.; Tae, G.; Kundu, S.C. Silk fibroin nanoparticles for cellular uptake and control release. *Int. J. Pharm.* **2010**, *388*, 242–250. [CrossRef]
50. Mathur, A.B.; Gupta, V. Silk fibroin-derived nanoparticles for biomedical applications. *Nanomedicine* **2010**, *5*, 807–820. [CrossRef] [PubMed]
51. Xu, Z.; Shi, L.; Yang, M.; Zhu, L. Preparation and biomedical applications of silk fibroin-nanoparticles composites with enhanced properties—A review. *Mater. Sci. Eng. C* **2019**, *95*, 302–311. [CrossRef]
52. Montalbán, M.G.; Coburn, J.M.; Lozano-Pérez, A.A.; Cenis, J.L.; Víllora, G.; Kaplan, D.L. Production of Curcumin-Loaded Silk Fibroin Nanoparticles for Cancer Therapy. *Nanomaterials* **2018**, *8*, 126. [CrossRef]
53. Shen, Y.; Wang, X.; Li, B.; Guo, Y.; Dong, K. Development of silk fibroin-sodium alginate scaffold loaded silk fibroin nanoparticles for hemostasis and cell adhesion. *Int. J. Biol. Macromol.* **2022**, *211*, 514–523. [CrossRef]
54. Fatemeh, M.; Hamidreza, M.; Ali, S.M.; Shahrokh, S.; Mehdi, F. Fabri-cation of Silk Scaffold Containing Simvastatin-Loaded Silk Fibroin Nanoparticles for Regenerating Bone Defects. *Iran. Biomed. J.* **2022**, *26*, 116–123.
55. Rahmani, H.; Fattahi, A.; Sadrjavadi, K.; Khaledian, S.; Shokoohinia, Y. Preparation and Characterization of Silk Fibroin Nanoparticles as a Potential Drug Delivery System for 5-Fluorouracil. *Adv. Pharm. Bull.* **2019**, *9*, 601–608. [CrossRef]
56. Li, Z.; Cheng, G.; Zhang, Q.; Wu, W.; Zhang, Y.; Wu, B.; Liu, Z.; Tong, X.; Xiao, B.; Cheng, L.; et al. PX478-loaded silk fibroin nanoparticles reverse multidrug resistance by inhibiting the hypoxia-inducible factor. *Int. J. Biol. Macromol.* **2022**, *222*, 2309–2317. [CrossRef]
57. Nidhin, M.; Vedhanayagam, M.; Sangeetha, S.; Kiran, M.S.; Nazeer, S.S.; Jayasree, R.S.; Sreeram, K.J.; Nair, B.U. Fluorescent nanonetworks: A novel bioalley for collagen scaffolds and Tissue Engineering. *Sci. Rep.* **2014**, *4*, 1–10. [CrossRef] [PubMed]
58. Liu, C.Z.; Czernuszka, J.T. Development of biodegradable scaffolds for tissue engineering: A perspective on emerging technology. *Mater. Sci. Technol.* **2007**, *23*, 379–391. [CrossRef]
59. Gómez-Guillén, M.; Giménez, B.; López-Caballero, M.; Montero, M. Functional and bioactive properties of collagen and gelatin from alternative sources: A review. *Food Hydrocoll.* **2011**, *25*, 1813–1827. [CrossRef]
60. Grigore, M.E. Hydrogels for Cardiac Tissue Repair and Regeneration. *J. Cardiovasc. Med. Cardiol.* **2017**, *4*, 049–057. [CrossRef]
61. Lo, S.; Fauzi, M. Current Update of Collagen Nanomaterials—Fabrication, Characterisation and Its Applications: A Review. *Pharmaceutics* **2021**, *13*, 316. [CrossRef]
62. Arun, A.; Malrautu, P.; Laha, A.; Luo, H.; Ramakrishna, S. Collagen Nanoparticles in Drug Delivery Systems and Tissue Engineering. *Appl. Sci.* **2021**, *11*, 11369. [CrossRef]
63. Alarcon, E.I.; Udekwu, K.; Skog, M.; Pacioni, N.L.; Stamplecoskie, K.G.; González-Béjar, M.; Polisetti, N.; Wickham, A.; Richter-Dahlfors, A.; Griffith, M.; et al. The biocompatibility and antibacterial properties of collagen-stabilized, photochemically prepared silver nanoparticles. *Biomaterials* **2012**, *33*, 4947–4956. [CrossRef]
64. Nicklas, M.; Schatton, W.; Heinemann, S.; Hanke, T.; Kreuter, J. Preparation and characterization of marine sponge collagen nanoparticles and employment for the transdermal delivery of 17β-estradiol-hemihydrate. *Drug Dev. Ind. Pharm.* **2009**, *35*, 1035–1042. [CrossRef] [PubMed]
65. Foster, J.A.; Mecham, R.; Imberman, M.; Faris, B.; Franzblau, C. A High Molecular Weight Species of Soluble Elastin-Proelastin. *Adv. Exp. Med. Biol.* **1977**, *79*, 351–369. [CrossRef] [PubMed]
66. Seyoung, H.; Wook, C.D.; Nam, K.H.; Gwon, P.C. Protein-Based Nanoparticles as Drug Delivery Sys-tems. *Pharmaceutics* **2020**, *12*, 604.
67. Almine, J.F.; Bax, D.V.; Mithieux, S.M.; Nivison-Smith, L.; Rnjak, J.; Waterhouse, A.; Wise, S.G.; Weiss, A.S. Elastin-based materials. *Chem. Soc. Rev.* **2010**, *39*, 3371–3379. [CrossRef]
68. Wu, Y.; Mackay, J.A.; McDaniel, J.R.; Chilkoti, A.; Clark, R.L. Fabrication of Elastin-Like Polypeptide Nanoparticles for Drug Delivery by Electrospraying. *Biomacromolecules* **2008**, *10*, 19–24. [CrossRef] [PubMed]
69. Machado, R.; Bessa, P.C.; Reis, R.L.; Rodriguez-Cabello, J.C.; Casal, M. Elastin-Based Nanoparticles for Delivery of Bone Morphogenetic Proteins. In *Nanoparticles in Biology and Medicine*; Humana Press: Totowa, NJ, USA, 2012; pp. 353–363. [CrossRef]
70. Kim, J.D.; Jung, Y.J.; Woo, C.H.; Choi, Y.C.; Choi, J.S.; Cho, Y.W. Thermo-responsive human α-elastin self-assembled nanoparticles for protein delivery. *Colloids Surf. B Biointerfaces* **2017**, *149*, 122–129. [CrossRef] [PubMed]
71. Hanafy, N.A.; El-Kemary, M.A. Silymarin/curcumin loaded albumin nanoparticles coated by chitosan as muco-inhalable delivery system observing anti-inflammatory and anti COVID-19 characterizations in oleic acid triggered lung injury and in vitro COVID-19 experiment. *Int. J. Biol. Macromol.* **2022**, *198*, 101–110. [CrossRef]
72. Liu, F.; Xue, L.; Xu, L.; Liu, J.; Xie, C.; Chen, C.; Liu, Y. Preparation and characterization of bovine serum albumin nanoparticles modified by Poly-l-lysine functionalized graphene oxide for BMP-2 delivery. *Mater. Des.* **2022**, *215*, 110479. [CrossRef]
73. Vaghasiya, K.; Ray, E.; Singh, R.; Jadhav, K.; Sharma, A.; Khan, R.; Katare, O.P.; Verma, R.K. Efficient, enzyme responsive and tumor receptor targeting gelatin nanoparticles decorated with concanavalin-A for site-specific and controlled drug delivery for cancer therapy. *Mater. Sci. Eng. C* **2021**, *123*, 112027. [CrossRef]

74. Xue, Y.; Lee, J.; Kim, H.-J.; Cho, H.-J.; Zhou, X.; Liu, Y.; Tebon, P.; Hoffman, T.; Qu, M.; Ling, H.; et al. Rhodamine Conjugated Gelatin Methacryloyl Nanoparticles for Stable Cell Imaging. *ACS Appl. Bio Mater.* **2020**, *3*, 6908–6918. [CrossRef]
75. Ansari, M.; Ahmad, A.; Kumar, A.; Alam, P.; Khan, T.H.; Jayamurugan, G.; Raza, S.S.; Khan, R. Aminocellulose-grafted-polycaprolactone coated gelatin nanoparticles alleviate inflammation in rheumatoid arthritis: A combinational therapeutic approach. *Carbohydr. Polym.* **2021**, *258*, 117600. [CrossRef]
76. Chen, X.; Zou, J.; Zhang, K.; Zhu, J.; Zhang, Y.; Zhu, Z.; Zheng, H.; Li, F.; Piao, J.-G. Photothermal/matrix metalloproteinase-2 dual-responsive gelatin nanoparticles for breast cancer treatment. *Acta Pharm. Sin. B* **2021**, *11*, 271–282. [CrossRef]
77. Moin, A.; Wani, S.U.D.; Osmani, R.A.; Abu Lila, A.S.; Khafagy, E.-S.; Arab, H.H.; Gangadharappa, H.V.; Allam, A.N. Formulation, characterization, and cellular toxicity assessment of tamoxifen-loaded silk fibroin nanoparticles in breast cancer. *Drug Deliv.* **2021**, *28*, 1626–1636. [CrossRef]
78. Vallejo, R.; Gonzalez-Valdivieso, J.; Santos, M.; Rodriguez-Rojo, S.; Arias, F. Production of elastin-like recombinamer-based nanoparticles for docetaxel encapsulation and use as smart drug-delivery systems using a supercritical anti-solvent process. *J. Ind. Eng. Chem.* **2021**, *93*, 361–374. [CrossRef]
79. Roldo, M.; Hornof, M.; Caliceti, P.; Bernkop-Schnürch, A. Mucoadhesive thiolated chitosans as platforms for oral controlled drug delivery: Synthesis and in vitro evaluation. *Eur. J. Pharm. Biopharm.* **2004**, *57*, 115–121. [CrossRef]
80. Mohammed, M.A.; Syeda, J.T.M.; Wasan, K.M.; Wasan, E.K. An Overview of Chitosan Nanoparticles and Its Application in Non-Parenteral Drug Delivery. *Pharmaceutics* **2017**, *9*, 53. [CrossRef] [PubMed]
81. Sharifi-Rad, J.; Quispe, C.; Butnariu, M.; Rotariu, L.S.; Sytar, O.; Sestito, S.; Rapposelli, S.; Akram, M.; Iqbal, M.; Krishna, A.; et al. Chitosan nanoparticles as a promising tool in nanomedicine with particular emphasis on oncological treatment. *Cancer Cell Int.* **2021**, *21*, 1–21. [CrossRef]
82. Yanat, M.; Schroën, K. Preparation methods and applications of chitosan nanoparticles; with an outlook toward reinforcement of biodegradable packaging. *React. Funct. Polym.* **2021**, *161*, 104849. [CrossRef]
83. Nagpal, K.; Singh, S.K.; Mishra, D.N. Chitosan Nanoparticles: A Promising System in Novel Drug Delivery. *Chem. Pharm. Bull.* **2010**, *58*, 1423–1430. [CrossRef] [PubMed]
84. Dev, A.; Binulal, N.; Anitha, A.; Nair, S.; Furuike, T.; Tamura, H.; Jayakumar, R. Preparation of poly(lactic acid)/chitosan nanoparticles for anti-HIV drug delivery applications. *Carbohydr. Polym.* **2010**, *80*, 833–838. [CrossRef]
85. Li, P.; Wang, Y.; Peng, Z.; She, F.; Kong, L. Development of chitosan nanoparticles as drug delivery systems for 5-fluorouracil and leucovorin blends. *Carbohydr. Polym.* **2011**, *85*, 698–704. [CrossRef]
86. Li, Y.; Liu, Y.; Guo, Q. Silk fibroin hydrogel scaffolds incorporated with chitosan nanoparticles repair articular cartilage defects by regulating TGF-β1 and BMP-2. *Arthritis Res. Therapy.* **2021**, *23*, 50. [CrossRef]
87. Zahiri, M.; Khanmohammadi, M.; Goodarzi, A.; Ababzadeh, S.; Farahani, M.S.; Mohandesnezhad, S.; Bahrami, N.; Nabipour, I.; Ai, J. Encapsulation of curcumin loaded chitosan nanoparticle within poly (ε-caprolactone) and gelatin fiber mat for wound healing and layered dermal reconstitution. *Int. J. Biol. Macromol.* **2020**, *153*, 1241–1250. [CrossRef] [PubMed]
88. Severino, P.; Da Silva, C.F.; Andrade, L.N.; de Lima Oliveira, D.; Campos, J.; Souto, E.B. Alginate Nanoparticles for Drug Delivery and Targeting. *Curr. Pharm. Des.* **2019**, *25*, 1312–1334. [CrossRef]
89. Almutairi, F.M. Biopolymer Nanoparticles: A Review of Prospects for Application as Carrier for Therapeutics and Diagnostics. *Int. J. Pharm. Res. Allied Sci.* **2019**, *8*, 25–35.
90. Pawar, S.N.; Edgar, K.J. Alginate derivatization: A review of chemistry, properties and applications. *Biomaterials* **2012**, *33*, 3279–3305. [CrossRef]
91. Oliveira, D.M.L.; Rezende, P.S.; Barbosa, T.C.; Andrade, L.N.; Bani, C.; Tavares, D.S.; da Silva, C.F.; Chaud, M.V.; Padilha, F.; Cano, A.; et al. Double membrane based on lidocaine-coated polymyxin-alginate nanoparticles for wound healing: In vitro characterization and in vivo tissue repair. *Int. J. Pharm.* **2020**, *591*, 120001. [CrossRef] [PubMed]
92. Li, P.; Dai, Y.N.; Wang, A.-Q.; Wei, Q. Chitosan-Alginate Nanoparticles as a Novel Drug Delivery System for Nifedi-pine. *Int. J. Biomed. Sci.* **2008**, *4*, 221. [PubMed]
93. Sorasitthiyanukarn, F.N.; Muangnoi, C.; Rojsitthisak, P.; Rojsitthisak, P. Chitosan-alginate nanoparticles as effective oral carriers to improve the stability, bioavailability, and cytotoxicity of curcumin diethyl disuccinate. *Carbohydr. Polym.* **2021**, *256*, 117426. [CrossRef]
94. Zohri, M.; Arefian, E.; Javar, H.A.; Gazori, T.; Aghaee-Bakhtiari, S.H.; Taheri, M.; Fatahi, Y.; Azadi, A.; Khoshayand, M.R.; Ghahremani, M.H. Potential of chitosan/alginate nanoparticles as a non-viral vector for gene delivery: Formulation and optimization using D-optimal design. *Mater. Sci. Eng. C* **2021**, *128*, 112262. [CrossRef]
95. Zhu, Z.; He, F.; Shao, H.; Shao, J.; Li, Q.; Wang, X.; Ren, H.; You, C.; Zhang, Z.; Han, C. Chitosan/Alginate Nanoparticles with Sustained Release of Esculentoside A for Burn Wound Healing. *ACS Appl. Nano Mater.* **2023**, *6*, 573–587. [CrossRef]
96. Caldonazo, A.; Almeida, S.L.; Bonetti, A.F.; Lazo, R.E.L.; Mengarda, M.; Murakami, F.S. Pharmaceutical applications of starch nanoparticles: A scoping review. *Int. J. Biol. Macromol.* **2021**, *181*, 697–704. [CrossRef]
97. Campelo, P.H.; Sant'Ana, A.S.; Clerici, M.T.P.S. Starch nanoparticles: Production methods, structure, and properties for food applications. *Curr. Opin. Food Sci.* **2020**, *33*, 136–140. [CrossRef]
98. Le Corre, D.; Angellier-Coussy, H. Preparation and application of starch nanoparticles for nanocomposites: A review. *React. Funct. Polym.* **2014**, *85*, 97–120. [CrossRef]
99. Le Corre, D.; Bras, J.; Dufresne, A. Starch Nanoparticles: A Review. *Biomacromolecules* **2010**, *11*, 1139–1153. [CrossRef]

100. Alp, E.; Damkaci, F.; Güven, E.; Tenniswood, M. Starch nanoparticles for delivery of the histone deacetylase inhibitor CG-1521 in breast cancer treatment. *Int. J. Nanomed.* **2019**, *14*, 1335–1346. [CrossRef] [PubMed]
101. Acevedo-Guevara, L.; Nieto-Suaza, L.; Sanchez, L.T.; Pinzon, M.I.; Villa, C.C. Development of native and modified banana starch nanoparticles as vehicles for curcumin. *Int. J. Biol. Macromol.* **2018**, *111*, 498–504. [CrossRef] [PubMed]
102. Li, L.; He, S.; Yu, L.; Elshazly, E.H.; Wang, H.; Chen, K.; Zhang, S.; Ke, L.; Gong, R. Codelivery of DOX and siRNA by folate-biotin-quaternized starch nanoparticles for promoting synergistic suppression of human lung cancer cells. *Drug Deliv.* **2019**, *26*, 499–508. [CrossRef]
103. Mariadoss, A.V.A.; Saravanakumar, K.; Sathiyaseelan, A.; Karthikkumar, V.; Wang, M.-H. Smart drug delivery of p-Coumaric acid loaded aptamer conjugated starch nanoparticles for effective triple-negative breast cancer therapy. *Int. J. Biol. Macromol.* **2022**, *195*, 22–29. [CrossRef]
104. Nallasamy, P.; Ramalingam, T.; Nooruddin, T.; Shanmuganathan, R.; Arivalagan, P.; Natarajan, S. Polyherbal drug loaded starch nanoparticles as promising drug delivery system: Antimicrobial, antibiofilm and neuroprotective studies. *Process. Biochem.* **2020**, *92*, 355–364. [CrossRef]
105. Majcher, M.J.; McInnis, C.L.; Himbert, S.; Alsop, R.J.; Kinio, D.; Bleuel, M.; Rheinstadter, M.; Smeets, N.M.; Hoare, T. Photopolymerized Starchstarch Nanoparticle (SNP) network hydrogels. *Carbohydr. Polym.* **2020**, *236*, 115998. [CrossRef]
106. Díaz-Montes, E. Dextran: Sources, Structures, and Properties. *Polysaccharides* **2021**, *2*, 554–565. [CrossRef]
107. Wasiak, I.; Kulikowska, A.; Janczewska, M.; Michalak, M.; Cymerman, I.A.; Nagalski, A.; Kallinger, P.; Szymanski, W.W.; Ciach, T. Dextran Nanoparticle Synthesis and Properties. *PLoS ONE* **2016**, *11*, e0146237. [CrossRef]
108. Thambi, T.; Gil You, D.; Han, H.S.; Deepagan, V.G.; Jeon, S.M.; Suh, Y.D.; Choi, K.Y.; Kim, K.; Kwon, I.C.; Yi, G.-R.; et al. Bioreducible Carboxymethyl Dextran Nanoparticles for Tumor-Targeted Drug Delivery. *Adv. Heal. Mater.* **2014**, *3*, 1829–1838. [CrossRef] [PubMed]
109. Keliher, E.J.; Yoo, J.; Nahrendorf, M.; Lewis, J.S.; Marinelli, B.; Newton, A.; Pittet, M.J.; Weissleder, R. [89]Zr-Labeled Dextran Nanoparticles Allow in Vivo Macrophage Imaging. *Bioconjugate Chem.* **2011**, *22*, 2383–2389. [CrossRef] [PubMed]
110. Jamwal, S.; Ram, B.; Ranote, S.; Dharela, R.; Chauhan, G.S. New glucose oxidase-immobilized stimuli-responsive dextran nanoparticles for insulin delivery. *Int. J. Biol. Macromol.* **2019**, *123*, 968–978. [CrossRef]
111. Butzbach, K.; Konhäuser, M.; Fach, M.; Bamberger, D.N.; Breitenbach, B.; Epe, B.; Wich, P.R. Receptor-mediated Uptake of Folic Acid-functionalized Dextran Nanoparticles for Applications in Photodynamic Therapy. *Polymers* **2019**, *11*, 896. [CrossRef] [PubMed]
112. Lan, J.; Li, Y.; Wen, J.; Chen, Y.; Yang, J.; Zhao, L.; Xia, Y.; Du, H.; Tao, J.; Li, Y.; et al. Acitretin-Conjugated Dextran Nanoparticles Ameliorate Psoriasis-like Skin Disease at Low Dosages. *Front. Bioeng. Biotechnol.* **2022**, *9*, 1430. [CrossRef]
113. Naha, P.C.; Hsu, J.C.; Kim, J.; Shah, S.; Bouché, M.; Si-Mohamed, S.; Rosario-Berrios, D.N.; Douek, P.; Hajfathalian, M.; Yasini, P.; et al. Dextran-Coated Cerium Oxide Nanoparticles: A Computed Tomography Contrast Agent for Imaging the Gastrointestinal Tract and Inflammatory Bowel Disease. *ACS Nano* **2020**, *14*, 10187–10197. [CrossRef]
114. Abid, M.; Naveed, M.; Azeem, I.; Faisal, A.; Nazar, M.F.; Yameen, B. Colon specific enzyme responsive oligoester crosslinked dextran nanoparticles for controlled release of 5-fluorouracil. *Int. J. Pharm.* **2020**, *586*, 119605. [CrossRef] [PubMed]
115. Arulmozhi, V.; Pandian, K.; Mirunalini, S. Ellagic acid encapsulated chitosan nanoparticles for drug delivery system in human oral cancer cell line (KB). *Colloids Surf. B Biointerfaces* **2013**, *110*, 313–320. [CrossRef]
116. El-Alfy, E.A.; El-Bisi, M.K.; Taha, G.M.; Ibrahim, H.M. Preparation of biocompatible chitosan nanoparticles loaded by tetracycline, gentamycin and ciprofloxacin as novel drug delivery system for improvement the antibacterial properties of cellulose based fabrics. *Int. J. Biol. Macromol.* **2020**, *161*, 1247–1260. [CrossRef] [PubMed]
117. Rahbar, M.; Morsali, A.; Bozorgmehr, M.R.; Beyramabadi, S.A. Quantum chemical studies of chitosan nanoparticles as effective drug delivery systems for 5-fluorouracil anticancer drug. *J. Mol. Liq.* **2020**, *302*, 112495. [CrossRef]
118. Yu, A.; Shi, H.; Liu, H.; Bao, Z.; Dai, M.; Lin, D.; Lin, D.; Xu, X.; Li, X.; Wang, Y. Mucoadhesive dexamethasone-glycol chitosan nanoparticles for ophthalmic drug delivery. *Int. J. Pharm.* **2020**, *575*, 118943. [CrossRef]
119. Afshar, M.; Dini, G.; Vaezifar, S.; Mehdikhani, M.; Movahedi, B. Preparation and characterization of sodium alginate/polyvinyl alcohol hydrogel containing drug-loaded chitosan nanoparticles as a drug delivery system. *J. Drug Deliv. Sci. Technol.* **2020**, *56*, 101530. [CrossRef]
120. Zahoor, A.; Rajesh, P.; Sadhna, S.; Khuller, G.K. Alginate Nanoparticles as Antituberculosis Drug Carriers: Formulation Development, Pharmacokinetics and Therapeutic Potential. *Indian J. Chest. Dis. Allied. Sci.* **2005**, *48*, 171–176. [CrossRef]
121. Sorasitthiyanukarn, F.N.; Muangnoi, C.; Rojsitthisak, P.; Rojsitthisak, P. Chitosan oligosaccharide/alginate nanoparticles as an effective carrier for astaxanthin with improving stability, in vitro oral bioaccessibility, and bioavailability. *Food Hydrocoll.* **2022**, *124*, 107246. [CrossRef]
122. Ma, X.; Jian, R.; Chang, P.R.; Yu, J. Fabrication and Characterization of Citric Acid-Modified Starch Nanoparticles/Plasticized-Starch Composites. *Biomacromolecules* **2008**, *9*, 3314–3320. [CrossRef] [PubMed]
123. Delrish, E.; Ghassemi, F.; Jabbarvand, M.; Lashay, A.; Atyabi, F.; Soleimani, M.; Dinarvand, R. Biodistribution of Cy5-labeled Thiolated and Methylated Chitosan-Carboxymethyl Dextran Nanoparticles in an Animal Model of Retinoblastoma. *J. Ophthalmic Vis. Res.* **2022**, *17*, 58. [CrossRef]
124. Jafar, M.M.A.; Heather, S.; Al, A.A. *Functional Biopolymers*; Springer International Publishing: Cham, Switzerland, 2019. [CrossRef]

125. Espinoza, S.M.; Patil, H.I.; San Martin Martinez, E.; Casañas Pimentel, R.; Ige, P.P. Poly-ε-caprolactone (PCL), a promising polymer for pharmaceutical and biomedical applications: Focus on nanomedicine in cancer. *Int. J. Polym. Mater. Polym. Biomater.* **2020**, *69*, 85–126. [CrossRef]
126. Vert, M.; Li, S.M.; Spenlehauer, G.; Guerin, P. Bioresorbability and biocompatibility of aliphatic polyesters. *J. Mater. Sci. Mater. Med.* **1992**, *3*, 432–446. [CrossRef]
127. Woodruff, M.A.; Hutmacher, D.W. The return of a forgotten polymer—Polycaprolactone in the 21st century. *Prog. Polym. Sci.* **2010**, *35*, 1217–1256. [CrossRef]
128. Sinha, V.R.; Bansal, K.; Kaushik, R.; Kumria, R.; Trehan, A. Poly-ε-caprolactone microspheres and nanospheres: An overview. *Int. J. Pharm.* **2004**, *278*, 1–23. [CrossRef]
129. Łukasiewicz, S.; Mikołajczyk, A.; Błasiak, E.; Fic, E.; Dziedzicka-Wasylewska, M. Polycaprolactone Nanoparticles as Promising Candidates for Nanocarriers in Novel Nanomedicines. *Pharmaceutics* **2021**, *13*, 191. [CrossRef] [PubMed]
130. Alex, A.; Joseph, A.; Shavi, G.; Rao, J.V.; Udupa, N. Development and evaluation of carboplatin-loaded PCL nanoparticles for intranasal delivery. *Drug Deliv.* **2016**, *23*, 2144–2153. [CrossRef] [PubMed]
131. Mei, L.; Zeng, X.; Chen, H.; Zheng, Y.; Song; Huang, L.; Ma, Y. Novel docetaxel-loaded nanoparticles based on PCL-Tween 80 copolymer for cancer treatment. *Int. J. Nanomed.* **2011**, *6*, 2679–2688. [CrossRef]
132. Abriata, J.P.; Turatti, R.C.; Luiz, M.T.; Raspantini, G.L.; Tofani, L.B.; Amaral, R.L.F.D.; Swiech, K.; Marcato, P.D.; Marchetti, J.M. Development, characterization and biological in vitro assays of paclitaxel-loaded PCL polymeric nanoparticles. *Mater. Sci. Eng. C* **2019**, *96*, 347–355. [CrossRef]
133. El-Habashy, S.E.; Eltaher, H.M.; Gaballah, A.; Zaki, E.I.; Mehanna, R.A.; El-Kamel, A.H. Hybrid bioactive hydroxyapatite/polycaprolactone nanoparticles for enhanced osteogenesis. *Mater. Sci. Eng. C* **2021**, *119*, 111599. [CrossRef]
134. Hao, Y.; Chen, Y.; He, X.; Yang, F.; Han, R.; Yang, C.; Li, W.; Qian, Z. Near-infrared responsive 5-fluorouracil and indocyanine green loaded MPEG-PCL nanoparticle integrated with dissolvable microneedle for skin cancer therapy. *Bioact. Mater.* **2020**, *5*, 542–552. [CrossRef] [PubMed]
135. Shahab, M.S.; Rizwanullah, M.; Alshehri, S.; Imam, S.S. Optimization to development of chitosan decorated polycaprolactone nanoparticles for improved ocular delivery of dorzolamide: In vitro, ex vivo and toxicity assessments. *Int. J. Biol. Macromol.* **2020**, *163*, 2392–2404. [CrossRef]
136. Chen, Y.; Lu, Y.; Hu, D.; Peng, J.; Xiao, Y.; Hao, Y.; Pan, M.; Yuan, L.; Qian, Z. Cabazitaxel-loaded MPEG-PCL copolymeric nanoparticles for enhanced colorectal cancer therapy. *Appl. Mater. Today* **2021**, *25*, 101210. [CrossRef]
137. Pan, Q.; Tian, J.; Zhu, H.; Hong, L.; Mao, Z.; Oliveira, J.M.; Reis, R.L.; Li, X. Tumor-Targeting Polycaprolactone Nanoparticles with Codelivery of Paclitaxel and IR780 for Combinational Therapy of Drug-Resistant Ovarian Cancer. *ACS Biomater. Sci. Eng.* **2020**, *6*, 2175–2185. [CrossRef] [PubMed]
138. Kumari, A.; Yadav, S.K.; Pakade, Y.B.; Kumar, V.; Singh, B.; Chaudhary, A.; Yadav, S.C. Nanoencapsulation and characterization of Albizia chinensis isolated antioxidant quercitrin on PLA nanoparticles. *Colloids Surf. B Biointerfaces* **2011**, *82*, 224–232. [CrossRef] [PubMed]
139. Da Costa, D.; Exbrayat-Héritier, C.; Rambaud, B.; Megy, S.; Terreux, R.; Verrier, B.; Primard, C. Surface charge modulation of rifampicin-loaded PLA nanoparticles to improve antibiotic delivery in Staphylococcus aureus biofilms. *J. Nanobiotechnology* **2021**, *19*, 1–17. [CrossRef] [PubMed]
140. Niza, E.; Božik, M.; Bravo, I.; Clemente-Casares, P.; Sánchez, A.L.; Juan, A.; Klouček, P.; Alonso-Moreno, C. PEI-coated PLA nanoparticles to enhance the antimicrobial activity of carvacrol. *Food Chem.* **2020**, *328*, 127131. [CrossRef]
141. Ghaffarzadegan, R.; Khoee, S.; Rezazadeh, S. Fabrication, characterization and optimization of berberine-loaded PLA nanoparticles using coaxial electrospray for sustained drug release. *DARU J. Pharm. Sci.* **2020**, *28*, 237–252. [CrossRef] [PubMed]
142. Zhang, L.; Zhu, H.; Gu, Y.; Wang, X.; Wu, P. Dual drug-loaded PLA nanoparticles bypassing drug resistance for improved leukemia therapy. *J. Nanoparticle Res.* **2019**, *21*, 83. [CrossRef]
143. Xie, Z.; Su, Y.; Kim, G.B.; Selvi, E.; Ma, C.; Aragon-Sanabria, V.; Hsieh, J.; Dong, C.; Yang, J. Immune Cell-Mediated Biodegradable Theranostic Nanoparticles for Melanoma Targeting and Drug Delivery. *Small* **2017**, *13*, 1603121. [CrossRef]
144. Kumar, A.; Han, S.S. PVA-based hydrogels for tissue engineering: A review. *Int. J. Polym. Mater. Polym. Biomater.* **2017**, *66*, 159–182. [CrossRef]
145. Kumar, A.; Negi, Y.S.; Choudhary, V.; Bhardwaj, N.K. Microstructural and mechanical properties of porous biocomposite scaffolds based on polyvinyl alcohol, nano-hydroxyapatite and cellulose nanocrystals. *Cellulose* **2014**, *21*, 3409–3426. [CrossRef]
146. Rivera-Hernández, G.; Antunes-Ricardo, M.; Martínez-Morales, P.; Sánchez, M.L. Polyvinyl alcohol based-drug delivery systems for cancer treatment. *Int. J. Pharm.* **2021**, *600*, 120478. [CrossRef]
147. El-Dafrawy, S.M.; Tarek, M.; Samra, S.; Hassan, S.M. Synthesis, photocatalytic and antidiabetic properties of ZnO/PVA nanoparticles. *Sci. Rep.* **2021**, *11*, 11404. [CrossRef] [PubMed]
148. Li, J.K.; Wang, N.; Wu, X.S. Poly(vinyl alcohol) nanoparticles prepared by freezing–thawing process for protein/peptide drug delivery. *J. Control. Release* **1998**, *56*, 117–126. [CrossRef]
149. Byun, Y.; Hwang, J.B.; Bang, S.H.; Darby, D.; Cooksey, K.; Dawson, P.L.; Park, H.J.; Whiteside, S. Formulation and characterization of α-tocopherol loaded poly ε-caprolactone (PCL) nanoparticles. *LWT-Food Sci. Technol.* **2011**, *44*, 24–28. [CrossRef]
150. Joye, I.J.; McClements, D.J. Biopolymer-based nanoparticles and microparticles: Fabrication, characterization, and application. *Curr. Opin. Colloid Interface Sci.* **2014**, *19*, 417–427. [CrossRef]

151. Kakran, M.; Antipina, M.N. Emulsion-based techniques for encapsulation in biomedicine, food and personal care. *Curr. Opin. Pharmacol.* **2014**, *18*, 47–55. [CrossRef] [PubMed]
152. Jin, H.Y.; Xia, F.; Zhao, Y.P. Preparation of hydroxypropyl methyl cellulose phthalate nanoparticles with mixed solvent using supercritical antisolvent process and its application in co-precipitation of insulin. *Adv. Powder Technol.* **2012**, *23*, 157–163. [CrossRef]
153. Jones, O.G.; McClements, D.J. Biopolymer Nanoparticles from Heat-Treated Electrostatic Protein-Polysaccharide Complexes: Factors Affecting Particle Characteristics. *J. Food Sci.* **2010**, *75*, N36–N43. [CrossRef]
154. Vecchione, D.; Grimaldi, A.M.; Forte, E.; Bevilacqua, P.; Netti, P.A.; Torino, E. Hybrid Core-Shell (HyCoS) Nanoparticles produced by Complex Coacervation for Multimodal Applications. *Sci. Rep.* **2017**, *7*, 45121. [CrossRef]
155. Schmitt, C.; Turgeon, S.L. Protein/polysaccharide complexes and coacervates in food systems. *Adv. Colloid Interface Sci.* **2011**, *167*, 63–70. [CrossRef]
156. Bock, N.; Woodruff, M.A.; Hutmacher, D.W.; Dargaville, T.R. Electrospraying, a Reproducible Method for Production of Polymeric Microspheres for Biomedical Applications. *Polymers* **2011**, *3*, 131–149. [CrossRef]
157. Ma, J.; Lee, S.M.-Y.; Yi, C.; Li, C.-W. Controllable synthesis of functional nanoparticles by microfluidic platforms for biomedical applications—A review. *Lab Chip* **2017**, *17*, 209–226. [CrossRef] [PubMed]
158. Xu, S.; Nie, Z.; Seo, M.; Lewis, P.; Kumacheva, E.; Stone, H.A.; Garstecki, P.; Weibel, D.B.; Gitlin, I.; Whitesides, G.M. Generation of Monodisperse Particles by Using Microfluidics: Control over Size, Shape, and Composition. *Angew. Chem. Int. Ed.* **2005**, *44*, 724–728. [CrossRef] [PubMed]
159. Krishnadasan, S.; Brown, R.J.C.; Demello, A.J.; Demello, J.C. Intelligent routes to the controlled synthesis of nanoparticles. *Lab Chip* **2007**, *7*, 1434–1441. [CrossRef]
160. Gómez-Guillén, M.C.; Montero, M.P. Enhancement of oral bioavailability of natural compounds and probiotics by mucoadhesive tailored biopolymer-based nanoparticles: A review. *Food Hydrocoll.* **2021**, *118*, 106772. [CrossRef]
161. Lynch, C.R.; Kondiah, P.P.D.; Choonara, Y.E. Advanced Strategies for Tissue Engineering in Regenerative Medicine: A Biofabrication and Biopolymer Perspective. *Molecules* **2021**, *26*, 2518. [CrossRef] [PubMed]
162. Banerjee, A.; Bandopadhyay, R. Use of dextran nanoparticle: A paradigm shift in bacterial exopolysaccharide based biomedical applications. *Int. J. Biol. Macromol.* **2016**, *87*, 295–301. [CrossRef]

Disclaimer/Publisher's Note: The statements, opinions and data contained in all publications are solely those of the individual author(s) and contributor(s) and not of MDPI and/or the editor(s). MDPI and/or the editor(s) disclaim responsibility for any injury to people or property resulting from any ideas, methods, instructions or products referred to in the content.

Article

Enhanced Electrical and Thermal Conductivities of Polymer Composites with a Segregated Network of Graphene Nanoplatelets

Ki Hoon Kim [1,†], Ji-Un Jang [2,†], Gyun Young Yoo [3], Seong Hun Kim [2], Myung Jun Oh [1,*] and Seong Yun Kim [3,*]

1. Department of Carbon Composites Convergence Materials Engineering, Jeonbuk National University, 567 Baekje-daero, Deokjin-gu, Jeonju-si 54896, Jeonbuk, Republic of Korea; kihoon2376@jbnu.ac.kr
2. Research Institute of Industrial Science, Hanyang University, 222 Wangsimni-ro, Haengdang-dong, Seongdong-gu, Seoul 04763, Republic of Korea; jju204@hanyang.ac.kr (J.-U.J.); kimsh@hanyang.ac.kr (S.H.K.)
3. Department of Organic Materials and Textile Engineering, Jeonbuk National University, 567 Baekje-daero, Deokjin-gu, Jeonju-si 54896, Jeonbuk, Republic of Korea; ky5932@gmail.com
* Correspondence: mjoh@jbnu.ac.kr (M.J.O.); sykim82@jbnu.ac.kr (S.Y.K.); Tel.: +82-63-270-2387 (M.J.O.); +82-63-270-2336 (S.Y.K.)
† These authors contributed equally to this work.

Abstract: Introducing a segregated network constructed through the selective localization of small amounts of fillers can be a solution to overcome the limitations of the practical use of graphene-based conductive composites due to the high cost of fillers. In this study, polypropylene composites filled with randomly dispersed GNPs and a segregated GNP network were prepared, and their conductive properties were investigated according to the formation of the segregated structure. Due to the GNP clusters induced by the segregated structure, the electrical percolation threshold was 2.9 wt% lower than that of the composite incorporating randomly dispersed GNPs. The fully interconnected GNP cluster network inside the composite contributed to achieving the thermal conductivity of 4.05 W/m·K at 10 wt% filler content. Therefore, the introduction of a segregated filler network was suitable to simultaneously achieve excellent electrical and thermal conductivities at a low content of GNPs.

Keywords: composites; segregated network; electrical conductivity; thermal conductivity; graphene

Citation: Kim, K.H.; Jang, J.-U.; Yoo, G.Y.; Kim, S.H.; Oh, M.J.; Kim, S.Y. Enhanced Electrical and Thermal Conductivities of Polymer Composites with a Segregated Network of Graphene Nanoplatelets. *Materials* **2023**, *16*, 5329. https://doi.org/10.3390/ma16155329

Academic Editor: Georgios C. Psarras

Received: 23 June 2023
Revised: 24 July 2023
Accepted: 26 July 2023
Published: 29 July 2023

Copyright: © 2023 by the authors. Licensee MDPI, Basel, Switzerland. This article is an open access article distributed under the terms and conditions of the Creative Commons Attribution (CC BY) license (https://creativecommons.org/licenses/by/4.0/).

1. Introduction

Graphene is well known for its excellent conductive properties, such as charge mobility (~200,000 cm^2/V·s [1,2]), electrical conductivity (~10^5 S/m [3]), and thermal conductivity (3000−6500 W/m·K [4,5]). In particular, graphene nanoplatelet (GNP)-filled conductive polymer composites (CPCs), which are lightweight, easy to process, and have excellent portability, are receiving a lot of attention because the GNPs manufactured via the top-down method of exfoliating graphite are advantageous in terms of price and mass production compared to bottom-up fabrication based graphene [6–10]. To maximize the electrical and thermal conductivities of CPCs, uniform filler dispersion has been identified as an important structural factor [11–14]. To achieve the uniform filler dispersion, various methods [15–19], such as covalent functionalization [15], noncovalent functionalization [16,18,19], and polymer wrapping [17], have been reported. However, despite these efforts, there is a need for a method for spreading the application of GNP-based CPCs by innovatively reducing the amounts of expensive GNPs incorporated into the composites.

Various strategies [20–26] have been proposed to achieve excellent electrical and thermal conductivities of polymer composites by incorporating smaller amounts of fillers. Double percolation can be generated by the selective localization of nanofillers based on thermodynamic (chemical affinity) and kinetic (melting point difference) factors between nanofillers and an immiscible matrix [20]. This strategy induces a predominant distribution of conductive fillers in one matrix phase. Hence, the amount of filler used to implement double-percolated CPCs with a certain conductive performance level is significantly smaller than that to implement randomly dispersed CPCs (R-CPCs) [21–23]. To further reduce the amount of conductive filler used, a segregated structure in which the filler is selectively localized at the interface has been proposed [24]. The segregated structure can minimize the proportions of fillers by forming a matrix region inside the network where fillers are not mixed [25–29]. The introduction of the segregated network can innovatively enhance the electrical and thermal conductivities of CPCs simultaneously. Therefore, there is a need to understand the simultaneous enhancement of electrical and thermal conductivities with respect to the structural development of GNP-based segregated composites.

In this study, the electrical and thermal conductivities of composites according to their network structures of conductive fillers were investigated experimentally and theoretically. GNP-based R-CPCs and segregated CPCs (S-CPCs) were prepared, and the electrical and thermal conductivities of the CPCs were analyzed. The percolation threshold in the electrical conductivity of the R-CPCs was 3 wt%, and the thermal conductivity of the R-CPC increased linearly according to the fitting of Nan's model. In contrast, the electrical percolation threshold of the S-CPC was observed at 0.1 wt% (0.04 vol%), and the thermal percolation behavior where the thermal conductivities of the S-CPCs rapidly increased following the thermal percolation model was confirmed.

2. Materials and Methods

2.1. Materials

GNPs (M25, XG Science, Lansing, MI, USA) with a lateral size of 25 μm, a thickness of 5 nm, and a density of 2.2 g/cm^3 [30], respectively, were used as fillers to improve the conductivities of the composites. Pelletized polypropylene (PP, Y-120A, Lotte Chemical, Daejeon, Republic of Korea), with an average diameter of 2–4 mm, was used as polymer matrix. The used PP exhibited a density of 0.9 g/cm^3 and a melting temperature (T_m) of 165 °C.

2.2. Composite Fabrication

Fabrication process of S-CPC is shown schematically in Figure 1. Before fabricating the composite, the raw materials were dried overnight at 85 °C to remove moisture. PP and GNP were weighed at the target content and mixed at 2000 rpm for 2 min using a mechanical mixer (ARE 310, Thinky Corp., Tokyo, Japan). The mixture was hot-compacted using a heating press (D3P-20J, Dae Heung Science, Incheon, Republic of Korea) at 15 MPa and 160 °C (below T_m) for 15 min. Fabrication of the segregated composite below T_m can lead to stable segregated networks that were fillers localized on the matrix interface and induces relatively superior conductive properties [26]. The composites with randomly dispersed GNP were fabricated by molding to size of 25 × 25 × 2 mm^3 using hot pressing after stirring (60 rpm) at the temperature (180 °C) that the matrix melted completely (Figure S1). The segregated composite and the composite with randomly dispersed GNPs were labeled S-CPCX and R-CPCX, respectively. X implied the weight fraction of GNP. The compositions of the prepared composites are presented in Table 1.

Table 1. Composition of the fabricated composites.

Sample Code	PP, wt% (vol%)	GNP, wt% (vol%)
R-CPC0	100 (100)	0 (0)
S-CPC0		
R-CPC0.1	99.9 (99.96)	0.1 (0.04)
S-CPC0.1		
R-CPC0.3	99.7 (99.88)	0.3 (0.12)
S-CPC0.3		

Table 1. Cont.

Sample Code	PP, wt% (vol%)	GNP, wt% (vol%)
R-CPC0.5	99.5 (99.80)	0.5 (0.20)
S-CPC0.5		
R-CPC1	99 (99.59)	1 (0.41)
S-CPC1		
R-CPC2	98 (99.17)	2 (0.83)
S-CPC2		
R-CPC3	97 (98.75)	3 (1.25)
S-CPC3		
R-CPC5	95 (97.89)	5 (2.11)
S-CPC5		
R-CPC7	93 (97.01)	7 (2.99)
S-CPC7		
R-CPC10	90 (95.65)	10 (4.35)
S-CPC10		

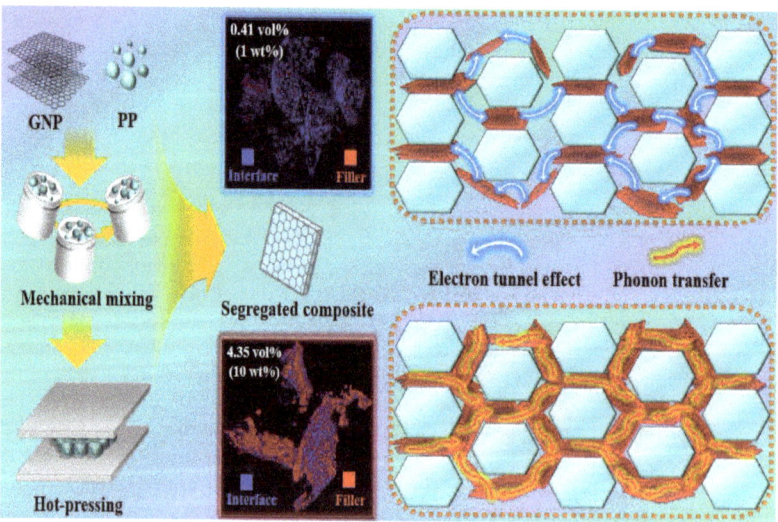

Figure 1. Schematic for fabrication process of S-CPC.

2.3. Characterization

The morphologies of the composites were observed via a field emission electron microscope (FE-SEM, GeminiSEM 500, Zeiss, Oberkrochen, Germany). Before observation, the specimens were dipped in liquid nitrogen for 5 min and mechanically fractured. The surfaces of specimens were Pt-coated for 140 sec in vacuum via a sputtering machine (Ion Sputter E-1030, Hitachi High Technologies Co., Tokyo, Japan). The Pt-coated specimens were observed under a nitrogen condition with an applied voltage of 2 kV. Nondestructive three-dimensional (3D) analysis was conducted with micro-computed tomography (μ-CT, Skyscan 1172, Bruker Co., Billerica, MA, USA) to analyze the 3D internal structure of the specimen. The 3D image of the specimen was attained via X-ray irradiation and reconstructed using a software program (NRecon, Version 1.6.10.2). The high electrical resistance of the composite was measured via an ultrahigh resistance meter (SM-8220,

HIOKI E. E. Corporation, Nagano, Japan). The low electrical resistance of the composite was measured based on direct current resistance (DC) using a Keithley 2400 Source Meter. Before the measurement, the specimens were coated with silver paste to reduce the contact resistance levels between the specimens and electrodes. The electrical conductivity (σ) of the specimen was calculated via the following equation:

$$\sigma = \frac{L}{Rwt} \quad (1)$$

where R, L, w, and t are the measured electrical resistance, length, width, and thickness of the specimen, respectively. The isotropic thermal conductivity of the sample with $25 \times 25 \times 2$ mm^3 was analyzed using a hot-disk method (TPS 2500S, Hot disk AB, Gothenburg, Sweden) according to ISO 22007-2 [31].

2.4. Electrical Percolation Model

The theoretical electrical conductivities of the fabricated specimens are calculated via the electrical percolation equation [32]. The electrical conductivity of the composite (σ_c) is enhanced due to the electron tunnel effect induced by conductive particles (GNP fillers) located in the insulating matrix (PP). Prior to the filler content where the tunnel effect is maximized (ϕ_c, percolation threshold), the electrical conductivity of the composite is expressed with the electrical conductivity of the matrix (σ_m) and the exponent (s). In addition, the σ_c after ϕ_c is expressed with the electrical conductivity of the filler (σ_f) and another exponent (v). The critical percolation exponents—s and v—are governed by the size and shape of the conductive particle and the thickness of the insulating layer (=distance between GNP fillers), respectively. The slope reaching the saturation region of electrical conductivity is determined. In this study, the electrical conductivities of the segregated composites (σ_{sc}) are expressed as follows:

$$\begin{aligned}\sigma_{sc} &= \sigma_m \left[(\phi_{ec})/(\phi_{ec} - \phi_f) \right]^s \quad (if,\ \phi_f < \phi_{ec}) \\ \sigma_{sc} &= \sigma_f \left[(\phi_f - \phi_{ec})/(1 - \phi_{ec}) \right]^v \quad (if,\ \phi_f > \phi_{ec}) \end{aligned} \quad (2)$$

where σ_m, ϕ_{ec}, ϕ_f, and σ_f denote the electrical conductivity of the PP matrix (1.04×10^{-13} S/m), electrical percolation threshold of the segregated composite (0.04 vol%), GNP filler content and electrical conductivity of the GNP filler (10^4 S/m [33]), respectively. In addition, s and v indicate the percolation exponents before and after the critical volume fraction, respectively.

2.5. Nan's Model and Thermal Percolation Model

The thermal conductivity (TC_c) of the composite incorporating the nanofiller is lower than expected for the rule of mixtures due to the interfacial thermal resistance (ITR, $\approx$ Kapitza radius) generated at the interface between the nanocarbon filler and the polymer matrix. Nan's model is an effective tool for evaluating the thermal conductivity of composite because TC_c can be predicted according to the size, shape, and content of the nanocarbon filler by assuming the uniform dispersion of fillers [34]. The theoretical thermal conductivities of the R-CPC and S-CPC fabricated in this study are described using Nan's model as follows:

$$TC_{Nan} = TC_m \times \left(\frac{3 + \phi_f \times (\beta_x + \beta_z)}{3 - \phi_f \times \beta_x} \right) \quad (3)$$

where,

$$\beta_x = \frac{2(K_{11}^C - TC_m)}{K_{11}^C + TC_m},\ \beta_z = \frac{K_{33}^C}{TC_m} - 1 \quad (4)$$

TC_m is the thermal conductivity of the PP (0.30 W/m·K), ϕ_f is the volume fraction of GNP fillers, and K_{11}^C and K_{33}^C are the equivalent thermal conductivities of GNPs surrounded with parallel and perpendicular interfacial barrier layers of the unit cell, respectively, which can be described as following equation:

$$K_{11}^C = \frac{TC_f}{1 + \frac{2a_k TC_f}{hTC_m}}, \quad K_{33}^C = \frac{TC_f}{1 + \frac{2a_k TC_f}{dTC_m}}, \quad a_k = R_{ITR} \times TC_m \tag{5}$$

where TC_f, a_k, and R_{ITR} are the thermal conductivity of the GNP (3000 W/m·K), Kapitza radius (25.1 nm), and ITR (8.40×10^{-8} m^2 K/W [35]), respectively, and h and d are the thickness (5 nm) and lateral size (25 μm) of the GNP, respectively.

$$TC_P = TC_m \times (1 - \phi_f) + TC_o \left(\frac{\phi_f - \phi_{tc}}{1 - \phi_{tc}}\right)^z \tag{6}$$

where TC_P, TC_m, and ϕ_f are the theoretical conductivity of a composite filled with a thermally percolated filler network, the thermal conductivity of the matrix (0.30 W/m·K), and the volume fraction of GNP fillers, respectively. In this study, ϕ_{tc} is 0.0045 (0.45 vol%). TC_o and z are the pre-exponential factor (91 W/m·K in this work) and critical exponent (0.98 in this work).

2.6. Applications

The improved electrical and thermal conductivities of the manufactured composites were applied as humidity sensors and thermal interface management (TIM) materials, respectively. After the prepared composite was placed on an electrode in an acrylic chamber connected to an external multimeter (Fluke 17B+ MAX Digital Multimeter, Fluke Corporation, Everett, WA, USA), surface resistance was measured according to humidity control (30–80% relative humidity (RH)) using a humidifier. Humidity-sensing sensitivity (HS) was calculated based on the measured resistance and humidity as following equation:

$$HS\ (\%) = \frac{\Delta R}{R_{30}} \times 100 \tag{7}$$

where, R_{30} is the resistance at 30% RH and ΔR represents the difference between the resistances at 80% RH and R_{30}. In addition, the thermal images for the application of the TIM material were obtained by measuring the average surface temperature of the fabricated composite at 10 sec using a thermal camera (Testo 875 infrared thermal imager, Testo Ltd., Lenzkirch, Germany) after being placed on the hot plate at 100 °C.

3. Results and Discussion

Cross-sectional FE-SEM images of the prepared specimens are placed in Figure 2a–l. A uniform dispersion of GNPs was confirmed in R-CPC0.3 (Figure 2a). In contrast, the selective localization of GNP in the interface of the segregated composite was confirmed in S-CPC0.3 (Figure 2b). In the magnified image of S-CPC0.3 (Figure 2c), GNPs were compacted at the interface, indicating that the segregated structure was formed in the S-CPC even at low filler contents. R-CPC1 and S-CPC1 showed more obvious differences in filler distribution (Figure 2d–f). A uniform GNP dispersion of R-CPC was observed, while S-CPC obviously exhibited a segregated structure. In addition, R-CPC showed an insulating gap (PP matrix) between GNPs due to uniform dispersion, and GNP clusters formed by the contact of incorporated fillers were observed in S-CPC. This morphological difference could affect the conductivities of the R-CPC and S-CPC. Despite the successful formation of GNP clusters in S-CPC, the partial disconnection between adjacent GNP clusters observed at the interface of the segregated structure indicated that an interconnected network was not formed at the 1 wt% filler content.

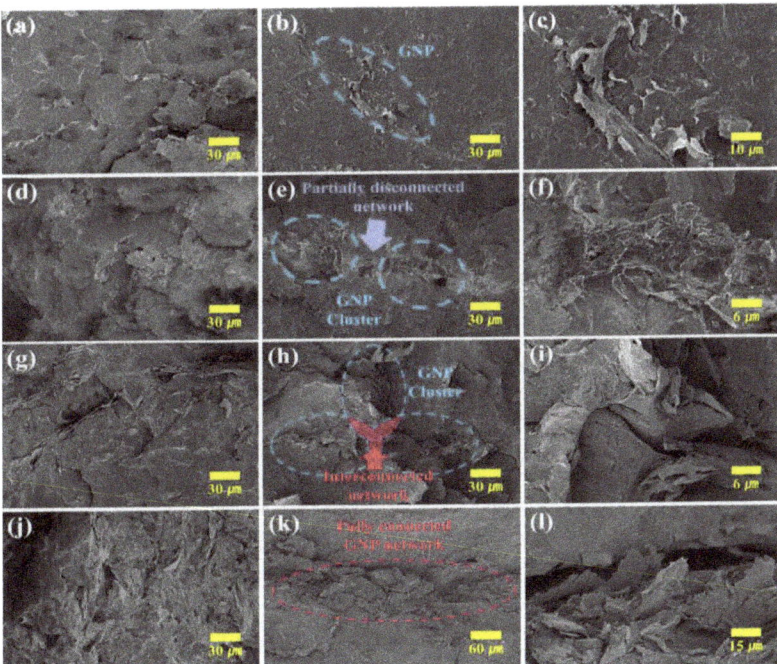

Figure 2. FE-SEM images of (**a**) R-CPC0.3 and S-CPC0.3 at (**b**) low and (**c**) high magnifications, (**d**) R-CPC1 and S-CPC1 at (**e**) low and (**f**) high magnifications, (**g**) R-CPC2 and S-CPC2 at (**h**) low and (**i**) high magnifications, and (**j**) R-CPC10 and S-CPC10 at (**k**) low and (**l**) high magnifications.

R-CPC2 still exhibited an insulating gap between the GNPs (Figure 2g). A few interconnected networks in which GNP clusters contacted each other were observed in S-CPC2 (Figure 2h,i). The presence of interconnected networks within the composite can greatly enhance tunneling conductivity and reduce phonon scattering, resulting in improved phonon transfer [36,37]. This indicates that fabricating segregated structures is an efficient strategy for forming a compact conductive network via selective localization of fillers. A uniform GNP dispersion of R-CPC10 was observed despite the maximum filler loading (Figure 2j). The reduced distances between fillers due to the high content could improve the electron tunnel effect and effective phonon transfer. From the observation of GNPs well localized at the matrix interface in S-CPC10, the applied process was suitable for fabricating segregated composites up to 10 wt% GNP loading (Figure 2k,l). In addition, a fully connected network (Figure 2k) was formed by GNP clusters consisting of fillers in contact (Figure 2l) with each other in the segregated composite. The fully connected filler network could contribute to the dramatic enhancements in conductivities.

Nondestructive observation of the 3D segregated structure using μ-CT was performed. Figure 3a–c shows the μ-CT images of R-CPC0.3 and S-CPC0.3. In Figure 3a, uniformly dispersed GNPs within the PP matrix were observed. The obvious distances between the fillers as insulating gaps could result in the low conductivities of the composite. On the other hand, GNPs were selectively located at the matrix interface in S-CPC0.3 (Figure 3b,c) and formed a conductive pathway (Figure 3c). Clusters of GNPs were formed with the selective localization in S-CPC1 (Figure 3e,f). In addition, partial disconnection between adjacent GNP clusters was observed, as discussed in Figure 2e. In the case of S-CPC2, multiple GNP clusters in the segregated structure were observed (Figure 3h,i). The interconnections between the clusters identified in the high-magnification image could be beneficial for both electron tunneling and phonon transfer. Figure 3j–l shows the μ-CT images of the

R-CPC10 and S-CPC10. Despite the high filler content, a uniform dispersion and segregated structures were induced. In particular, a fully connected network of S-CPC10 was clearly confirmed via nondestructive 3D structural analysis based on μ-CT and 2D observation using FE-SEM.

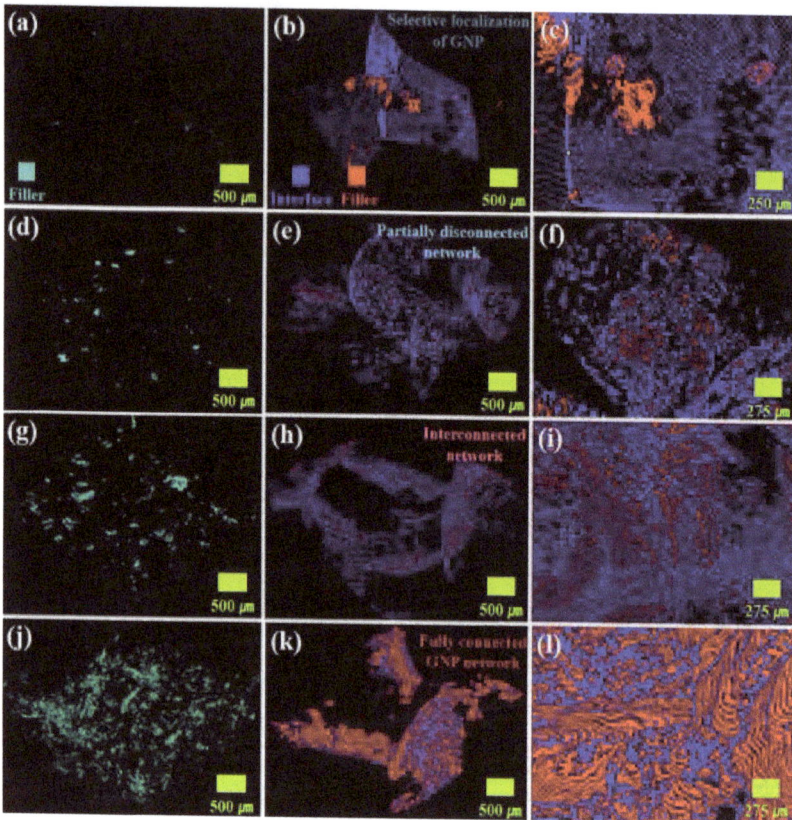

Figure 3. μ-CT images of (**a**) R-CPC0.3 and S-CPC0.3 at (**b**) low and (**c**) high magnifications, (**d**) R-CPC1 and S-CPC1 at (**e**) low and (**f**) high magnifications, (**g**) R-CPC2 and S-CPC2 at (**h**) low and (**i**) high magnifications and (**j**) R-CPC10 and S-CPC10 at (**k**) low and (**l**) high magnifications.

The electrical conductivity of composite could be described by percolation theory based on the tunnel effects of electrons [38,39]. Electrical percolation behavior is represented by a dramatic improvement in the electrical conductivities of composites when fillers are located at a specific distance where the tunnel effect is generated [40]. The segregated structure could form a compact conductive network derived from the selective localization of fillers, achieving superior electrical conductivities of S-CPCs at low filler contents relative to R-CPCs [41,42]. A significant difference in the electrical conductivities of the R-CPC and S-CPC was observed in Figure 4a. For example, R-CPC0.3 and S-CPC0.3 achieved 7.90×10^{-13} S/m and 2.06×10^{-2} S/m, respectively, indicating that the electron tunnel effect was generated by the selective localization of the fillers based on the segregated structure. In addition, R-CPCs showed a moderate increase of the electrical conductivity up to 10 wt% (~5.81 S/m), whereas the electrical conductivity of S-CPC1 was 20.79 S/m, where GNP clusters were identified. Furthermore, the electrical conductivity of the S-CPC2 showed 192.87 S/m before saturation, where interconnections between adjacent GNP clusters were confirmed. Therefore, based on the internal structure analysis and electrical conductivity results, the segregated structure of the fabricated composite was effective for increasing the electrical conductivities and reducing the filler contents required for generating the electron tunnel effect. The theoretically evaluated electrical conductivity using the electrical percolation equation showed that the segregated composites were advantageous in inducing a reduced percolation threshold (ϕ_{ec} at 0.04 vol% (0.1 wt%)) compared to that of R-CPCs (ϕ_{ec} at 1.25 vol% (3 wt%)). Comparisons for the electrical percolation thresholds and maximum electrical conductivities of previously reported GNP-segregated composites are shown in Figure S2 and Table S1 in the Supplementary Materials [43–51]. Therefore, it was experimentally and theoretically confirmed that the interconnected filler networks between GNP clusters generated by the segregated structures contributed to the improved electrical conductivities.

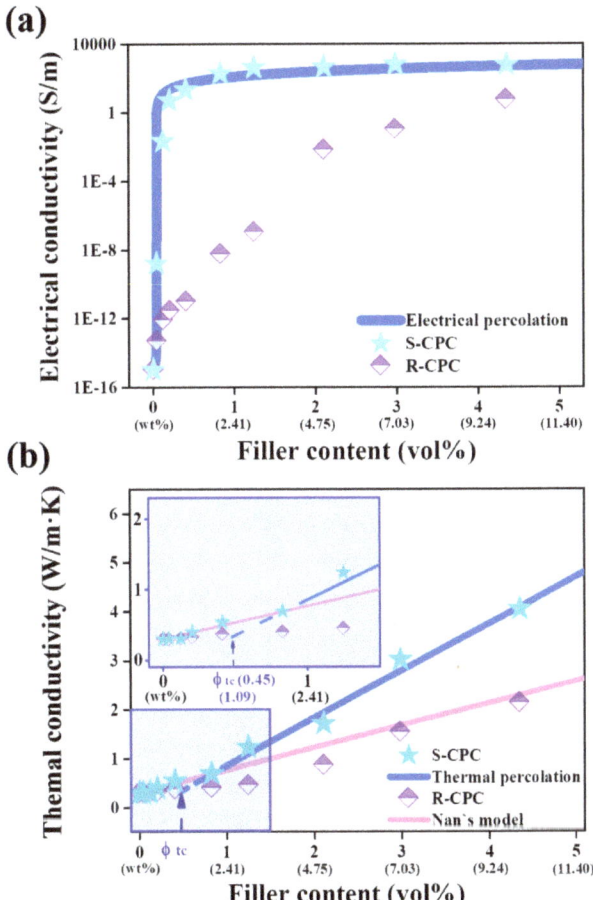

Figure 4. Experimentally and theoretically obtained (**a**) electrical and (**b**) thermal conductivities of the fabricated specimens.

The thermal conductivities of the prepared specimens are displayed in Figure 4b. The thermal conductivity of R-CPC increased linearly with increasing filler content, and the thermal conductivity of R-CPC10 (2.13 W/m·K) was improved by 610% compared to that of raw PP (0.30 W/m·K). These results were similar to the thermal conductivity trends of the R-CPCs prepared using the uniform filler dispersion method [52]. Thus, the theoretical thermal conductivity evaluated via Nan's model based on ITR was in good agreement with the thermal conductivity of R-CPC. In contrast, the thermal conductivity of the segregated composite enhanced linearly to 2 wt% GNP (S-CPC2, 0.69 W/m·K), then rapidly improved up to 10 wt% (S-CPC10, 4.05 W/m·K). The thermal conductivity of S-CPCs was evaluated via the percolation model because the experimental thermal conductivity of S-CPC at higher than 3 wt% exceeded that theoretically calculated with Nan's model. Thermal percolation is a behavior in which the thermal conductivity rapidly improves with increasing contact between the thermally conductive particles [53]. The thermal conductivity of S-CPC evaluated via a thermal percolation threshold ($\varnothing_{tc}$) of 0.45 vol% was in good agreement with the measured value, indicating that the phonon transfer system dominated by the ITR between the PP and GNP in the S-CPC was converted to a system based on direct contact between the GNPs. These results were in good agreement with the partially disconnected network between GNP clusters that contributed to the formation of multiple interfaces (≈ITR) and the fully connected filler networks induced by interconnected GNP clusters, as discussed in Figures 2 and 3, respectively. The observed GNP clusters within the segregated structure of the composite formed by the selective localization of fillers during the process enhanced the electrical and thermal conductivities by inducing effective electron tunneling and phonon transfer. In particular, the fully connected filler network induced the excellent thermal conductivity of the S-CPC by reducing the contribution of ITR. Therefore, it was experimentally and theoretically confirmed that the applied strategy was advantageous for the dramatic improvements in the electrical and thermal conductivities of the composites. In addition, as shown in the humidity sensing sensitivities and thermal images of the fabricated composites (Figure 5), it was confirmed that the sensitivity of the humidity sensor and the heat dissipation performance as a TIM material were improved by the enhanced electrical and thermal conductivity of S-CPC.

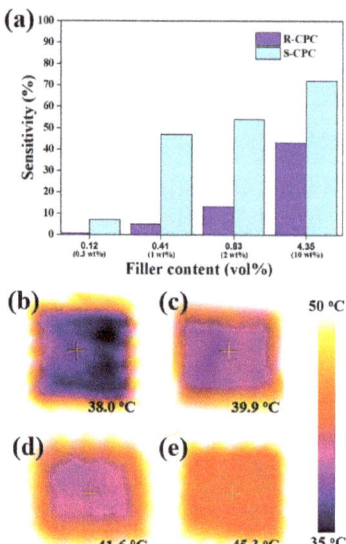

Figure 5. (a) Humidity-sensing sensitivity results of the fabricated composites and thermal images of the (b) R-CPC2, (c) S-CPC2, (d) R-CPC10, and (e) S-CPC10.

4. Conclusions

Introducing a segregated network inside the composite is a useful method for achieving excellent conductivities at low contents of conductive fillers. In this study, theoretical and experimental investigations were conducted to discover the improvements in the electrical and thermal conductivities of composites according to the selective localization of GNP filler using the segregated structure and the generation of the conductive network. In the internal structures of the composites observed using FE-SEM and μ-CT, the GNP clusters located at the interfaces of the PP and the fully connected GNP networks were obviously observed, indicating that the applied process was suitable for the fabrication of segregated composites. S-CPC achieved an improved electrical conductivity of 20.79 S/m at a low filler content (1 wt%) compared to R-CPC by achieving the electron tunnel effect generated by GNP clusters on the interfaces of the PP particles. The enhancement trend of thermal conductivity in S-CPC, before the incorporation of filler of 3 wt%, was calculated via Nan's equation considering ITR; however, the dramatic increase in the measured thermal conductivity at high filler contents (>3 wt%) was evaluated via the thermal percolation equation. This thermal behavior was determined by the formation of a filler network using interconnected GNP clusters, and the enhanced thermal conductivity of S-CPC10 (4.05 W/m·K) due to a fully connected filler network in the segregated structure was observed. It was confirmed from the experimentally and theoretically evaluated electrical and thermal conductivities of the composites that the segregated structure induced using the applied process was useful for effective electron tunneling and phonon transfer, providing potential options for achieving excellent sensing property and TIM in composites with low filler content.

Supplementary Materials: The following supporting information can be downloaded at: https://www.mdpi.com/article/10.3390/ma16155329/s1, Figure S1: Schematic for fabrication process of R-CPC; Figure S2: Comparisons of percolation threshold (ϕ_{ec}) and maximum electrical conductivity of segregated composites incorporating GNP; Table S1: Comparisons of materials, fabrication methods and percolation thresholds of segregated composites.

Author Contributions: Methodology, S.Y.K.; writing—original draft preparation, K.H.K. and J.-U.J.; visualization, K.H.K.; investigation, G.Y.Y. and S.H.K.; writing—review and editing, J.-U.J. and S.Y.K.; supervision, M.J.O. and S.Y.K.; project administration, M.J.O. and S.Y.K.; funding acquisition: S.H.K. and M.J.O. All authors have read and agreed to the published version of the manuscript.

Funding: This paper was supported by research funds for newly appointed professors of Jeonbuk National University in 2020. This study was also partially supported by "HRD Program for Industrial Innovation (Carbon Composite Professional Human Resources Training Program)" grant funded by the Korea Government (MOTIE). (P0017002, 2022), the National Research Foundation of Korea (NRF) grant funded by the Korea government (MSIT) (No. 2021R1A2C1093839), and Basic Science Research Program through the National Research Foundation of Korea (NRF) funded by the Ministry of Education (2016R1A6A1A03013422).

Institutional Review Board Statement: Not applicable.

Informed Consent Statement: Not applicable.

Data Availability Statement: Not applicable.

Acknowledgments: R-CPC and S-CPC specimens were observed by GeminiSEM 500 (Zeiss, Oberkrochen, Germany) installed in the Center for University Wide Research Facilities (CURF) at Jeonbuk National University. The authors would like to thank Han Gyeol Jang, Ph.D. Candidate from Korea Institute of Science and Technology (KIST) for the thermal conductivity analysis and provision.

Conflicts of Interest: The authors declare no conflict of interest.

References

1. Novoselov, K.S.; Fal'ko, V.I.; Colombo, L.; Gellert, P.R.; Schwab, M.G.; Kim, K. A roadmap for graphene. *Nature* **2012**, *490*, 192–200. [CrossRef] [PubMed]
2. Morozov, S.V.; Novoselov, K.S.; Katsnelson, M.I.; Schedin, F.; Elias, D.C.; Jaszczak, J.A.; Geim, A.K. Giant intrinsic carrier mobilities in graphene and its bilayer. *Phys. Rev. Lett.* **2008**, *100*, 016602. [CrossRef] [PubMed]
3. Stankovich, S.; Dikin, D.A.; Dommett, G.H.B.; Kohlhaas, K.M.; Zimney, E.J.; Stach, E.A.; Piner, R.D.; Nguyen, S.T.; Ruoff, R.S. Graphene-based composite materials. *Nature* **2006**, *442*, 282–286. [CrossRef] [PubMed]
4. Balandin, A.A.; Ghosh, S.; Bao, W.; Calizo, I.; Teweldebrhan, D.; Miao, F.; Lau, C.N. Superior thermal conductivity of single-layer graphene. *Nano Lett.* **2008**, *8*, 902–907. [CrossRef] [PubMed]
5. Balandin, A.A. Thermal properties of graphene and nanostructured carbon materials. *Nat. Mater.* **2011**, *10*, 569–581.

6. Li, J.; Liu, X.; Feng, Y.; Yin, J. Recent progress in polymer/two-dimensional nanosheets composites with novel performances. *Prog. Polym. Sci.* **2022**, *126*, 101505.
7. Punetha, V.D.; Rana, S.; Yoo, H.J.; Chaurasia, A.; McLeskey, J.T., Jr.; Ramasamy, M.S.; Sahoo, N.G.; Cho, J.W. Functionalization of carbon nanomaterials for advanced polymer nanocomposites: A comparison study between CNT and graphene. *Prog. Polym. Sci.* **2017**, *67*, 1–47.
8. Masarra, N.-A.; Batistella, M.; Quantin, J.-C.; Regazzi, A.; Pucci, M.F.; Hage, R.E.; Lopez-Cuesta, J.M. Fabrication of PLA/PCL/Graphene Nanoplatelet electrically conductive circuit using the fused filament fabrication 3d printing technique. *Materials* **2022**, *15*, 762. [CrossRef]
9. Lee, H.; Kim, M.N.; Jang, H.G.; Jang, J.-U.; Kim, J.; Kim, S.Y. Phenyl glycidyl ether-based non-covalent functionalization of nano-carbon fillers for improving conductive properties of polymer composites. *Compos. Commun.* **2022**, *33*, 101237. [CrossRef]
10. Chen, X.; Zhang, X.; Xiang, D.; Wu, Y.; Zhao, C.; Li, H.; Li, Z.; Wang, P.; Li, Y. 3D printed high-performance spider web-like flexible strain sensors with directional strain recognition based on conductive polymer composites. *Mater. Lett.* **2022**, *306*, 130935. [CrossRef]
11. Efros, A.L.; Shklovskii, B.I. Critical behaviour of conductivity and dielectric constant near the metal-non-metal transition threshold. *Phys. Status Solidi B Basic Res.* **1976**, *76*, 475–485. [CrossRef]
12. Jang, J.-U.; Cha, J.E.; Lee, S.H.; Kim, J.; Yang, B.; Kim, S.Y.; Kim, S.H. Enhanced electrical and electromagnetic interference shielding properties of uniformly dispersed carbon nanotubes filled composite films via solvent-free process using ring-opening polymerization of cyclic butylene terephthalate. *Polymer* **2020**, *186*, 122030.
13. Noh, Y.J.; Joh, H.-I.; Yu, J.; Hwang, S.H.; Lee, S.; Lee, C.H.; Kim, S.Y.; Youn, J.R. Ultra-high dispersion of graphene in polymer composite via solvent free fabrication and fuctionalization. *Sci. Rep.* **2015**, *5*, 9141. [CrossRef] [PubMed]
14. Xiang, D.; Zhang, X.; Harkin-Jones, E.; Zhu, W.; Zhou, Z.; Shen, Y.; Li, Y.; Zhao, C.; Wang, P. Synergistic effects of hybrid conductive nanofillers on the performance 3D printed highly elastic strain sensors. *Compos. Part A Appl. Sci. Manuf.* **2020**, *129*, 105730. [CrossRef]
15. Vryonis, O.; Andritsch, T.; Vaughan, A.S.; Lewin, P.L. Effect of surfactant molecular structure on the electrical and thermal performance of epoxy/functionalized-graphene nanocomposites. *Polym. Compos.* **2020**, *41*, 2753–2767. [CrossRef]
16. Park, M.; Lee, H.; Jang, J.-U.; Park, J.H.; Kim, C.H.; Kim, S.Y.; Kim, J. Phenyl glycidyl ether as an effective noncovalent functionalization agent for multiwalled carbon nanotube reinforced polyamide 6 nanocomposite fibers. *Compos. Sci. Technol.* **2019**, *177*, 96–102. [CrossRef]
17. Jang, J.-U.; Lee, H.S.; Kim, J.W.; Kim, S.Y.; Kim, S.H.; Hwang, I.; Kang, B.J.; Kang, M.K. Facile and cost-effective strategy for fabrication of polyamide 6 wrapped multi-walled carbon nanotube via anionic melt polymerization of ε-caprolactam. *J. Chem. Eng.* **2019**, *373*, 251–258. [CrossRef]
18. Xiang, D.; Zhang, X.; Li, Y.; Harkin-Jones, E.; Zheng, Y.; Wang, L.; Zhao, C.; Wang, P. Enhanced performance of 3D printed highly elastic strain sensors of carbon nanotube/thermoplastic polyurethane nanocomposites via non-covalent interactions. *Compos. B Eng.* **2019**, *176*, 107250. [CrossRef]
19. Xiang, D.; Zhang, Z.; Han, Z.; Zhang, X.; Zhou, Z.; Zhang, J. Effects of non-covalent interactions on the properties of 3D printed flexible piezoresistive strain sensors of conductive polymer composites. *Compos. Interface* **2021**, *28*, 577–591. [CrossRef]
20. Li, T.-T.; Wang, Y.; Wang, Y.; Sun, F.; Xu, J.; Lou, C.-W.; Lin, J.-H. Preparation of flexible, highly conductive polymer composite films based on double percolation structures and synergistic dispersion effect. *Polym. Compos.* **2021**, *42*, 5159–5167. [CrossRef]
21. Liebscher, M.; Domurath, J.; Saphiannikova, M.; Müller, M.T.; Heinrich, G.; Pötschke, P. Dispersion of graphite nanoplates in melt mixed PC/SAN polymer blends and its influence on rheological and electrical properties. *Polymer* **2020**, *200*, 122577. [CrossRef]
22. Guo, M.; Kashfipour, M.A.; Li, Y.; Dent, R.S.; Zhu, J.; Maia, J.M. Structure-rheology-property relationships in double-percolated polypropylene/poly(methylmethacrylate)/boron nitride polymer composites. *Compos. Sci. Technol.* **2020**, *198*, 108306. [CrossRef]
23. Xiang, D.; Liu, L.; Chen, X.; Wu, Y.; Wang, M.; Zhang, J.; Zhao, C.; Li, H.; Li, Z.; Wang, P.; et al. High-performance fiber strain sensor of carbon nanotube/thermoplastic polyurethane@styrene butadiene styrene with a double percolated structure. *Front. Mater. Sci.* **2022**, *16*, 220586. [CrossRef]
24. Malliaris, A.; Turner, D.T. Influence of particle size on electrical resistivity of compacted mixture of polymeric and metallic powders. *J. Appl. Phys.* **1971**, *42*, 614–618. [CrossRef]
25. Qi, X.-D.; Yang, J.-H.; Zhang, N.; Huang, T.; Zhou, Z.-W.; Kühnert, I.; Pötschke, P.; Wang, Y. Selective localization of carbon nanotubes and its effect on the structure and properties of polymer blends. *Prog. Polym. Sci.* **2021**, *123*, 101471. [CrossRef]
26. Pang, H.; Xu, L.; Yan, D.-X.; Li, Z.-M. Conductive polymer composites with segregated structures. *Prog. Polym. Sci.* **2014**, *39*, 1908–1933. [CrossRef]
27. Xiang, D.; Wang, L.; Tang, Y.; Zhao, C.; Harkin-Jones, E.; Li, Y. Effect of phase transition on electrical properties of polymer/carbon nanotube and polymer/graphene nanoplatelet composites with different conductive network structures. *Polym. Int.* **2017**, *67*, 227–235. [CrossRef]
28. Liu, L.; Zhang, X.; Xiang, D.; Wu, Y.; Sun, D.; Shen, J.; Wang, M.; Zhao, C.; Li, H.; Li, Z.; et al. Highly stretchable, sensitive and wide linear responsive fabric-based strain sensors with a self-segregated carbon nanotube (CNT)/polydimethylsiloxane (PDMS) coating. *Prog. Nat. Sci.* **2022**, *32*, 34–42. [CrossRef]
29. Xiang, D.; Wang, L.; Tang, Y.; Harkin-Jones, E.; Zhao, C.; Wang, P.; Li, Y. Damage self-sensing behavior of carbon nanofiller reinforced polymer composites with different conductive network structures. *Polymer* **2018**, *158*, 308–319. [CrossRef]

30. Um, J.G.; Jun, Y.-S.; Alhumade, H.; Krithivasan, H.; Lui, G.; Yu, A. Investigation of the size effect of graphene nano-platelets (GnPs) on the anti-corrosion performance of polyurethane/GnP composites. *RSC Adv.* **2018**, *8*, 17091–17100. [CrossRef]
31. ISO 22007-2:2015; Plastic—Determination of Thermal Conductivity and Thermal Diffusivity—Part 2: Transient Plane Heat Source (Hot Disc) Method. Swedish Standards Institute: Stockholm, Sweden, 2015.
32. McLachalan, D.S. Equation for the conductivity of metal-insulator mixtures. *J. Phys. C Solid State Phys.* **1985**, *18*, 1891–1897. [CrossRef]
33. Becerril, H.A.; Mao, J.; Liu, Z.; Stoltenberg, R.M.; Bao, Z.; Chen, Y. Evaluation of solution processed reduced graphene oxide films as transparent conductors. *ACS Nano* **2008**, *2*, 463–470. [CrossRef] [PubMed]
34. Nan, C.-W.; Liu, G.; Lin, Y.; Li, M. Interface effect on thermal conductivity of carbon nanotube composites. *Appl. Phys. Lett.* **2004**, *85*, 3549–3551. [CrossRef]
35. Huxtable, S.T.; Cahill, D.G.; Shenogin, S.; Xue, L.; Ozisik, R.; Barone, P.; Usrey, M.; Strano, M.S.; Siddons, G.; Shim, M.; et al. Interfacial heat flow in carbon nanotube suspensions. *Nat. Mater.* **2003**, *2*, 731–734. [CrossRef]
36. Nigro, B.; Grimaldi, C.; Ryser, P. Tunneling and transport regimes in segregated composites. *Phys. Rev. E* **2012**, *85*, 011137. [CrossRef] [PubMed]
37. Zhang, F.; Feng, Y.; Feng, W. Three-dimensional interconnected networks for thermally conductive polymer composites: Design, preparation, properties and mechanism. *Mater. Sci. Eng. R Rep.* **2020**, *142*, 100580. [CrossRef]
38. Kim, S.Y.; Noh, Y.J.; Yu, J. Prediction and experimental validation of electrical percolation by applying a modified micromechanics model considering multiple heterogeneous inclusions. *Compos. Sci. Technol.* **2015**, *106*, 156–162. [CrossRef]
39. Kirkpatrick, S. Percolation and conduction. *Rev. Mod. Phys.* **1973**, *45*, 574. [CrossRef]
40. Kim, H.S.; Kim, J.H.; Kim, W.Y.; Lee, H.S.; Kim, S.Y.; Khil, M.-S. Volume control of expanded graphite based on inductively coupled plasma and enhanced thermal conductivity of epoxy composite by formation of the filler network. *Carbon* **2017**, *119*, 40–46. [CrossRef]
41. Park, S.-H.; Ha, J.-H. Improve electromagnetic interference shielding properties through the use of segregate carbon nanotube network. *Materials* **2019**, *12*, 1395. [CrossRef]
42. Vovchenko, L.; Matzui, L.; Oliynyk, V.; Milovanov, Y.; Mamunya, Y.; Volynets, N.; Plyushch, A.; Kuzhir, P. Polyethylene composite with segregated carbon nanotubes network: Low frequency plasmons and high electromagnetic interference shielding efficiency. *Materials* **2020**, *13*, 1118. [CrossRef]
43. Wang, D.; Zhang, X.; Zha, J.-W.; Zhao, J.; Dang, Z.-M.; Hu, G.-H. Dielectric properties of reduced graphene oxide/polypropylene composites with ultralow percolation threshold. *Polymer* **2013**, *54*, 1916–1922. [CrossRef]
44. Mamunya, Y.; Matzui, L.; Vovchenko, L.; Maruzhenko, O.; Oliynyk, V.; Pusz, S.; Kumanek, B.; Szeluga, U. Influence of conductive nano- and microfiller distribution on electrical conductivy and EMI shielding properties of polymer/carbon composites. *Compos. Sci. Technol.* **2019**, *170*, 51–59. [CrossRef]
45. Wu, C.; Huang, X.; Wang, G.; Lv, L.; Chen, G.; Li, G.; Jiang, P. Highly Conductive nanocomposites with three-dimensional compactly interconnected graphene networks via a self assembly process. *Adv. Funct. Mater.* **2012**, *28*, 506–513.
46. Pang, H.; Yan, D.-X.; Bao, Y.; Chen, J.-B.; Chen, C.; Li, Z.-M. Super-tough conducting carbon nanotube/ultrahigh-molecular-weight polyethylene composites with segregated and double-percolated structure. *J. Mater. Chem.* **2012**, *22*, 23568–23575. [CrossRef]
47. Tu, Z.; Wang, J.; Yu, C.; Xiao, H.; Jiang, T.; Yang, Y.; Shi, D.; Mai, Y.-W.; Li, R.K.Y. A facile approach for preparation of polystyrene/graphene nanocomposites with ultra-low percolation threshold through an electrostatic assembly process. *Compos. Sci. Technol.* **2016**, *134*, 49–56. [CrossRef]
48. Pang, H.; Bao, Y.; Lei, J.; Tang, J.-H.; Ji, X.; Zhang, W.-Q.; Chen, C. Segregated conductive ultrahigh-molecular-weight polyethylene composites containing high-density polyethylene as carrier polymer of graphene nanosheets. *Polym. Plast. Technol. Eng.* **2012**, *51*, 1483–1486. [CrossRef]
49. Wang, B.; Li, H.; Li, L.; Chen, P.; Wang, Z.; Gu, Q. Electrostatic adsorption method for preparing electrically conducting ultrahigh molecular weight polyethylene/graphene nanosheets composites with a segregated network. *Compos. Sci. Technol.* **2013**, *89*, 180–185. [CrossRef]
50. Hu, H.; Zhang, G.; Xiao, L.; Wang, H.; Zhang, Q.; Zhao, Z. Preparation and electrical conductivity of graphene/ultrahigh molecular weight polyethylene composites with a segregated structure. *Carbon* **2012**, *50*, 4596–4599. [CrossRef]
51. Yang, J.-C.; Wang, X.-J.; Zhang, G.; Wei, Z.-M.; Long, S.-R.; Yang, J. Segregated poly(arylene sulfide sulfone)/graphene nanoplatelet composites for electromagnetic interference shielding prepared by the partial dissolution method. *RSC Adv.* **2020**, *10*, 20817–20826. [CrossRef]
52. Jang, J.-U.; Nam, H.E.; So, S.O.; Lee, H.; Kim, G.S.; Kim, S.Y.; Kim, S.H. Thermal percolation behavior in thermal conductivity of polymer nanocomposite with lateral size of graphene nanoplatelet. *Polymers* **2022**, *14*, 323. [CrossRef] [PubMed]
53. Shtein, M.; Nadiv, R.; Buzagio, M.; Kahil, K.; Regev, O. Thermally conductive graphene-polymer composites: Size, percolation, and synergy effects. *Chem. Mater.* **2015**, *27*, 2100–2106. [CrossRef]

Disclaimer/Publisher's Note: The statements, opinions and data contained in all publications are solely those of the individual author(s) and contributor(s) and not of MDPI and/or the editor(s). MDPI and/or the editor(s) disclaim responsibility for any injury to people or property resulting from any ideas, methods, instructions or products referred to in the content.

Article

Fabrication of Multifunctional Silylated GO/FeSiAl Epoxy Composites: A Heat Conducting Microwave Absorber for 5G Base Station Packaging

Zhuyun Xie [1], Dehai Xiao [1], Qin Yu [1], Yuefeng Wang [1], Hanyi Liao [1], Tianzhan Zhang [2], Peijiang Liu [3,*] and Liguo Xu [4,*]

1. Centre of Chip Chemistry, Huangpu Institution of Materials, Changchun Institute of Applied Chemistry, Chinese Academy of Sciences, Guangzhou 510663, China; zx263@cornell.edu (Z.X.); dhxiao@ciac.ac.cn (D.X.); y187459762379@163.com (Q.Y.); wyf@mail.ipc.ac.cn (Y.W.); liaohanyi@ciac.ac.cn (H.L.)
2. College of Material Science and Engineering, Jilin Jianzhu University, Changchun 130119, China; zhangtz048@mail.ipc.ac.cn
3. Reliability Physics and Application Technology of Electronic Component Key Laboratory, The Fifth Electronics Research Institute of the Ministry of Information Industry, Guangzhou 510610, China
4. College of Light Chemical Industry and Materials Engineering, Shunde Polytechnic, Foshan 528333, China
* Correspondence: cz2343222@163.com (P.L.); 21099@sdpt.edu.cn (L.X.)

Abstract: A multifunctional microwave absorber with high thermal conductivity for 5G base station packaging comprising silylated GO/FeSiAl epoxy composites were fabricated by a simple solvent-handling method, and its microwave absorption properties and thermal conductivity were presented. It could act as an applicable microwave absorber for highly integrated 5G base station packaging with 5G antennas within a range of operating frequency of 2.575–2.645 GHz at a small thickness (2 mm), as evident from reflection loss with a maximum of −48.28 dB and an effective range of 3.6 GHz. Such a prominent microwave absorbing performance results from interfacial polarization resonance attributed to a nicely formed GO/FeSiAl interface through silylation. It also exhibits a significant enhanced thermal conductivity of 1.6 W/(mK) by constructing successive thermal channels.

Keywords: wave absorption; thermal conduction; 5G base station; packaging materials; epoxy resin

1. Introduction

With the popularization of 5G communication, 5G base stations are increasingly distributed. The 5G base station has a higher frequency band, an ultra large bandwidth, more transmitting and receiving antennas, and more complex beamforming working modes compared to 4G applications, which requires increasing power density and integration [1]. This may cause a series of problems. On one hand, a number of digital parts inside the 5G base station, such as high-frequency signal lines, pins of integrated circuits, and various types of connectors, may emit mass microwaves, affecting the normal operation of microwave-sensitive elements inside and outside the 5G station [2,3]. It may also have an interaction with living species [4]. On the other hand, higher power density generates more heat. This excess heat has difficulty exporting through thermal conduction between electronic components, causing the temperature inside the base station to rise, significantly reducing its operating life [5]. Therefore, the demand for heat-conducting and wave-absorbing materials is growing rapidly. To solve this problem, thermally conductive materials are applied on the surface of electronic components. However, since thermal conductive material already occupies limited space inside of the device gap, there is no space for additional wave-absorbing material [6]. Also, the continuous thermal path required for heat conduction will be drastically reduced once it is blocked by wave-absorbing materials with low thermal conductivity [7]. Therefore, the demand for a material with both heat-conducting and wave-absorbing properties is growing rapidly.

Figure 1 shows the typical application and working mechanism of existing wave-absorbing materials in 5G base stations. There are two widely used solutions—one is microwave-absorbing thermal interface material (TIM) [8] by attaching a functional polymer pad directly to the surface of the microwave or heat source so that the heat can be directly transmitted to the shielding and dissipated. The second method is the form-in-place (FIP) sealing gasket [9], which acts in the gaps at bulkhead joints and the gaps at shielding joints to avoid the leakage of electromagnetic waves and the blocking of heat conduction. Leading product providers include Laird Co., Nolato Co., FRD Co., etc. However, existing products either have a relatively low reflection loss of −10 dB with a low thermal conductivity of 1 W/(mK) or have a high thermal conductivity but with no wave-absorbing ability [10]. A material that combines both properties has yet to be developed. The most widely used packaging materials for the 5G base station are polymer materials, among which epoxy resin attracts attention for its excellent adhesive and mechanical properties. However, epoxy resins have poor thermal conductivity and microwave-absorbing properties [11,12]. Normally, functional fillers are used in polymer materials to achieve better thermal and magnetic properties. Recently, carbon materials have received extensive attention thanks to their excellent dielectric properties as well as high electrical conductivity, such as carbon nanotubes, graphene, etc. [13–18]. Among them, graphene oxide has residual defects and an amount of epoxy, hydroxyl, and carboxyl groups on its surface. These factors may result in the transition from contiguous states to the Fermi level, thus proposing an impedance match performance. Moreover, defect polarization relaxation and groups' electronic dipole polarization relaxation are also beneficial for enhancing their wave-absorbing performance [19]. Last but not least, materials with 2D structures, such as graphene and MXene, exhibit excellent thermal conducting and wave-absorbing properties. Sun et al. [20] reported a self-assembly anchored MXene nanosheet loaded with CMWCNTs with a maximum reflection loss of −46 dB at a thickness of 1.5 mm. Graphene has a thermal conductivity of 5000 W/(mK) in the plane direction and 30 W/(mK) in the longitudinal direction [21].

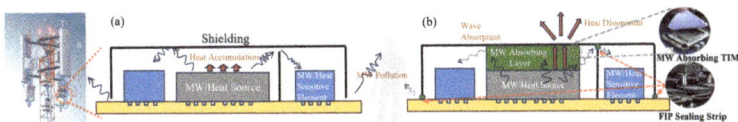

Figure 1. 5G base station (**a**) without wave-absorbing and heat conducting measure, (**b**) with wave absorbing layers and FIP sealing strips.

Yet, simply using graphene oxide as a thermally conductive and wave-absorbing filler for packaging material is impractical for several reasons. Firstly, due to its large specific surface area, a very small amount of graphene oxide will have a huge impact on the fluidity of the compounds [22]. It has been reported that pure graphene filling can only achieve a maximum of 7 dB of reflection loss [19]. Secondly, graphene oxide, working as a dielectric loss absorber and exhibiting a high reflection of microwaves, may emit microwave pollution to the surroundings [22]. To improve this, many attempts have been made to combine GO with magnetic particles. Zou et al. [23] developed Fe/GO nanocomposites by inserting Fe^{3+} into GO followed by a reduced reaction in H_2. The maximum reflection loss of 9 dB was at 11–18 GHz. Li et al. [24] synthesized GO/Fe_3O_4/ iron phthalocyanine composites using a facile one-step solvothermal method. They observed the maximum microwave absorption of −27.92 dB at 10.8 GHz. Ghosh et al. [25] fabricated n-doped GO/$MnCo_2O_4$ nanocomposites using a facile hydrothermal method followed by an annealing process, and the reflection loss was observed to fall in the range of −90 to −77 dB.

Nevertheless, the synthesis processes of the above studies are not practical for scaling up production. Selecting a suitable magnetic particle and finding a feasible way to produce the compound is important for rapid application. FeSiAl, as a type of soft magnetic alloy, has become widely used in the microwave absorption field for its excellent magnetic properties such as high saturation magnetization and high eddy current loss causing by its high permeability in the low-frequency range [26–28]. However, its high conductivity, along with a low Snoek limit at a high frequency, limit its application [29–31].

In this study, a novel epoxy-based 5G base station packaging material with comprehensive properties of high wave-absorbing ability and high thermal conductivity ability was introduced, in which GO/FeSiAl particles were made and modified by γ-aminopropyl triethoxysilane in gentle solvent conditions. By combining these two particles, an impedance match can be formed, improving its high-frequency range wave-absorbing ability [30]. Meanwhile, graphene has a large specific surface area and can act as an anti-settling agent [32], providing another application advantage. Our work provides a simple and practical way to produce multifunctional 5G base station packaging materials for application at a large scale, which has an advanced reflection loss (RL) of −48.28 dB and a wide effective range of 3.6 GHz within the range of operating frequency of 5G antennas of 2.575–2.645 GHz at a small thickness (2 mm).

2. Materials and Methods

2.1. Materials

Graphene powder was purchased from XFnano Inc. (Nanjing, China) and N-butylamine (99.7%) was purchased from Macklin (Shanghai, China). γ-aminopropyl triethoxysilane was purchased from Changhe Chemical Co., (Hangzhou, China), FeSiAl was purchased from Mana New Material Co., (Changsha, China), Epoxy resin (BE 188EL) was purchased from Changchun Chemical Co., (Taipei, Taiwan), and Polyetheramine (Jeffamine D-2000) was purchased from Huntsman Corp., (The Woodlands, TX, USA).

2.2. Preparation of GO/Ethanol Suspension

The GO was prepared by oxidizing graphene powder using the Hummers' method [33]. After oxidation, the obtained graphene oxide powder was washed several times using deionized water through a centrifuge to remove the residual salts and acids. The GO was then silylated according to Mastsuo's method [34]. The washed GO was mixed with a certain amount of butylamine for exfoliation and then subjected to ultrasonication for 30 min. The dispersion was refluxed at 60 °C for 60 min. Then, the exfoliated GO was centrifuged with ethanol several times to remove the residual butylamine and was dispersed into ethanol via ultrasonication for 30 min.

2.3. Preparation of Silylated GO@FeSiAl Nanoparticles

A certain amount of FeSiAl powder was added into GO/ethanol suspension at different ratios. The weight ratio of FeSiAl and GO is listed in Table 1, and varied from 1000:1 to 10:1. Then, the GO/FeSiAl was dispersed into an ethanol/deionized water solution (0.1 g/mL), in which the weight ratio of ethanol and deionized water was 9:1. The mixture was stirred and refluxed for 3 h at 80 °C, while slowly dripping γ-aminopropyl triethoxysilane (0.2 mg/mL). The silylated GO/FeSiAl was suction-filtrated and vacuum dried at 80 °C for 24 h. A group of pure GO without FeSiAl was created as the control group compared to the GF groups. GO-1 was the GO having gone through all the processes but without FeSiAl.

Table 1. The composition of the samples.

Samples	Coupling Agent Content (mg/mL)	Mass Ratio of FeSiAl (%)	Mass Ratio of GO (%)
GO-1	0.2	0	1
GF-0	0	99	1
GF-1	0.2	100	0
GF-2	0.2	99.9	0.1
GF-3	0.2	99	1
GF-4	0.2	90	10

2.4. Preparation of GO/FeSiAl Epoxy Compounds

The dried GO/FeSiAl powder was then mixed with epoxy resin and polyetheramine using a planetary stir at a speed of 2000 rpm for 5 min, in which the active hydrogen equivalent of epoxy resin and polyetheramine was 1:1. Then, the epoxy resin compound was cured at 80 °C for 24 h. The resulting sample was called a GO/FeSiAl epoxy compound.

2.5. Characterization Techniques

The scanning electron microscope (SEM) images of particles were taken on a JSM-IT800 instrument, JEOL, Akishima-shi, Tokyo, Japan operating with 5 kV. The X-ray diffraction (XRD) patterns of the compounds were acquired in the range of 10–90° on a D8 ADVANCE instrument, Bruker, Billerica, MA, USA with CuKα radiation and a scanning rate of 3°/min. The Fourier transform infrared spectroscopy (FTIR) analysis of particles was recorded in KBr in the form of a compressed pellet in the range of 400–4000 cm^{-1} on Vertex 70 V. Bruker, Billerica, MA, USA.

Microwave absorption properties were characterized by measuring the magnetic and dielectric properties of the compounds in the frequency range from 0.1 GHz to 18 GHz, covering most of the microwave pollution frequency in our daily life, such as communication devices, satellite communications, radar, etc. [35]. The sample was prepared in the shape of concentric circles with a thickness of 3 mm, an inner diameter of 3 mm, and an outer diameter of 7 mm, and was tested using the coaxial method based on the ASTM D5568-2a standard [36], on an ENA Series Network Analyzer, N5080a Agilent, Santa Clara, CA, USA. As shown in Figure 2, the concentric circle sample was put on the coaxial airline (sample holder) and then accessed the coaxial transmission line through a connector to the Network Analyzer. Then, the reflection loss was calculated using transmission line theory. A detailed illustration of the fabrication and the characterization method is presented in Figure 2.

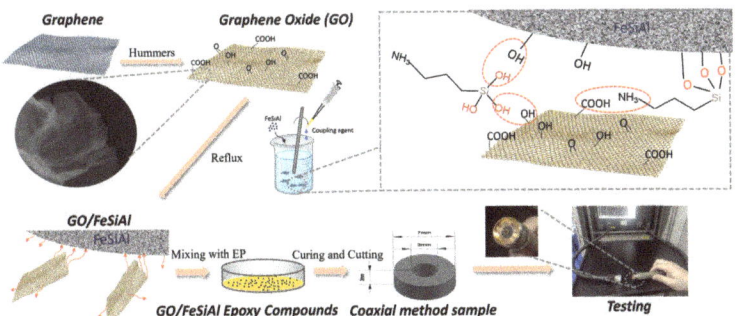

Figure 2. Schematic illustration for the fabrication of GO/FeSiAl epoxy compounds and testing methods.

3. Results and Discussions

3.1. Morphology and Structural Analysis

SEM images of GO/FeSiAl composites with or without a coupling agent are displayed in Figure 3. GF-3 presents a blurred interface of GO and FeSiAl, suggesting a good compatibility, whereas GF-0 shows a clear interface, indicating poor compatibility. It can be concluded that the use of coupling agents enhances the compatibility of GO and FeSiAl particles, which is important for the generation of heterogeneous interfaces and a stronger interfacial polarization loss.

Figure 4 shows the XRD spectra of GO, GO-1, FeSiAl, and GF-3. The diffraction peak at $2\theta = 11°$ derived from GO shifted to the lower region in GO-1, indicating that the interlayer space increased during the silylating process. It might be caused by long coupling agent molecules attaching to the GO sheet surface and exfoliating the GO sheets, causing the interlayer spacing to increase. In Figure 4 (c), there are three peaks associated with (220),

(400), and (211) planes, consisting of the standard diffraction spectra of the body-centered cubic structure, FeSiAl. There is a fairly gentle peak within the range of $2\theta = 20°$ to $2\theta = 35°$ of GF-3 (Figure 4 (d)) compared to FeSiAl, which is the characteristic peak of GO, indicating that a small amount of GO was grafted onto FeSiAl.

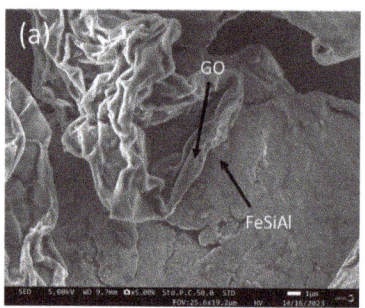 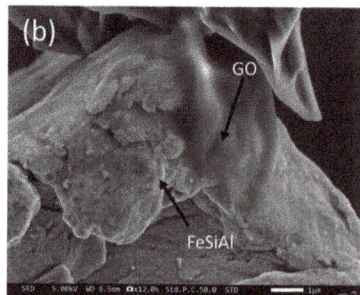

Figure 3. SEM images of (**a**) GF-0 and (**b**) GF-3.

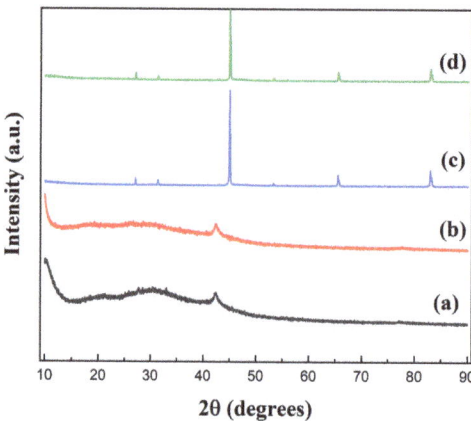

Figure 4. XRD spectra of (a) GO, (b) silylated GO, (c) FeSiAl, and (d) GF-3.

Figure 5 shows the FTIR spectra of the GF-0 and GF-3 samples. The absorption peak of GF-0 at 3450 cm^{-1} is attributed to -OH from absorbing water and hydrogen bonds between the layered structure of GO. In addition, the absorption peak of GF-3 at around 3450 cm^{-1} shifts to 3425 cm^{-1} and is lower than that of GO, indicating that the force between GO layers is weakened and chemical bonding between GO and FeSiAl is generated instead. The relative intensity of the peaks at 1637 cm^{-1} is derived from the C=O group, which represents functional groups on the GO surface and becomes wider in GF-3, resulting from several peaks overlapping each other. Those peaks are generated from the amide groups by the carboxyl group on the GO surface reacting with the amino group on the coupling agent. The peak at 1250 cm^{-1} represents the stretching and bending vibration of C-N groups in the amide group or coupling agent. Moreover, the peaks around 1120 cm^{-1} and 1250 cm^{-1} are owing to the Si-O bond and the stretching and bending vibration of the C-N groups overlapping each other, which may be from the amide group, products of the coupling agent and epoxy groups, or the coupling agent itself. The peak at 658 cm^{-1} is owing to the stretching vibration of N-H groups from amide or amino groups. Since FeSiAl is a metallic compound and shows little or no transmittance, these results suggest that the coupling agent was successfully grafted on GO.

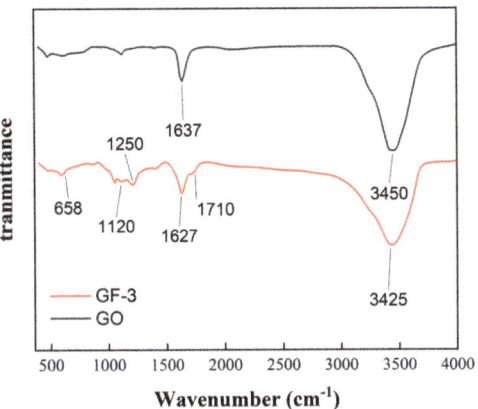

Figure 5. FTIR spectra of GO and GF-3.

3.2. Microwave Absorption Properties

To relieve the microwave absorption mechanism of the GO/FeSiAl/epoxy compound, the influence of GO amounts on reflection loss, impedance matching, and the attenuation constant was studied.

The electromagnetic properties were tested through the coaxial method within the frequency range of 0 GHz to 18 GHz to reveal the relationship between the complex permeability real part (μ'), the complex permeability imaginary part (μ''), the relative complex permittivity real part (ε'), the relative complex permittivity imaginary part (ε''), and the amount of GO in the compounds, as shown in Figure 6.

Figure 6a shows a decreasing trend of ε' with increasing GO amounts, responding to the ability of the material to store charge decreases. This can be explained through the micro-capacitance model [37]. Neighboring FeSiAl presented in the matrix can be seen as two flat plates of micro-capacitor. While GO conducts electricity very well and has poor ability to stores electrons, therefor cannot form a micro-capacitor structure within itself or with FeSiAl [38]. Thus, the addition of GO will destroy the micro-capacitor between the FeSiAl and leads to a decrease in charge storing ability. In Figure 6b, the ε'' values exhibit an obvious peak in sample GF-1 and sample GF-2 with fewer or no GO. As the amount of GO increases, more peaks appear at higher frequencies. These peaks can be ascribed to the interfacial polarization resonance, indicating more GO-FeSiAl interface in the system as well as defects and functional groups on the surface of GO resulting in higher interfacial polarization. This contributes to higher microwave absorption at higher frequencies [39,40]. As the frequency rises, μ' shows a collectively decreasing trend (Figure 6c). This may be explained by high-frequency fields resulting in polarization hysteresis and dielectric relaxation. This trend slows down as more GO is added. The complex permeability imaginary part (ε'') of GF-1 and GF-2 with a lower amount of GO presents a pronounced and sharp peak at 0.3 GHz, which is attributed to the magnetic capacity of the FeSiAl consuming electromagnetic energy, as shown in Figure 6d. The high-frequency shift to 2 GHz and widening of peaks for samples with a higher GO amount may be ascribed to the suppression of eddy current resulting from the lower electric conductivity of GO-FeSiAl compared to pure FeSiAl [41].

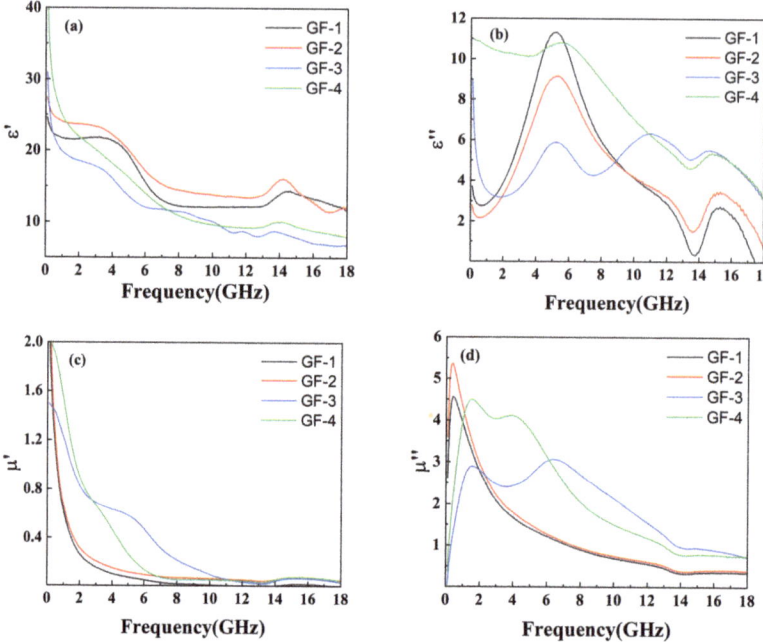

Figure 6. Plots of (**a**) ε', (**b**) ε'', (**c**) μ', and (**d**) μ'' versus frequency for GO/FeSiAl epoxy compounds with different GO amounts.

According to the transmission line theory, ε_r ($\varepsilon_r = \varepsilon' - j\varepsilon''$) represents relative complex permittivity and μ_r ($\mu_r = \mu' - j\mu''$) represents relative complex permeability. The reflection loss properties (RL) of the GO/FeSiAl epoxy compound can be calculated from ε_r and μ_r through the following equations:

$$RL = 20\lg\left|\frac{z_{in} - z_0}{z_{in} + z_0}\right| \quad (1)$$

$$Z_{in} = Z_0\sqrt{\frac{\mu_r}{\varepsilon_r}}\tan\left(j\frac{2\pi f d}{c}\sqrt{\mu_r \varepsilon_r}\right), \quad (2)$$

where Z_{in} is the normalized impedance of the sample, Z_0 is the free space impedance, f is the frequency of the incident microwave, c is the velocity of light in free space, and d is the thickness of the testing sample. The 3D projections of the reflection loss of the epoxy composites from 0 mm to 5 mm thickness and of the 0 GHz to 18 GHz frequency variation are shown in Figure 7. The black contour range of the effective bandwidth indicates an RL value below −10 dB (90% absorption). All the epoxy composites have a fairly wide area to within this standard. In the sample with pure silylated FeSiAl (Figure 7a), the maximum reflection loss is −30.75 dB at 0.5 GHz and at a 4.95 mm thickness, with an efficient bandwidth of 0.4 GHz. While, as shown in Figure 7b, with only 0.1% of GO amount in the FeSiAl system, the maximum reflection loss increases by 10% to −34.00 dB at 0.5 GHz, and the effective bandwidth increases by 20% to 0.5 GHz. However, epoxy compounds with pure magnetic particles or a low GO amount show fairly narrow effective bandwidths, and strong absorption peaks only fall in a very low frequency band. As shown in Figure 7c for the GF-3 sample with 1% of GO, the maximum reflection loss increases to −48.29 dB at 1.6 GHz and the effective bandwidth increases to 3.8 GHz, which exhibits a wider bandwidth (3.8 GHz vs. 3.12 GHz [42] and 3.52 GHz [43]) and a higher refection loss (−48.29 dB vs. −47 dB [44], −44.47 dB [42] and −48.08 dB [39]) in the most recent relevant literature, using FeSiAl as a microwave absorber. This provides a perfect match for 5G station application, as the effective bandwidth just falls in the range of the 5G antennas' operating frequency band in the 2.575–2.645 GHz range [45], as shown in Figure 8. It can be concluded that, in most cases, the microwave reflection loss increased greatly as the GO amount increased. However, it decreases as the amount of GO reaches 10% in GF-4, as shown in Figure 7d. This might result from the decrease of complex permittivity. Overall, controlling the amount of GO in GO/SiFeAl epoxy compounds can facilely adjust the dissociation and frequency of the absorption peak.

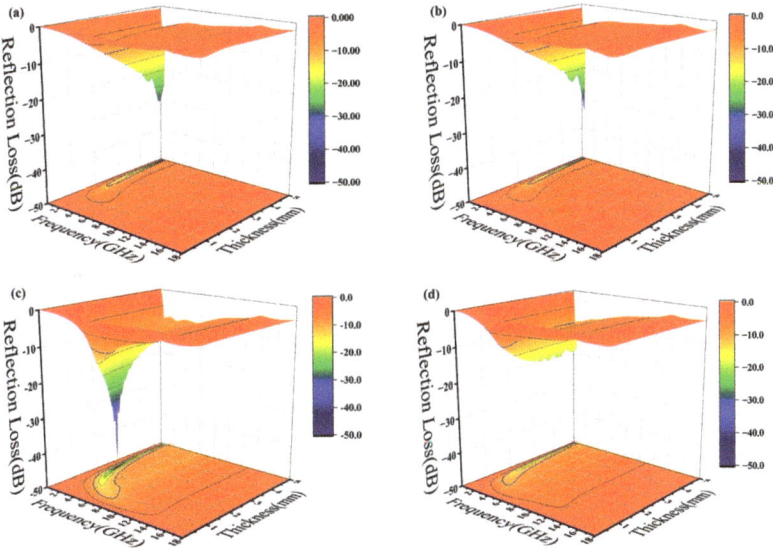

Figure 7. Three-dimensional (3D) projections of reflection loss of epoxy composites with (**a**) GF-1, (**b**) GF-2, (**c**) GF-3, and (**d**) GF-4.

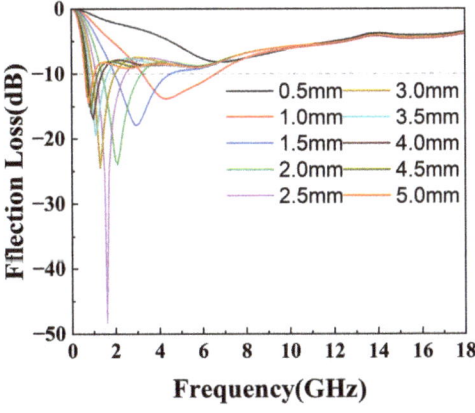

Figure 8. Reflection loss of the epoxy compound with GF-3 at different thicknesses.

Factors affecting the performance of wave absorption also include impedance matching and the attenuation constant. Generally, the closer the value of impedance matching Z ($Z = Z_{in}/Z_0$) is to 1, the more the incident electromagnetic wave can enter the material without being emitted. Another parameter reflecting the attenuation characteristics is the attenuation constant (α), which can be calculated through the following equation:

$$\alpha = \frac{\sqrt{2}\pi f}{c}\sqrt{(\mu''\varepsilon'' - \mu'\varepsilon') + \sqrt{(\mu''\varepsilon'' - \mu'\varepsilon')^2 + (\mu'\varepsilon'' + \mu''\varepsilon')^2}}. \tag{3}$$

The attenuation constant reflects the material's ability to dissipate electromagnetic waves. However, a strong attenuation constant often leads to low impedance matching, so it is necessary to balance the effects of both. The attenuation constant of the compounds is shown in Figure 9 while the impedance matching is shown in Figure 10. As shown in Figure 9, the attenuation constant increases as the amount of GO grows, indicating more electromagnetic energy loss resulting from the porous structure and defects on the surface of GO. However, GF-4 with the largest GO amount and the highest attenuation constant shows a low reflection loss due to its low impedance matching as shown in Figure 10d, which is consistent with the conclusion from the reflection loss results. In conclusion, 1% of GO is a reasonable amount among others that reaches a balance between magnetic loss and conductive loss.

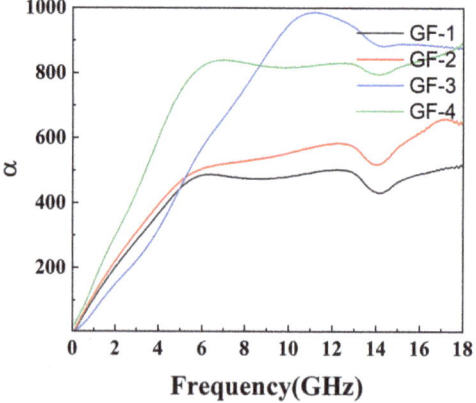

Figure 9. Attenuation constant (α) of epoxy compounds.

Figure 11 is the schematic of the GO/FeSiAl microwave absorption mechanism. To begin with, GO and FeSiAl can perform as wave-absorbing agents separately. GO has a high dielectric loss from defect polarization and dipole polarization, while FeSiAl has a high magnetic loss from eddy current loss at a low frequency and nature resonance at a higher frequency. Then, since GO and FeSiAl have different polarity and conductivity, interfacial polarization occurs at the interface between GO and FeSiAl, which contributes greatly to the microwave dielectric loss. There is also conductive loss between FeSiAl interfaces. Finally, GO and FeSiAl have a synergistic effect as, at a high frequency, GO can lower FeSiAl's permeability and prevent a skin effect, therefore forming a better impedance matching. It can be concluded that the GO/FeSiAl epoxy compounds have strong absorption and a broad effective bandwidth covering the operating frequency of 5G antennas with a thin coating so that it could provide a potential means for efficient microwave absorption for a 5G base station.

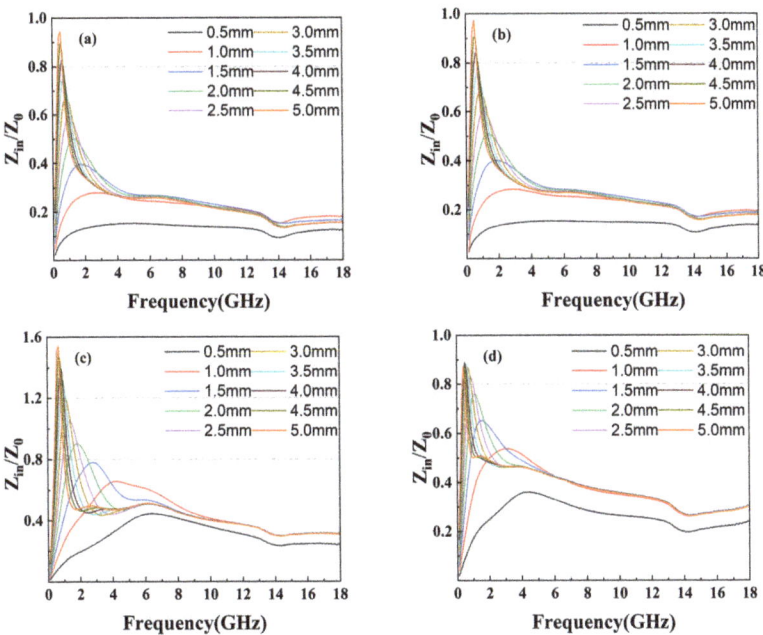

Figure 10. Impedance matching Z of epoxy composites with (**a**) GF-1, (**b**) GF-2, (**c**) GF-3, and (**d**) GF-4.

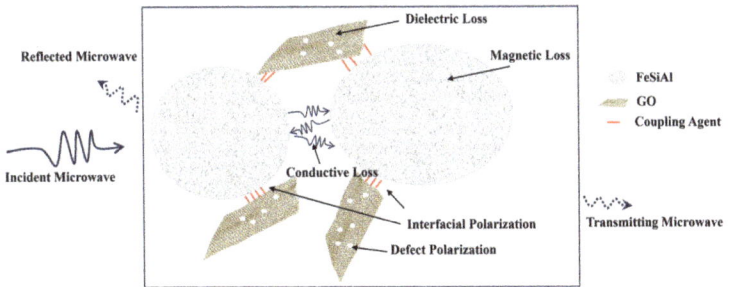

Figure 11. Schematic of the GO/FeSiAl epoxy compounds microwave absorption mechanism.

3.3. Thermal Conductivity Analysis

Figure 12 shows the thermal conductivity of GF 3/epoxy compounds in different weight fractions. The thermal conductivity grows as the weight fraction increases, and a sudden acceleration of growth occurs at 60%. This is because the epoxy matrix's thermal conductivity is very low compared to the filler itself. When the fraction of filler is low, the epoxy matrix acts as a thermal barrier positioned between fillers [46]. When the filler fraction reaches a certain point, a thermal channel is formed. The maximum result of 1.6 W/(mK) was obtained when the weight fraction of GF-3 in epoxy compounds was 90%, which was 605% higher than that of pure epoxy, respectively.

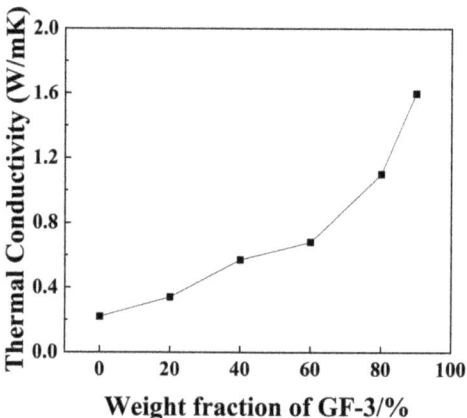

Figure 12. Thermal conductivity analysis of epoxy compounds with different ratios of GF-3.

4. Conclusions

Our work provides a simple and practical way to fabricate multifunctional 5G base station packaging materials comprising silylated GO/FeSiAl/epoxy, which can be realized by using a simple solvent-handling method. It could act as a small thickness microwave absorber within a range of operating frequency for the 5G antennas of 2.575–2.645 GHz, as evidenced by a reflection loss with a maximum of −48.28 dB and an effective range of 3.6 GHz. Such a prominent microwave absorbing performance results not only from the interfacial polarization reasonance attributed to the nicely formed GO/FeSiAl interface but also the synergistic effect from excellent impedance matching by GO compromising FeSiAl's high conductivity at a high frequency. It also exhibits a significantly enhanced thermal conductivity of 1.6 W/(mK), performing a remarkable heat conducting ability by constructing successive thermal channels. All in all, the GO/FeSiAl epoxy compounds show promising results of strong absorption and a broad effective bandwidth covering the operating frequency of 5G antennas with a thin coating so that it could provide potential efficient microwave absorption for a 5G base station.

Author Contributions: Conceptualization, Z.X.; Data curation, T.Z.; Formal analysis, Y.W.; Funding acquisition, D.X.; Investigation, Q.Y.; Methodology, Z.X.; Project administration, D.X.; Software, H.L.; Supervision, L.X.; Validation, Z.X.; Writing—original draft, Z.X.; Writing—review and editing, P.L. All authors have read and agreed to the published version of the manuscript.

Funding: This research received no external funding.

Institutional Review Board Statement: Not applicable.

Informed Consent Statement: Not applicable.

Data Availability Statement: Data are contained within the article.

Conflicts of Interest: The authors declare no conflict of interest.

References

1. Tao, H.; Landy, N.I.; Bingham, C.M.; Zhang, X.; Averitt, R.D.; Padilla, W.J. A metamaterial absorber for the terahertz regime: design, fabrication and characterization. *Opt. Express* **2008**, *16*, 7181–7188. [CrossRef]
2. Bai, M. Elctromagnetic Compatibility Analysis of Electronic Circuit. *Mod. Electron. Tech.* **2009**, *32*, 191–194.
3. Li, F.; Tang, B.; Qi, D.; Liu, X.; Kuang, Y.; Xie, H. Electromagnetic Interference from 5G Base Station Antenna in Substation on Secondary Equipment. *South. Power Syst. Technol.* **2021**, *15*, 111–117.
4. Gultekin, D.H.; Siegel, P.H. Absorption of 5G radiation in brain tissue as a function of frequency, power and time. *IEEE Access* **2020**, *8*, 115593–115612. [CrossRef]

5. Zhou, W.Y.; Qi, S.H.; Zhao, H.Z.; Liu, N.L. Thermally conductive silicone rubber reinforced with boron nitride particle. *Polym. Compos.* **2007**, *28*, 23–28. [CrossRef]
6. Tao, L.; Li, C.; Ren, Y.; Li, H.; Chen, J.; Yang, Q. Synthesis of polymer/CNTs composites for the heterogeneous asymmetric hydrogenation of quinolines. *Chin. J. Catal.* **2019**, *40*, 1548–1556. [CrossRef]
7. Wong, C.; Bollampally, R.S. Comparative study of thermally conductive fillers for use in liquid encapsulants for electronic packaging. *IEEE Trans. Adv. Packag.* **1999**, *22*, 54–59. [CrossRef]
8. Zhou, Y.; Wu, S.; Long, Y.; Zhu, P.; Wu, F.; Liu, F.; Murugadoss, V.; Winchester, W.; Nautiyal, A.; Wang, Z.; et al. Recent advances in thermal interface materials. *ES Mater. Manuf.* **2020**, *7*, 4–24. [CrossRef]
9. Ji, J.; Li, W.; Liu, Y.; Du, H.; Guo, W.; Su, Y. A method to determine an electrical parameter of form-in-place shielding material with two other parameters known by calculation and simulation. In Proceedings of the 2022 Asia-Pacific International Symposium on Electromagnetic Compatibility (APEMC), Beijing, China, 1–4 September 2022; pp. 807–809.
10. Turunen, J. 3W SFP Interface Development. 2022. Available online: https://urn.fi/URN:NBN:fi:amk-2022092920595 (accessed on 12 August 2022).
11. Gu, J.; Zhang, Q.; Dang, J.; Xie, C. Thermal conductivity epoxy resin composites filled with boron nitride. *Polym. Adv. Technol.* **2012**, *23*, 1025–1028. [CrossRef]
12. Fu, Y.X.; He, Z.X.; Mo, D.C.; Lu, S.S. Thermal conductivity enhancement with different fillers for epoxy resin adhesives. *Appl. Therm. Eng.* **2014**, *66*, 493–498. [CrossRef]
13. Guo, X.; Liu, G. Electromagnetic Shielding Enhancement of Butyl Rubber/Single-Walled Carbon Nanotube Composites via Water-Induced Modification. *Polymers* **2023**, *15*, 2101. [CrossRef] [PubMed]
14. Guo, X.; Liu, L.; Ding, N.; Liu, G. Transformation from Electromagnetic Inflection to Absorption of Silicone Rubber and Accordion-Shaped Ti3C2MXene Composites by Highly Electric Conductive Multi-Walled Carbon Nanotubes. *Polymers* **2023**, *15*, 2332. [CrossRef] [PubMed]
15. Han, B.; Wang, Y. Performance Simulation and Fused Filament Fabrication Modeling of the Wave-Absorbing Structure of Conductive Multi-Walled Carbon Nanotube/Polyamide 12 Composite. *Polymers* **2023**, *15*, 804. [CrossRef] [PubMed]
16. Qian, Y.; Luo, Y.; Li, Y.; Xiong, T.; Wang, L.; Zhang, W.; Gang, S.; Li, X.; Jiang, Q.; Yang, J. Enhanced electromagnetic wave absorption, thermal conductivity and flame retardancy of BCN@ LDH/EP for advanced electronic packing materials. *Chem. Eng. J.* **2023**, *467*, 143433. [CrossRef]
17. Zhao, Y.; Long, A.; Zhao, P.; Liao, L.; Wang, R.; Li, G.; Wang, B.; Liao, X.; Yu, R.; Liao, J. Natural Hollow Fiber-Derived Carbon Microtube with Broadband Microwave Attenuation Capacity. *Polymers* **2022**, *14*, 4501. [CrossRef] [PubMed]
18. Yang, Z.; Wu, H.; Zhang, R.; Deng, K.; Li, Y.; Liu, Z.; Zhong, Q.; Kang, Y. Effect of graphene/spherical graphite ratio on the properties of PLA/TPU composites. *Polymers* **2022**, *14*, 2538. [CrossRef] [PubMed]
19. Wang, C.; Han, X.; Xu, P.; Zhang, X.; Du, Y.; Hu, S.; Wang, J.; Wang, X. The electromagnetic property of chemically reduced graphene oxide and its application as microwave absorbing material. *Appl. Phys. Lett.* **2011**, *98*, 072906. [CrossRef]
20. Sun, C.; Li, Q.; Jia, Z.; Wu, G.; Yin, P. Hierarchically flower-like structure assembled with porous nanosheet-supported MXene for ultrathin electromagnetic wave absorption. *Chem. Eng. J.* **2023**, *454*, 140277. [CrossRef]
21. Zhao, W.; Fu, R.; Gu, X.; Wang, X.; Fang, J. Interface structure and thermal conductivity of polymer matrix composite. *Mater. Rep.* **2013**, *3*, 76–79.
22. Cui, G.; Lu, Y.; Zhou, W.; Lv, X.; Hu, J.; Zhang, G.; Gu, G. Excellent microwave absorption properties derived from the synthesis of hollow Fe3O4@ reduced graphite oxide (RGO) nanocomposites. *Nanomaterials* **2019**, *9*, 141. [CrossRef]
23. Zou, Y.H.; Liu, H.B.; Yang, L.; Chen, Z.Z. The influence of temperature on magnetic and microwave absorption properties of Fe/graphite oxide nanocomposites. *J. Magn. Magn. Mater.* **2006**, *302*, 343–347. [CrossRef]
24. Li, J.; Wei, J.; Pu, Z.; Xu, M.; Jia, K.; Liu, X. Influence of Fe3O4/Fe-phthalocyanine decorated graphene oxide on the microwave absorbing performance. *J. Magn. Magn. Mater.* **2016**, *399*, 81–87. [CrossRef]
25. Ghosh, K.; Srivastava, S.K. Fabrication of N-doped reduced graphite Oxide/MnCo2O4 nanocomposites for enhanced microwave absorption performance. *Langmuir* **2021**, *37*, 2213–2226. [CrossRef] [PubMed]
26. Li, Z.; Li, Z.; Yang, H.; Li, H.; Liu, X. Soft magnetic properties of gas-atomized FeSiAl microparticles with a triple phosphoric acid-sodium silicate-silicone resin insulation treatment. *J. Electron. Mater.* **2022**, *51*, 2142–2155. [CrossRef]
27. He, J.; Liu, X.; Deng, Y.; Peng, Y.; Deng, L.; Luo, H.; Cheng, C.; Yan, S. Improved magnetic loss and impedance matching of the FeNi-decorated Ti3C2Tx MXene composite toward the broadband microwave absorption performance. *J. Alloys Compd.* **2021**, *862*, 158684. [CrossRef]
28. Zhi, D.; Li, T.; Qi, Z.; Li, J.; Tian, Y.; Deng, W.; Meng, F. Core-shell heterogeneous graphene-based aerogel microspheres for high-performance broadband microwave absorption via resonance loss and sequential attenuation. *Chem. Eng. J.* **2022**, *433*, 134496. [CrossRef]
29. Li, H.; Cheng, H.; Liu, H.; Long, L.; Liu, X. Carbon nanotubes/FeSiAl hybrid flake for enhanced microwave absorption properties. *J. Electron. Mater.* **2022**, *51*, 6986–6994. [CrossRef]
30. Li, Z.; Wang, J.; Zhao, F. Study on the electromagnetic properties and microwave absorbing mechanism of flaky FeSiAl alloy based on annealing and phosphate coating. *Mater. Res. Express* **2021**, *8*, 066526. [CrossRef]
31. Sun, J.; Xu, H.; Shen, Y.; Bi, H.; Liang, W.; Yang, R.B. Enhanced microwave absorption properties of the milled flake-shaped FeSiAl/graphite composites. *J. Alloys Compd.* **2013**, *548*, 18–22. [CrossRef]

32. Xie, L. Synthesis and Rheology of Magnetorheological Fluids Based on High Viscosity Linear Polysiloxane with Focus on Sedimentation. Ph.D. Thesis, Chongqing University, Chongqing, China, 2016.
33. Hummers, W.S., Jr.; Offeman, R.E. Preparation of graphitic oxide. *J. Am. Chem. Soc.* **1958**, *80*, 1339. [CrossRef]
34. Matsuo, Y.; Tabata, T.; Fukunaga, T.; Fukutsuka, T.; Sugie, Y. Preparation and characterization of silylated graphite oxide. *Carbon* **2005**, *43*, 2875–2882. [CrossRef]
35. Kitchen, R. *RF and Microwave Radiation Safety*; Newnes: Oxford, UK, 2001.
36. *ASTM D5568-22a*; Standard Test Method for Measuring Relative Complex Permittivity and Relative Magnetic Permeability of Solid Materials at Microwave Frequencies Using Waveguide. ASTM: West Conshohocken, PA, USA, 2022.
37. Tian, G.; Deng, W.; Xiong, D.; Yang, T.; Zhang, B.; Ren, X.; Lan, B.; Zhong, S.; Jin, L.; Zhang, H.; et al. Dielectric micro-capacitance for enhancing piezoelectricity via aligning MXene sheets in composites. *Cell Rep. Phys. Sci.* **2022**, *3*, 100814. [CrossRef]
38. Gusynin, V.; Sharapov, S.; Carbotte, J. Magneto-optical conductivity in graphene. *J. Phys. Condens. Matter* **2006**, *19*, 026222. [CrossRef]
39. He, E.; Yan, T.; Ye, X.; Gao, Q.; Yang, C.; Yang, P.; Ye, Y.; Wu, H. Preparation of FeSiAl–Fe$_3$O$_4$ reinforced graphene/polylactic acid composites and their microwave absorption properties. *J. Mater. Sci.* **2023**, *58*, 11647–11665. [CrossRef]
40. Wang, X.; Gong, R.; Li, P.; Liu, L.; Cheng, W. Effects of aspect ratio and particle size on the microwave properties of Fe–Cr–Si–Al alloy flakes. *Mater. Sci. Eng. A* **2007**, *466*, 178–182. [CrossRef]
41. Zhang, C.; Jiang, J.; Bie, S.; Zhang, L.; Miao, L.; Xu, X. Electromagnetic and microwave absorption properties of surface modified Fe–Si–Al flakes with nylon. *J. Alloys Compd.* **2012**, *527*, 71–75. [CrossRef]
42. Shi, S.; Liu, H.; Cheng, H.; Zhang, L.; Liu, X. Tailored microwave absorption performance through interface evolution by in-situ reduced Fe nanoparticles on the surface of FeSiAl microflake. *Phys. Scr.* **2023**, *98*, 105020. [CrossRef]
43. He, S.; Wang, G.S.; Lu, C.; Liu, J.; Wen, B.; Liu, H.; Guo, L.; Cao, M.S. Enhanced wave absorption of nanocomposites based on the synthesized complex symmetrical CuS nanostructure and poly (vinylidene fluoride). *J. Mater. Chem. A* **2013**, *1*, 4685–4692. [CrossRef]
44. Cao, Z.; Liu, W.; Li, M.; Zhu, X.; Su, H.; Wang, J.; Zhang, X. Flaky FeSiAl powders with high permeability towards broadband microwave absorption through tuning aspect ratio. *J. Mater. Sci. Mater. Electron.* **2023**, *34*, 1249. [CrossRef]
45. Luo, Y. Testing and research on electromagnetic environment of 5G base station. Ph.D. Thesis, Beijing University of Posts and Telecommunications, Beijing, China, 2022.
46. Nagai, Y.; Lai, G.C. Thermal conductivity of epoxy resin filled with particulate aluminum nitride powder. *J. Ceram. Soc. Jpn.* **1997**, *105*, 197–200. [CrossRef]

Disclaimer/Publisher's Note: The statements, opinions and data contained in all publications are solely those of the individual author(s) and contributor(s) and not of MDPI and/or the editor(s). MDPI and/or the editor(s) disclaim responsibility for any injury to people or property resulting from any ideas, methods, instructions or products referred to in the content.

Article

Comparative Studies of the Dielectric Properties of Polyester Imide Composite Membranes Containing Hydrophilic and Hydrophobic Mesoporous Silica Particles

Kuan-Ying Chen [1], Minsi Yan [1], Kun-Hao Luo [1], Yen Wei [2] and Jui-Ming Yeh [1,*]

[1] Department of Chemistry and Center for Nanotechnology, Chung Yuan Christian University, Chung Li District, Taoyuan City 32023, Taiwan
[2] Department of Chemistry, Tsinghua University, Beijing 100084, China
* Correspondence: juiming@cycu.edu.tw; Tel.: +886-3-265-3341

Abstract: In this paper, comparative studies of hydrophilic and hydrophobic mesoporous silica particles (MSPs) on the dielectric properties of their derivative polyester imide (PEI) composite membranes were investigated. A series of hydrophilic and hydrophobic MSPs were synthesized with the base-catalyzed sol-gel process of TEOS, MTMS, and APTES at a distinctive feeding ratio with a non-surfactant template of D-(-)-Fructose as the pore-forming agent. Subsequently, the MSPs were blended with the diamine of APAB, followed by introducing the dianhydride of TAHQ with mechanical stirring for 24 h. The obtained viscous solution was subsequently coated onto a copper foil, 36 μm in thickness, followed by performing thermal imidization at specifically programmed heating. The dielectric constant of the prepared membranes was found to show an obvious trend: PEI containing hydrophilic MSPs > PEI > PEI containing hydrophobic MSPs. Moreover, the higher the loading of hydrophilic MSPs, the higher the value of the dielectric constant and loss tangent. On the contrary, the higher the loading of hydrophobic MSPs, the lower the value of the dielectric constant with an almost unchanged loss tangent.

Citation: Chen, K.-Y.; Yan, M.; Luo, K.-H.; Wei, Y.; Yeh, J.-M. Comparative Studies of the Dielectric Properties of Polyester Imide Composite Membranes Containing Hydrophilic and Hydrophobic Mesoporous Silica Particles. *Materials* **2023**, *16*, 140. https://doi.org/10.3390/ma16010140

Academic Editors: Liguo Xu and Peijiang Liu

Received: 23 November 2022
Revised: 12 December 2022
Accepted: 20 December 2022
Published: 23 December 2022

Copyright: © 2022 by the authors. Licensee MDPI, Basel, Switzerland. This article is an open access article distributed under the terms and conditions of the Creative Commons Attribution (CC BY) license (https://creativecommons.org/licenses/by/4.0/).

Keywords: poly(ester imide); hydrophobic; mesoporous; silica; dielectric

1. Introduction

In a fully data-centric world populated with telecommunication devices of autonomous vehicles, sensors, robots, and cloud-connected resources, networks need to transfer a greater amount of data than the currently used 4G and 5G systems are capable of handling. To meet the requirements, such as an extraordinary level of data reliability, low latency, low power consumption and massive capacity, devices of the next generation of 5G and 6G wireless communication technologies are expected to use millimeter wave bands (30–300 GHz) [1–5]. These high frequencies imply extreme device densities with small sizes and immunity to interference, thus urging the development of new materials with unique properties and structures with unorthodox designs [6,7].

The polyimide membrane is a high-performance polymer with good dielectric properties, thermal stability, and chemical resistance [8,9]. It is widely used in the electronic industry. In recent years, with the vigorous development of mobile communication devices, the demand for the size reduction in integrated circuits has driven the advanced process line width of 7, 5, and 3 nm [10]. Low dielectric interlayer materials not only reduce the current leakage of integrated circuits but also improve the capacitance effect between wires, RC time delay, cross-talk, heat generation, and power dissipation in the new generation of high-density integrated circuits [11,12]. Commercially available polyimide has a dielectric constant between 3.1 and 3.6 [13], which is not enough for future development. Hence, extensive research on polyimides with low dielectric constant materials was reported [14–17].

There are two main approaches to reducing the dielectric constant of the polyimide polymer, both aiming at reducing the polarizability of the material: (1) introducing a fluorine atom or lipid ring structure [18,19] and (2) increasing the free volume of the structural skeleton or introducing air [20,21]. The dielectric constant of air is 1 [22], which is intrinsically low in comparison to water, which is approximately 80 [23]. Introducing air into the polyimide structure effectively reduces the dielectric constant of the overall material. This porous structure could be achieved through the direct introduction of air, such as polyimide aerogels, which requires a time-consuming solvent exchange process and the precise conditioning of a drying method. Another method is compositing with other porous materials. Mesoporous silica has gained much attention in recent research as a molecular sieve and has already been explored in organic and inorganic composites to achieve low dielectric constant materials [24]. The preparation of mesoporous silica could be further differentiated into silica with regular porous structures and irregular porous structures. Stucky et al. and Kresge et al. both demonstrated the formation and modification of MCM-41 silica using a surfactant as a template, which requires high-temperature calcination to remove the temperate [25,26]. This method is industrially unfavorable due to its high energy requirement and the nature of processing complications. To form silica with irregular porosity, Yen et al. conducted extensive research on a nonsufactant template method forsol-gel process using a hydroxy-carboxylic acid as a template [27–29]. In the same period, Yen et al. proposed a novel one-step method for enzyme encapsulation in mesoporous material using D-glucose as a template [30]. Furthermore, according to the study [31], the ester segment functional group is the dominant factor for low water uptake; thus, the PEI polymer system is a promising candidate as a novel dielectric substrate material for use in the next generation of high-performance flexible printed circuit boards operating at higher frequency (10 GHz).

Therefore, in this study, we attempted to prepare a PEI membrane with lower dielectric properties (e.g., the dielectric constant and loss tangent) by incorporating hydrophobic MSPs, which could effectively decrease the dielectric constant of the PEI membrane by introducing a large amount of air into the porous channels and significantly decreasing the absorption amount of moisture by the MSPs. The hydrophobic MSPs were synthesized via a base-catalyzed sol-gel reaction with D-(-)-Fructose as a non-surfactant template. Subsequently, the hydrophobic inorganic MSPs were introduced into the organic PEI membrane with a lower loss tangent to prepare the organic-inorganic hybrid membranes with a lower dielectric constant and loss tangent simultaneously. Moreover, the hydrophilic MSPs were also synthesized as a control experiment. The characterizations of the hydrophilic and hydrophobic MSPs were investigated and compared using FTIR, ^{13}C-NMR, ^{29}Si-NMR spectra, and EDS. The physical properties of the hydrophilic and hydrophobic MSPs were studied and compared using SEM, TEM, CA, TGA, and BET. Subsequently, comparative studies for the effect of hydrophilic and hydrophobic MSPs on the dielectric properties of the PEI and its corresponding composite membranes were systematically investigated. The mechanical strength, thermal properties, and surface wettability of the PEI and its corresponding composite membranes were also investigated using the tensile test, TGA, and CA, respectively.

2. Experiment

2.1. Materials

p-Phenylene bis(trimellitate) dianhydride (TAHQ) and 4-Aminophenyl-4-aminobenzoate (APAB) were both obtained from SHIFENG TECHNOLOGY CO., LTD (Tainan City, Taiwan) and were dried in an oven overnight before use. Tetraethyl orthosilicate (TEOS), trimethoxymethylsilane (MTMS), and ammonium hydroxide, (ACS reagent, 28.0–30.0% NH_3 basis) were purchased from Sigma-Aldrich (St. Louis, MO, USA) and used as received. Triethoxysilane (3-Aminopropyl) (APTES) and D-(-)-Fructose, both of 99% purity, were purchased from Sigma-Aldrich and used as received. N-Methyl-2-pyrrolidone (NMP) was purchased from Macron and stored in a container with molecular sieves overnight before use. Ethanol was purchased from J.T. Baker and used as received.

2.2. Instrumentation

All samples were characterized through an attenuated total reflectance FT-IR in the form of potassium bromide (KBr) powder-pressed pellets. Both the ^{13}C and ^{29}Si MAS solid-state NMR experiments were performed on a 400 MHz solid-state NMR spectrometer. ^{13}C MAS NMR spectra were obtained at 100.63 MHz with 7 kHz applying 90° pulses and 2.0 s pulse delays. ^{29}Si MAS NMR spectra were recorded at 79.49 MHz applying 90° pulses, 300 s pulse delays, and 5.0 ms of contact time, with samples in 5.0 mm zirconia rotors spinning at 7 kHz. The surface morphologies of the superhydrophobic samples were observed by using SEM (JOEL JSM-7600F). The silica dispersion was detected using energy dispersive spectroscopy (EDS) (Oxford xmax 80). The crystallographic structure of the polymer was studied with the X-ray diffraction (XRD) pattern on a Bruker D8 Advance Eco instrument using Cu Kα radiation (λ = 1.5418 nm, 40 kV, 25 mA) at a scanning rate of 2° min^{-1}. Thermogravimetric analysis (TGA) was used to analyze the thermal stability and moisture content of the sample under N$_2$ at constant pressure, with a temperature elevation rate of 10 °C/min, from 30 °C to 800 °C. The tensile strength, elongation at break, and Young's modulus were tested and recorded using a Hung Ta HT-9102 tensile test machine. Brunauer-Emmett-Teller (BET) data were obtained by performing nitrogen adsorption/desorption isotherms and accelerated surface area porosity analysis on Micromeritics ASAP-2010. The hydrophilicity of the sample was tested based on the contact angle (CA) of the water droplets on the prepared sample. The results were measured with a contact angle FTA125 instrument. The microstructure and surface morphology were studied using a transmission electron microscope (TEM) and scanning electron microscope (SEM), respectively. Lastly, all prepared samples were tested under impedance spectroscopy at 10 GHz with an AET Microwave Dielectrometer to measure their loss tangent and capacitance. The dielectric constant κ was calculated from the equation listed below, where C is the capacitance, ε_0 is the absolute dielectric constant, A is the component electrode surface area, and d is the thickness.

$$C = \frac{\kappa \varepsilon_0 A}{d}$$

2.3. Synthesis of the Hydrophilic and Hydrophobic MSPs

The fundamental design concept to synthesize the MSPs was to perform the base-catalyzed sol-gel process and to create the worm-like porous channels through a non-surfactant template route, as shown in Scheme 1. To prepare the hydrophilic MSPs, the base-catalyzed sol-gel reactions of TEOS and APTES (with a hydrophilic -NH$_2$) at a specific feeding ratio in the presence of the non-surfactant template of D-(-)-Fructose as the pore-forming agent was used. Moreover, to prepare the hydrophobic MSPs, the sol-gel reactions of TEOS, APTES (with a hydrophilic -NH$_2$), and MTMS (with a hydrophobic -CH$_3$) at a specific feeding ratio in the presence of D-(-)-Fructose were also employed. The detailed formulations to synthesize the hydrophilic MSPs (A1, A2, and A3) and hydrophobic MSPs (M1 and M2) are shown in Table 1. The flowchart of synthesizing the hydrophilic MSPs (A1–A3) and hydrophobic MSPs (M1 and M2) via the base-catalyzed sol-gel reaction in the presence of D-(-)-Fructose is illustrated in Scheme 2. The typical representative procedure to synthesize the hydrophilic and hydrophobic MSPs was as follows: Based on the formulation presented in Table 1, an appropriate amount of D-(-)-Fructose was poured into a double-layer beaker containing 200 mL of ethanol and 100 mL of deionized water with magnetic stirring at 35 °C. Subsequently, 9.0 mL of concentrated ammonia was dropped into the previous beaker while stirring. Three sol-gel precursors (i.e., TEOS, APTES, or MTMS) at specific feeding ratios were injected into the beaker with a syringe with magnetic stirring for 48 h. The synthesized products were collected by centrifuging the solution and then ultrasonically washing the collected paste with ethanol and water 3 and 10 times, respectively, to remove the D-(-)-Fructose. The washing of the products was followed by the freeze-drying procedure to obtain the final hydrophilic MSPs (A1, A2, and A3) and hydrophobic MSPs (M1 and M2).

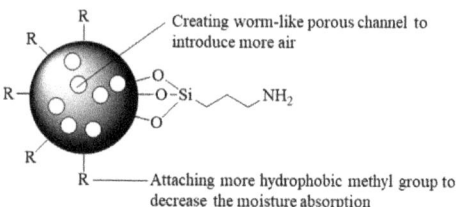

Scheme 1. Designed concept of synthesizing hydrophobic mesoporous silica particles (MSPs), with hydrophilic primary and hydrophobic methyl group as well as worm-like porous channels, functioning as fillers that decrease the dielectric properties of the polyester imide (PEI) composite membrane.

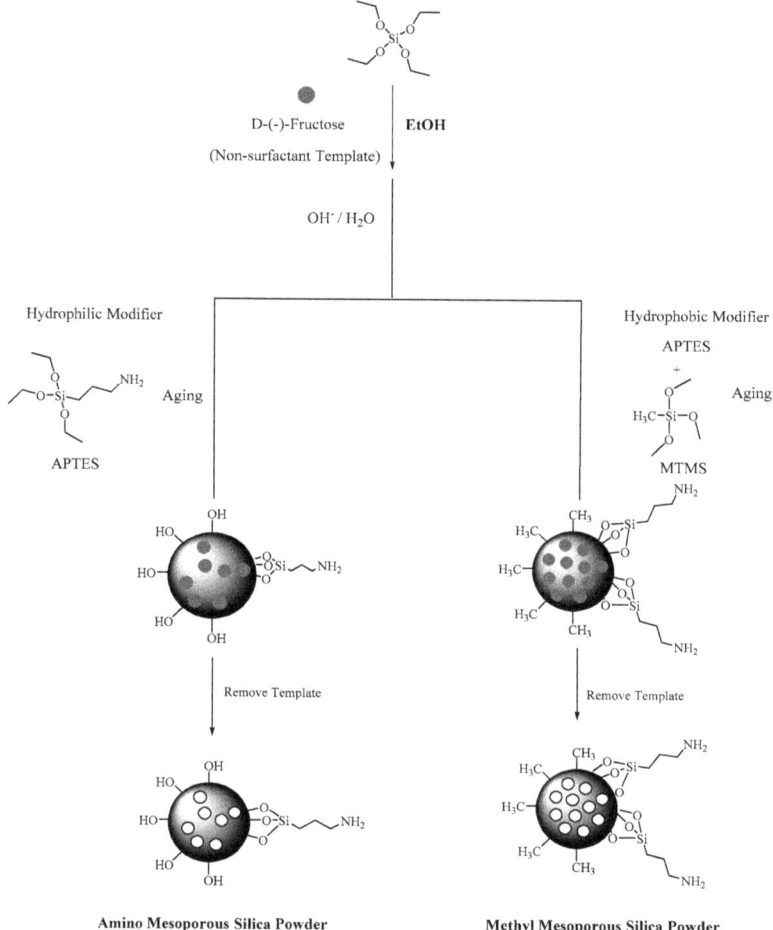

Scheme 2. Flowchart for the synthesis of the (**a**) hydrophilic amino-modified mesoporous silica particles (MSPs) and the (**b**) hydrophobic methyl-modified MSPs through the base-catalyzed sol-gel route with D-(-)-Fructose as a non-surfactant template.

2.4. Preparation of PEI Composite Membranes Containing Hydrophilic and Hydrophobic MSPs

To prepare the PEI composite membrane, distinctive MSPs were incorporated into the polyester imide at specific feeding ratios; the detailed formulations for the preparation of the composite membranes are shown in Table 2. A typical procedure to prepare the composite membrane was as follows: First, a suitable amount of diamine monomer was dissolved in a 3-neck rounded flask with 40 g of NMP at room temperature under a nitrogen atmosphere. Subsequently, different ratios of MSPs were then stirred into the mixture. A specific amount of dianhydride was then introduced with magnetic stirring for 24 h. As the reaction time increased, the viscosity of the MSP-containing mixing solution was gradually increased to produce the poly(ester amic acid) (PEAA)/MSP composites, followed by coating on top of a copper foil with a blade coater.

Subsequently, the prepared mixing solution was followed by performing a thermal imidization reaction to form a polyester imide (PEI). The programmed heating conditions were as follows: (150 °C, 5 min), (200 °C, 5 min), (250 °C, 5 min), (300 °C, 30 min), and (350 °C, 30 min). After immersing the PEI-coated copper foil in an acid etching solution for 30 min, the final PEI membrane was then obtained, which was 36 μm in thickness. The flowchart for the preparation of the PEI composite membranes containing hydrophilic and hydrophobic MSPs is shown in Scheme 3.

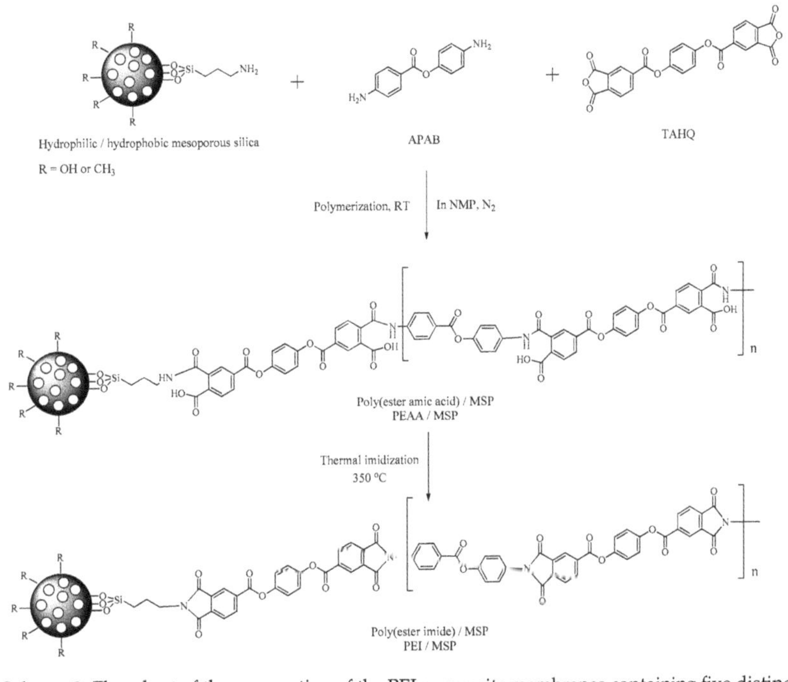

Scheme 3. Flowchart of the preparation of the PEI composite membranes containing five distinctive MSPs via thermal imidization.

3. Results and Discussion

In this section, the characterization of the prepared materials was classified into two parts: the inorganic MSPs and organic-inorganic PEI composite membranes.

3.1. Characterization of the Hydrophilic and Hydrophobic MSPs

3.1.1. FTIR Spectra

Figure 1 shows the representative FTIR spectra of the synthesized distinctive MSPs (i.e., A1, A2, A3, M1, and M2). First, the characteristic peak for the symmetrical absorption peak of Si-O-Si appeared at the wavenumber of 950 cm^{-1}. Moreover, the characteristic peak for the asymmetric absorption of Si-O-Si appeared at the position of 1036 cm^{-1} [32]. A broad-band peak of Si-OH was observed from 3000 to 3700 cm^{-1}. The characteristic peak that appeared at the position of 1638 cm^{-1} was assigned to be the bending of a hydrophilic primary amine -NH$_2$ [33], which confirmed the participation of APTES in the sol-gel reactions for the A1, A2, and A3 MSPs. On the other hand, for the characterization of the M1 and M2 hydrophobic MSPs, the characteristic peak for the asymmetric absorption of Si-CH$_3$ and hydrophobic -CH$_3$ was observed at the positions of 1279 and 2968 cm^{-1} [34], respectively, which confirmed the participation of MTMS in the sol-gel reactions of M1 and M2. Furthermore, it is worth noting that the bonding between MTMS and APTES on the surface is triple bonding. Certain studies [35–37] have mentioned that in the spectra of FT-IR, the peaks at 1050–1150 cm^{-1} are characteristic of C-O-C, Si-O-C, and Si-O-Si stretching. This was observed in this study in Figure 1.

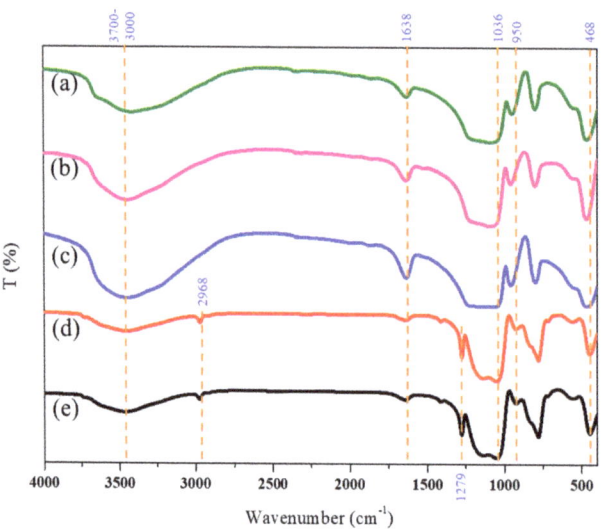

Figure 1. ATR-FTIR spectra of the hydrophilic (**a**) A1, (**b**) A2, and (**c**) A3 MSPs, as well as the hydrophobic (**d**) M1 and (**e**) M2 MSPs.

3.1.2. Solid-State ^{29}Si-NMR Spectra

Figure 2 shows the solid-state ^{29}Si-NMR spectra of the hydrophilic and hydrophobic MSPs. For example, Figure 2a–c exhibits the solid-state ^{29}Si-NMR spectra of the hydrophilic A1, A2, and A3 MSPs. The hydrophilic characteristics of A1~A3 were attributed to the hydrophilic group of the primary amine from APTES reacting with TEOS via the base-catalyzed sol-gel reactions. For the characterization of the hydrophilic MSPs, the T3 signal of APTES was observed at the chemical shift of ~−60 ppm. Moreover, the Q^3 and Q^4 signals of TEOS were observed at chemical shifts of −101 ppm and −110 ppm [38,39], respectively. However, the signal intensity and integral area of the Q series were found to be larger than those of the T series, which may be attributed to the higher feeding molar ratio of TEOS than that of APTES. On the other hand, Figure 2d,e are the solid-state silicon NMR spectra of the hydrophobic M1 and M2 MSPs. The hydrophobic characteristics of M1 and M2 were attributed to the hydrophobic -CH$_3$ from MTMS reacting with TEOS via the base-catalyzed sol-gel reactions. However, the signal intensity and integral area of the T series were found to be larger than those of the Q series, which may be attributed to the higher feeding molar ratio of APTES and MTMS than that of TEOS.

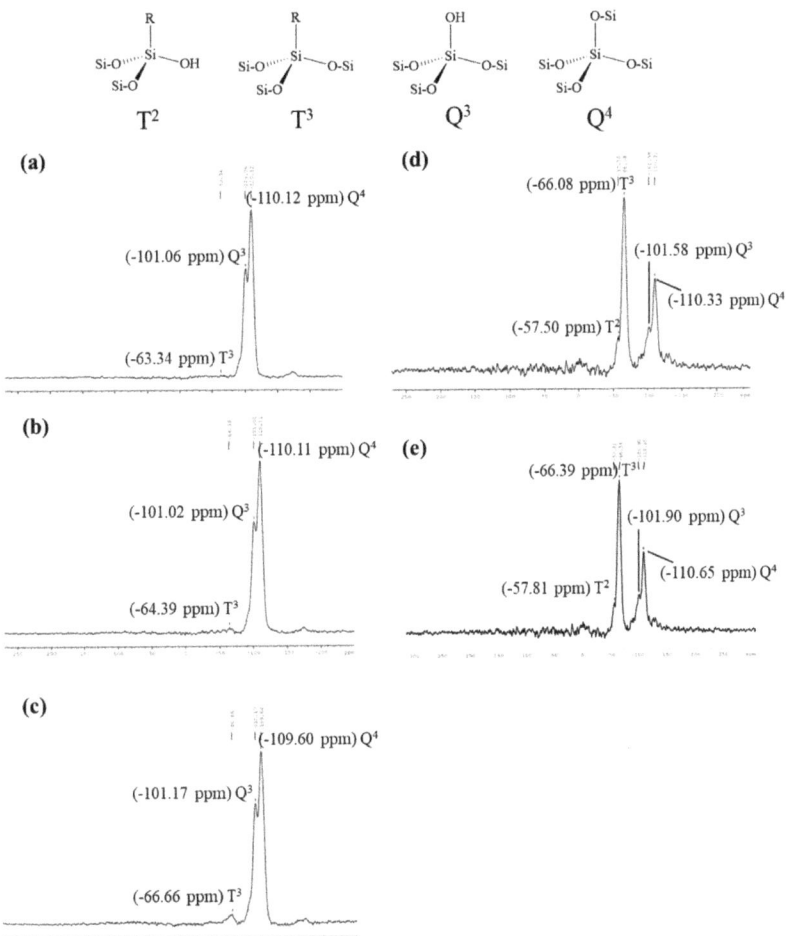

Figure 2. Solid-state 29Si-NMR spectra of the hydrophilic (**a**) A1, (**b**) A2, and (**c**) A3 MSPs, as well as the hydrophobic (**d**) M1 and (**e**) M2 MSPs.

3.1.3. Solid-State ^{13}C-NMR Spectra

Figure 3 reveals the solid-state ^{13}C-NMR spectra of the hydrophilic and hydrophobic MSPs. Figure 3a–c shows the hydrophilic A1, A2, and A3 MSPs modified by APTES. In these hydrophilic MSPs, the chemical shifts of carbon for **a, b** and **c** were found to appear at the positions at approximately 43 ppm, 22 ppm, and 10 ppm [40], respectively, which indicated the participation of APTES in the sol-gel reactions of A1, A2, and A3. On the other hand, the hydrophobic M1 and M2 MSPs modified by MTMS were prepared by incorporating a small amount of hydrophilic APTES and a large amount of hydrophobic MTMS. Figure 3d,e shows the solid-state ^{13}C-NMR spectra for the hydrophobic M1 and M2 MSPs modified by APTES and MTMS. In these MSPs, the chemical shifts of carbon for a, b and c were found to appear at positions at approximately 43 ppm, 22 ppm, and 10 ppm [40], respectively, which indicated the participation of APTES in the sol-gel reactions of M1 and M2. Moreover, the chemical shift of carbon for **d** was found to appear at the position of −3.5 ppm, which confirmed the participation of MTMS in the sol-gel reactions of M1 and M2.

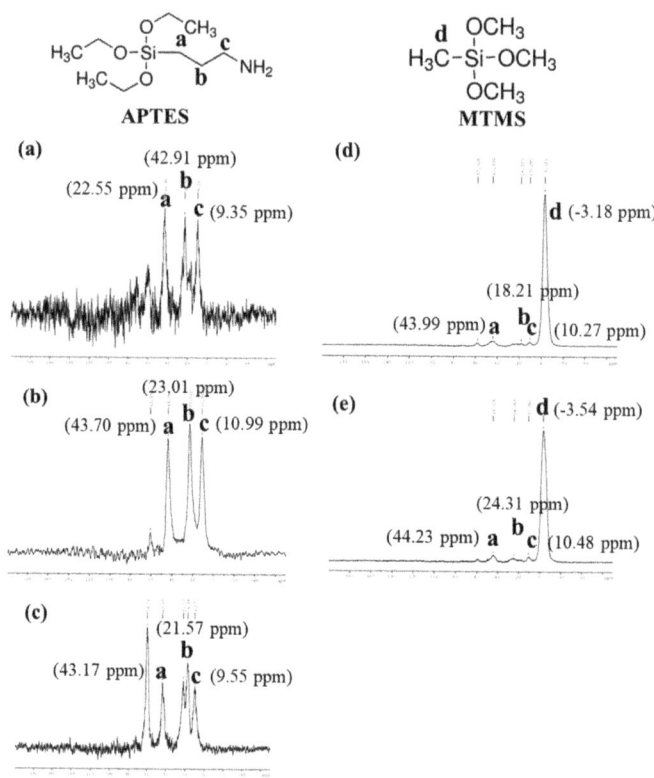

Figure 3. Solid-state 13C-NMR spectra of the hydrophilic (**a**) A1, (**b**) A2, and (**c**) A3 MSPs, as well as the hydrophobic (**d**) M1 and (**e**) M2 MSPs.

3.1.4. Energy Dispersive Spectroscopy (EDS)

The elemental analysis of the EDS spectra for the three hydrophilic MSPs (A1, A2, and A3) confirmed that the EDS data of A3 exhibited the highest N element content of 1.29 wt-%, and A1 revealed the lowest content of 0.72 wt-%, which was consistent with the highest and lowest initial feeding ratios of APTES in A3 and A1, respectively, as summarized in Table 1.

3.2. Physical Properties of MSPs

3.2.1. Surface Morphology of the Hydrophilic and Hydrophobic MSPs (SEM)

The SEM images of the surface morphologies for the hydrophilic MSPs (A1, A2, and A3) were found to be spherical, as shown in Figure 4a–c. It should be noted that with the increase in APTES content in the hydrophilic MSPs, an increase in the diameter of silica particles occurred. For example, the average hydrophilic MSP diameters of A1, A2, and A3 were ~800 nm, ~1000 nm, and ~1200 nm, respectively. It indicated that a higher loading of amino-silane may slightly increase the diameter of the corresponding hydrophilic MSPs. This conclusion is consistent with the previous publication reported by Krysztafkiewicz et al. [41]. At the same time, the surface morphologies of the M1 and M2 MSPs modified by the fixed feeding ratio of MTMS6 with different dosages of the pore-forming agent were found to be irregular in shape, as shown in Figure 4d,e.

Figure 4. SEM for (**a**) A1, (**b**) A2, (**c**) A3, (**d**) M1, and (**e**) M2, as well as the transmission images of the TEM for (**f**) A1, (**g**) A2, (**h**) A3, (**i**) M1, and (**j**) M2.

3.2.2. Transmission Morphology of the Hydrophilic and Hydrophobic MSPs (TEM)

This section may be divided into subheadings. It should provide a concise and precise description of the experimental results, their interpretation, as well as the experimental conclusions that could be drawn. For the TEM observation, the MSP samples were first dispersed in ethanol, followed by dropping on the copper mesh for the observation of the TEM image; the darker and brighter regions of the TEM imagery indicated that the thickness of the silica particle wall was thicker and thinner, respectively. For example, the TEM image of hydrophilic MSPs (A1, A2, and A3) is shown in Figure 4f, Figure 4g, and Figure 4h, respectively. Moreover, the TEM images of the M1 and M2 MSPs were found to exhibit a mesoporous structure, as shown in Figure 4i and Figure 4j, respectively.

3.2.3. Surface Wettability

In these studies, the hydrophilic MSPs (A1, A2, A3) and hydrophobic MSPs (M1 and M2) in powder form were first fabricated into the shape of a powder-pressed pellet before performing the CA measurements of the water droplets. It should be noted that the CA of three hydrophilic MSPs (i.e., A1, A2, and A3) could not be detected, as shown in Figure 5a, which might be attributed to the super-hydrophilic characteristics of the prepared MSPs modified with a primary amine group. Moreover, the CAs of the two hydrophobic MSPs (i.e., M1 and M2) were both found to be ~127°, as shown in Figure 5b. This indicated that the incorporation of MTMS into the MSPs may effectively increase the CA of the hydrophilic MSPs, which was attributed to the introduction of the hydrophobic -CH_3 group. All the data of the CA measurements of the water droplets for the MSPs are summarized in Table 1.

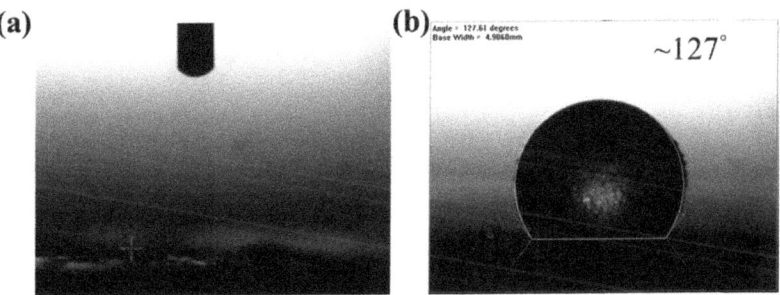

Figure 5. Contact angle of the water droplets for the (**a**) hydrophilic MSPs (A1, A2, and A3). (un-detectable) and the (**b**) hydrophobic MSPs (M1 and M2).

3.2.4. Determination of the Moisture Absorption of the MSPs with TGA

The TGA of the five distinctive hydrophilic and hydrophobic MSPs was operated at temperatures ranging between 30 °C and 800 °C in the air at a heating rate of 10 °C/min, as shown in Figure 6. In this study, the moisture absorption of MSPs was defined as the weight loss at 200 °C, and the data are summarized in Table 1. The moisture absorptions of the hydrophilic MSPs (A1, A2, and A3) were estimated at weight losses of 4.15, 4.24, and 7.68 wt-%, respectively, as shown in Figure 6. It indicated that the A3 sample revealed the highest hygroscopicity. The significant weight loss observed at 100 °C is the water content from the A3 MSPs. Compared to A1 and A2, A3 possessed the highest content of hydrophilic functional groups. Hence, it absorbed more moisture at RT.

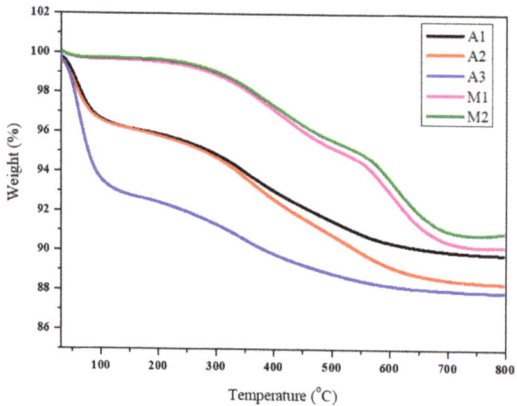

Figure 6. TGA curves of the hydrophilic MSPs (A1, A2, and A3) and the hydrophobic MSPs (M1 and M2).

On the other hand, the moisture absorptions of the hydrophobic MSPs (M1 and M2) were estimated at weight losses of 0.58 and 0.45 wt-%, respectively, as shown in Figure 6. It demonstrated that M2 exhibited the lowest hygroscopicity. These results confirmed that the MSP modified by MTMS could effectively improve the corresponding hydrophobicity compared to that of MSPs modified by APTES. Moreover, the lower hygroscopicity of M2 as compared to that of M1 may be attributed to the hydrophobic -CH_3 attached to the higher specific surface area of the MSP resulting from the higher dosage of the pore-forming agent. It should be noted that the trend of the MSP's moisture absorption determined with the TGA was consistent with the previous studies of the CA measurements of the water droplets. All the data on the moisture absorptions of the MSPs are summarized in Table 1.

3.2.5. BET Analysis of the MSPs

Figure 7a,b exhibit the N_2 adsorption-desorption isotherm and pore diameter distribution curves of the hydrophilic MSPs, respectively. It should be noted that a higher loading of APTES in the MSPs resulted in a smaller specific surface area. This indicated that the trend of the specific surface area for the hydrophilic MSPs was A1 (469 ± 12 m^2/g) > A2 (402 ± 21 m^2/g) > A3 (384 ± 14 m^2/g). The slightly decreasing surface area of the MSPs with higher APTES may be attributed to the longer chains of primary amines of APTES existing inside the mesoporous channel after the removal of D-(-)-Fructose. The average pore size distributions of the A1, A2, and A3 MSPs were 2.9, 3.0, and 2.7 nm, respectively.

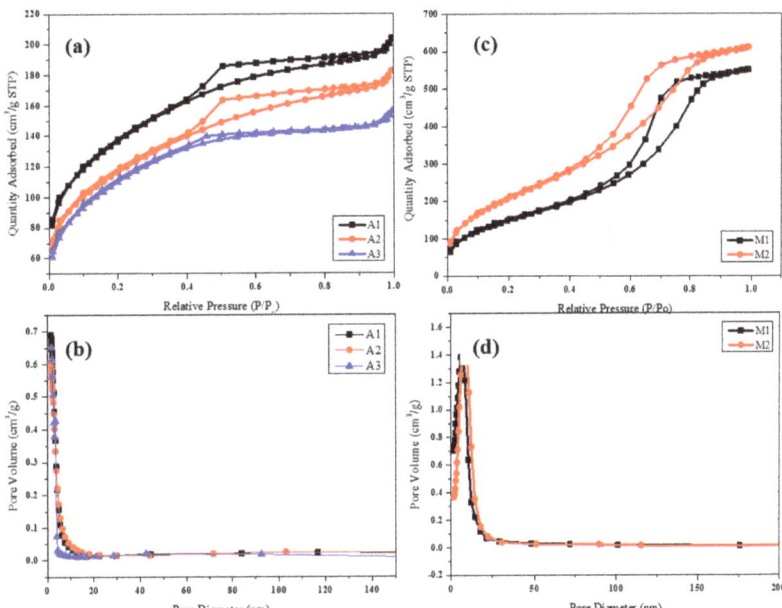

Figure 7. BET analysis of the series of MSPs. (**a**) The N_2 adsorption-desorption isotherm as well as (**b**) the pore size distribution curves for the hydrophilic MSPs (A1, A2 and A3). (**c**) The N_2 adsorption-desorption isotherm and (**d**) pore size distribution curves of the hydrophobic MSPs (M1 and M2).

Figure 7c,d exhibit the N_2 adsorption-desorption isotherm and pore diameter distribution curves of the hydrophobic MSPs, respectively. It should be noted that a higher loading of MTMS in the MSPs resulted in a larger specific surface area. This indicated that the specific surface area for the hydrophobic MSPs was M2 (771 ± 14 m^2/g) > M1 (548 ± 12 m^2/g). The specific surface area of the hydrophobic M2 MSP compared to that of M1 was attributed to the higher loading of the pore-forming agent, which is consistent with a previous report [42]. The average pore size distributions of the M1 and M2 MSPs were found to be ~4.9 and ~6.2 nm, respectively. All the data of the BET analysis of the surface areas, pore volumes, and average pore diameters for the five distinctive MSPs are summarized in Table 1.

3.3. Characterization of the MSP-Based PEI Composite Membranes

3.3.1. ATR-FTIR Spectra

In this study, the characterizations of the MSP-based PEI composite membranes were measured with ATR-FTIR, as shown in Figure 8. It should be noted that FTIR spectra of all composite membranes revealed the characteristic peaks of Si-O-Si groups except that of the neat PEI membrane. For example, the characteristic symmetric and asymmetric absorption peaks of Si-O-Si were found to appear at wavenumbers of 950 cm^{-1} and 1036 cm^{-1}, respectively [32]. On the other hand, the characteristic peak that appeared at the position of 735 cm^{-1} was attributed to the formation of the imide ring, which indicated the successful preparation of the PEI with thermal imidization [42]. However, a redshift was clearly observed for the PEI/MSP composite at this characteristic peak. According to Okada et al. [43], the reason for the possible redshift in the spectrum of the imide ring functional groups may be that the PEI is an ordered molecular chain, while silica has a network structure. When silica and imide ring are combined, the intervention of steric barriers deforms the orderly arrangement into an out-of-order one, thus producing a redshift phenomenon. Moreover, the characteristic peaks that appeared at the position of 1705 cm^{-1} and 1770 cm^{-1} were assigned to be the symmetric stretching of the imide ring and asymmetric C=O stretching of the imine ring, respectively [37].

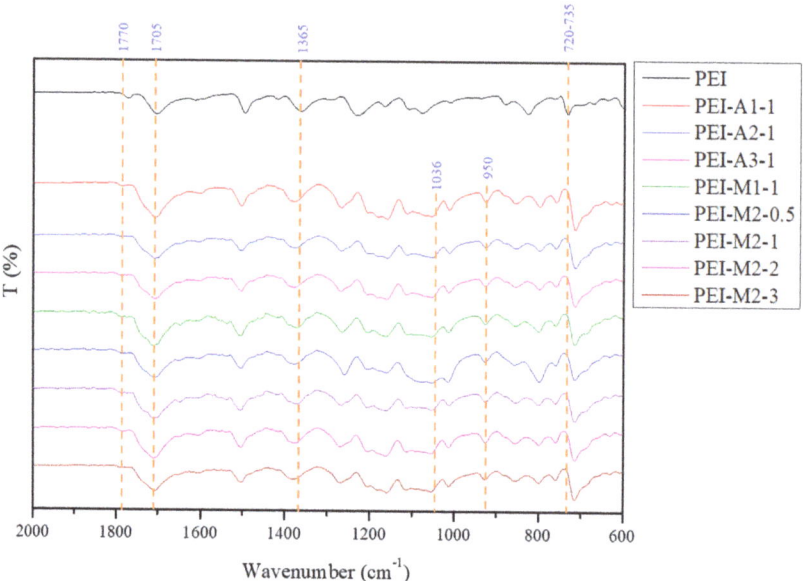

Figure 8. ATR-FTIR spectra of the polyester imide (PEI) and its corresponding composite membranes containing hydrophilic MSPs (A1, A2, and A3) and hydrophobic MSPs (M1 and M2).

3.3.2. XRD Pattern

In this characterization, the crystal behavior of the PEI membrane in the presence of the hydrophilic and hydrophobic MSPs was studied using the XRD pattern, as shown in Figure 9. The Bragg angle of 2θ = ~22° indicated that all prepared samples were of amorphous materials, attributed to the presence of the MSPs [44]. The orderly arrangement of the PEI molecules was disrupted in the presence of the MSPs, therefore leading to a decrease in the crystallinity of the PEI membranes. The full width at half maximum (denoted by FWHM) of the XRD profile was sensitive to the variation in the microstructure and stress-strain accumulation in the material. The FWHM value is inversely proportional to the crystallite size (Scherrer's formula), meaning the broader the peak, the smaller the crystallite dimension [45]. For example, by incorporating 1 wt-% A1, A2, and A3 MSPs, the FWHM value of the PEI increased from 3.2 to 3.4, 3.5, and 3.5, respectively, indicating that the incorporation of the A1, A2, and A3 MSPs into the PEI may decrease the crystallite size of the PEI membrane, as shown in Figure 9a and Table 2. On the other hand, incorporating 1 wt-% of the M1 and M2 MSPs, the FWHM value of the PEI increased from 3.2 to 3.3 and 3.3, respectively. It indicated that the PEI composite membranes incorporating 1 wt-% hydrophobic MSPs exhibited a lower FWHM value than that of 1 wt-% hydrophilic MSPs. Moreover, the higher loading of M2 in the PEI composite membranes may result in a higher FWHM value, reflecting a decrease in the crystallite size of the PEI membrane.

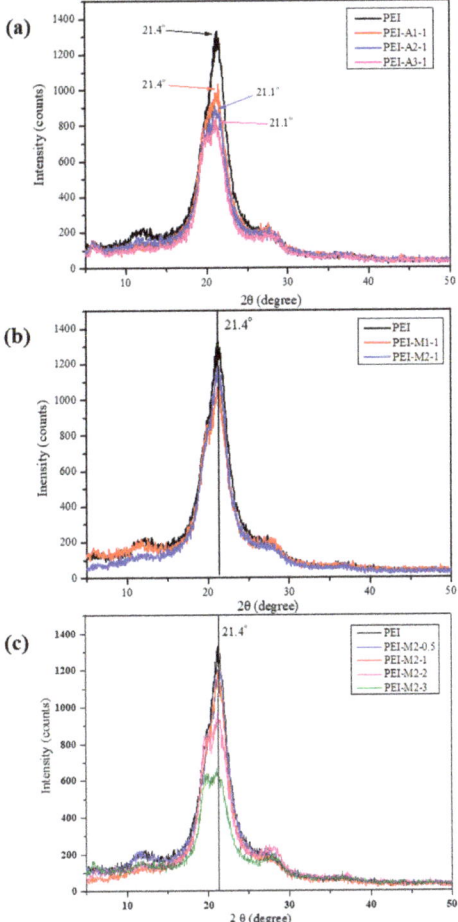

Figure 9. XRD curves of the PEI and its corresponding composite membranes containing the (**a**) hydrophobic MSPs (A1, A2, and A3), (**b**) hydrophobic M1 MSP, and (**c**) hydrophobic M2 MSP.

3.4. Physical Properties of MSP-Based PEI Composite Membranes

3.4.1. Dielectric Properties

In this study, the high-frequency dielectric measurement used the resonant cavity with impedance spectroscopy to measure the microwave dielectric of the substance, and the measured frequency used was 10 GHz. The dielectric constant and loss tangent of the PEI composite membranes containing the MSPs was measured in the 10 GHz frequency band, as summarized in Table 2. For example, the dielectric constant (D_k) and loss tangent (D_f) of the neat PEI were 3.27 and 0.007, respectively. By incorporating 1 wt-% hydrophilic A1, A2, and A3 MSPs, the D_k of the corresponding PEI composite membrane was found to increase up to 3.43, 3.46, and 3.92, respectively. The D_f increased up to 0.011, 0.012, and 0.015, respectively. The obvious increase in the dielectric constant and loss tangent in the PEI by incorporation of 1 wt-% hydrophilic MSPs may be attributed to two possible reasons: the higher moisture absorption and smaller crystallite size of the PEI membrane. The weight percentage of the moisture absorption of the PEI and corresponding composite membranes of PEI-A1-1, PEI-A2-1, and PEI-A3-1 were estimated to be 0.73, 0.95, 0.98, and 1.18 wt-%, respectively, based on Figure 6 and Table 2. According to the previous publications, the D_k

and D_f values of H_2O at room temperature tested under 10 GHz were reported to be ~80 and ~0.47 [46]. Therefore, the PEI composite membranes with higher moisture absorption capabilities may reveal higher D_k and D_f values.

On the other hand, by incorporating 1 wt-% hydrophobic M1 and M2 MSPs, the D_k of the corresponding PEI composite membrane was found to be suppressed to 3.24 and 3.18, respectively. In comparison, the D_f retained roughly the same values as the PEI alone, which were 0.009 and 0.008, respectively. The weight percentage of the moisture absorptions of the PEI-M1-1 and PEI-M2-1 PEI composite membranes were estimated to be 0.69 and 0.52 wt-%, respectively, based on the TGA investigation. The decrease of the D_k may be attributed to two possible reasons: (1) the low moisture absorption of the PEI or (2) the high surface area of silica with worm-like pore channels. First, the low moisture absorption of the PEI membrane may be related to the incorporation of the hydrophobic MSPs (M1 and M2). Second, the higher surface area may lead more air (D_k of air = ~1) into the porous channel of the hydrophobic MSP1s, reflecting a decrease in the D_k of the PEI membrane.

For the studies of the dielectric properties of the PEI composite membrane containing the M2 MSP at different loadings, the D_k of the PEI composite membrane containing 1 wt-%, 2 wt-%, and 3 wt-% of M2 were found to be 3.18, 2.97, and 2.93, respectively. On the other hand, the D_f of the PEI composite membrane containing 1 wt-%, 2 wt-%, and 3 wt-% of M2 were found to be 0.008, 0.008, and 0.009, respectively.

3.4.2. Mechanical Strength

The tensile modulus and elongation at the break of the PEI membrane were measured using the tensile test, according to ASTM D882 [47], and were found to be 11.26 MPa and 1.48%, respectively. By incorporating 1 wt-% hydrophilic MSPs, the tensile moduli of the PEI composite membranes containing 1 wt-% of A1, A2, and A3 were found to decrease to 4.03, 4.80, and 5.48 MPa, respectively. Moreover, the elongations at the break of the PEI composite membranes containing 1 wt-% of A1, A2, and A3 were found to increase to 2.15%, 2.52%, and 2.75%, respectively.

On the other hand, by incorporating 1 wt-% hydrophobic MSPs, the tensile moduli of the PEI composite membranes containing 1 wt-% of M1 and M2 were found to significantly increase to 10.22 MPa, and 16.23 MPa, respectively. At the same time, the elongations at the break of the PEI composite membranes containing 1 wt-% of M1 and M2 were found with noticeable reductions, which were 1.64% and 1.20%, respectively. It indicated that the incorporation of the hydrophobic M2 MSP with a higher surface area of 771 m^2/g promoted the mechanical strength and decreased the elongation at the break of the PEI membrane simultaneously as compared to the hydrophobic M1 MSP with the lower surface area of 548 m^2/g.

Eventually, by incorporating the M2 hydrophobic MSP at different feeding ratios of 1 wt-%, 2 wt-%, and 3 wt-% in the PEI membrane, the mechanical strengths of the neat PEI were found to be 16.23, 18.95, and 6.68 MPa of the PEI composites, respectively. Moreover, the elongations at the break of the neat PEI composites decreased from 1.20% to 1.19% and then escalated to 1.65%. It implied that, by comparing the 1 and 2 wt-% loading of M2, the higher the loading of M2 in the PEI membrane, the higher the mechanical strength and the lower the elongation at the break of the composite membranes. The decrease of the mechanical strength and increase in the elongation at the break of the PEI composite membrane at 3 wt-% of M2 loading may be attributed to the M2 particle aggregation that occurred at the higher loading of the MSPs. All test measurements are collectively listed in Table 2.

It is speculated that there was physical crosslinking caused by the strong interfacial interaction between the PEI molecules and MSPs. This may be the core reason for the strengthening effects observed. Within an appropriate range of crosslinking density, both a strengthening and toughening effect was observed. However, excessive incorporation would lead to agglomerates, which create defects within the polymetric structure, hence decreasing the overall mechanical properties. Although it is widely believed in a rubber/polymer system, the strengthening effect is a result of void formation in rubber particles under stress [48], it is unlikely to be what was found in this PEI/MSP composite system since the Si-O bond is too strong to be broken before the failure of the composite.

3.4.3. Moisture Absorption and Thermal Stability Determined with TGA

This section is divided into subheadings. It should provide a concise and precise description of the experimental results, their interpretation, as well as the experimental conclusions that could be drawn. The moisture absorption and decomposition temperature (T_{5d}) of the PEI membranes were investigated with TGA, as shown in Figure 10 and Table 2. The experimental conditions were as follows: the operational temperature of the TGA study for the prepared membrane samples was run from 30 °C to 800 °C under atmospheric conditions at a heating rate of 10 °C/min.

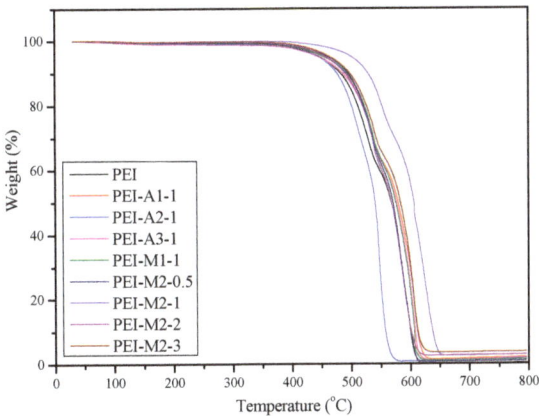

Figure 10. TGA curves for the PEI and its corresponding composite membranes containing five distinctive MSPs (A1, A1, A3, M1, and M2).

First, the moisture absorption amount of the PEI and corresponding composite membranes was defined as the weight loss of the TGA curve at 200 °C. Therefore, the moisture absorption of the neat PEI membrane was estimated at ~0.73 wt-%, which may be due to the appearance of oxygen and nitrogen atoms on the surface of the neat PEI membrane. By introducing the hydrophilic A1, A1, and A3 MSPs, the moisture absorption of the composite membranes increased up to 0.95, 0.98, and 1.18 wt-%, respectively. On the other hand, by introducing the hydrophobic MSPs (M1 and M2), the moisture absorption of the composite membranes decreased to 0.69 and 0.52 wt-%, respectively. Moreover, by introducing the hydrophobic M2 MSP at four different feeding ratios of 0.5 wt-%, 1 wt-%, 2 wt-%, and 3 wt-% in the PEI membrane, the moisture absorption of the composite membranes was further decreased to 0.61, 0.52, 0.38, and 0.45 wt-%, respectively.

Second, the T_{5d} of the PEI and its corresponding composite membranes were also determined with the weight loss from the TGA curve at 5 wt-%. It should be noted that all PEI membrane samples exhibited T_{5d} 400 °C. First, the T_{5d} of the neat PEI membrane appeared at 445.8 °C. Moreover, the T_{5d} of the PEI membrane containing the hydrophilic MSPs (A1, A2, and A3) were located at 453.8, 445.0, and 443.9 °C, respectively.

On the other hand, the T_{5d} of the PEI membrane containing the hydrophobic MSPs of M1 and M2 were 461.0 and 470.4 °C, respectively. It indicated that the hydrophobic M2 MSP with a higher surface area might have stronger chemical bonds between the primary amine groups inside and outside the mesoporous silica channels with the dianhydride groups of TAHQ, simultaneously leading to higher mechanical strength and thermal decomposition temperature.

The T_{5d} of the PEI membrane containing the hydrophobic M2 MSP at four different feeding ratios of 0.5 wt-%, 1 wt-%, 2 wt-%, and 3 wt-% were found to be 424.8, 470.0, 486.6, and 463.9 °C, respectively. It implied that, up to the 2 wt-% loading of M2, a higher loading of M2 in the PEI membrane resulted in a higher thermal decomposition temperature of the composite membranes. However, the increase in the T_{5d} of the PEI composite membrane at 3 wt-% of M2 loading may be attributed to the following two reasons: (1) the M2 particle aggregated at higher loading of MSPs, resulting in decreased chemical bond formation between the -NH_2 of M2 and the dianhydride of TAHQ or (2) the excessive amount of hydrophobic M2 particles existing in the solution of APAB and TAHQ retarded the polymerization reactions.

3.5. Discussion

The hydrophobic mesoporous silica powder prepared in this research experiment only used a methyl functional siloxane modifier for the hydrophobic modification. In future studies, non-fluorine-based modifier formulations could be considered as substitutions, such as siloxane with hydrophobic functional groups, including phenyl, vinyl, and ester groups could be adapted to modify the mesoporous silica. In addition, dispersibility could be studied and discussed by adjusting the particle size of the powder. The conditions of the sol-gel method could also be investigated in detail, such as varying the pH value of the synthesis environment. This could adjust the particle size of the powder, i.e., adjust the organic and inorganic phases, finding better compatibility. In addition, to achieve commercially practical applications, yield is also an important key factor because it is related to the cost of goods. The high specific surface area hydrophobic mesoporous silica powder M2 prepared in this study had a yield of approximately 25%, so it is bound to be optimized through conditions to improve the yield before it could be used in practical applications.

Table 1. Formulation and analytical data of the BET, CA, TGA, and EDS for the hydrophilic MSPs of A1, A2, and A3 as well as the hydrophobic MSPs of M1 and M2.

Sample Code	D-(-)-Fructose	TEOS	APTES (-NH_2)	MTMS (-CH_3)	BET			CA	TGA@200 °C	EDS (Atomic%)			
					Surface Area (m^2/g)	Pore Volume (cm^3/g)	Average Pore Diameter (nm)		Moisture				
Unit	g	mmole	mmole	mmole				(°)	(%)	C	N	O	Si
A1	5.2	27	22.6	-	469 ± 12	0.26 ± 0.02	2.9 ± 0.2	-	4.15	49.25	0.80	36.59	13.36
A2	5.2	27	45.2	-	402 ± 21	0.25 ± 0.02	3.0 ± 0.2	-	4.24	26.41	1.20	51.55	20.84
A3	5.2	27	90.4	-	384 ± 14	0.21 ± 0.01	2.7 ± 0.2	-	7.68	34.05	1.60	41.53	22.83
M1	5.2	9	0.45	18	548 ± 12	0.94 ± 0.02	4.9 ± 0.1	127.6	0.58	-	-	-	-
M2	7.2	9	0.45	18	771 ± 14	0.86 ± 0.01	6.2 ± 0.1	127.2	0.45	-	-	-	-

All samples were prepared with 9 mL of ammonia and 100 mL of distilled water.

Table 2. Formulation and analytical data of the dielectric properties, mechanical strengths, thermal properties, moisture absorptions, and crystallinities of the PEI and its corresponding composite membranes containing A1, A2, A3, M1, and M2.

Sample Code	PEI		Silica		10 GHz		Tensile Strength	Elongation at Break	T_{5d}	Moisture (TGA@ 200 °C)	Contact Angle	FWHM
	TAHQ	APAB										
Unit	(mmole)	(mmole)	Code	(wt-%)	D_k	D_f	MPa	%	°C	%	°	
PEI	20	20	-	-	3.27	0.007	11.26	1.48	445.8	0.73	74.6	3.2
PEI-A1-1	20	20	A1	1	3.43	0.011	4.03	2.15	453.8	0.95	83.7	3.4
PEI-A2-1	20	20	A2	1	3.46	0.012	4.80	2.52	445.0	0.98	80.9	3.5
PEI-A3-1	20	20	A3	1	3.92	0.015	5.48	2.75	443.9	1.18	61.5	3.5
PEI-M1-1	20	20	M1	1	3.24	0.009	-	-	461.0	0.69	83.5	3.3
PEI-M2-0.5	20	20	M2	0.5	3.24	0.007	15.91	1.37	424.8	0.61	81.1	3.3
PEI-M2-1	20	20	M2	1	3.18	0.008	16.23	1.20	470.4	0.52	84.3	3.3
PEI-M2-2	20	20	M2	2	2.97	0.008	18.95	1.19	486.6	0.38	88.5	3.4
PEI-M2-3	20	20	M2	3	2.93	0.009	6.68	1.65	463.9	0.45	93.8	3.5

4. Concluding Remarks

In this study, comparative studies of hydrophilic and hydrophobic MSPs on the dielectric properties of their derivative PEI composite membranes were performed. First of all, a series of hydrophilic and hydrophobic MSPs were synthesized via the base-catalyzed sol-gel process of TEOS, MTMS, and APTES at distinctive specific feeding ratios with the non-surfactant template of D-(-)-Fructose as the pore-forming agent. The prepared MSPs were characterized with FTIR, ^{29}Si-NMR, and ^{13}C NMR spectra, and EDS. The surface morphological images of the prepared hydrophilic MSPs modified with APTES and hydrophobic MSPs modified with MTMS were found to show spherical and irregular shapes, respectively, as identified by observing the SEM and TEM. For the study of BET, by increasing the APTES content of the hydrophilic MSPs, a trend of decreasing surface area and pore volume was found. On the other hand, by increasing the content of the pore-forming agent in the hydrophobic MSPs, the surface area and average pore diameter were both elevated. The contact angle of the hydrophilic MSPs (A1, A2, and A3) could not be detected, and the contact angle of the hydrophobic MSPs (M1 and M2) all measured at ~127°. The moisture absorptions of the A1, A2, A3, M1, and M2 MSPs were estimated at 4.15, 4.24, 7.68, 0.58, and 0.45 wt-%, respectively, as determined with TGA.

Subsequently, the MSPs were blended with the diamine of APAB, followed by the introduction of the dianhydride of TAHQ with mechanical stirring for 24 h. The obtained PEI2 composites were characterized with FTIR and XRD. It should be noted that the dielectric constant of the PEI composites was found to show an obvious trend: PEI containing hydrophilic MSP > PEI > PEI containing hydrophobic MSP. Moreover, the higher the loading of hydrophilic MSPs, the lower the dielectric properties. On the contrary, the higher the loading of the hydrophobic MSP, the better the dielectric properties. The mechanical strength and thermal stability of the PEI and its composite membranes were investigated with the tensile test and TGA, respectively. It should be noted that the PEI composite membranes containing hydrophilic MSPs and hydrophobic MSPs were found to exhibit weaker and stronger mechanical strengths, respectively. Moreover, the PEI composite containing hydrophilic and hydrophobic MSPs was also found to reveal lower and higher thermal decomposition temperatures, respectively. The PEI composite membranes containing hydrophilic and hydrophobic MSPs were found to exhibit higher and lower moisture absorption amounts, respectively.

Author Contributions: Conceptualization, investigation and validation, K.-Y.C.; Assistive analysis, K.-H.L.; writing—original draft preparation, writing—review and editing, M.Y.; supervision and methodology, Y.W., supervision, J.-M.Y. All authors have read and agreed to the published version of the manuscript.

Funding: This research received external funding from the The National Science and Technology Council, Taiwan, R.O.C. (grant number: NSTC107-2113-M-033-007-).

Institutional Review Board Statement: Not applicable.

Informed Consent Statement: Not applicable.

Data Availability Statement: The data that support the findings of this study are available from the corresponding author, but restrictions apply to the availability of these data, which were used under license for the current study, and so are not publicly available. The data are, however, available from the authors upon reasonable request and with permission of the funding party, Ministry of Science and Technology, Taiwan, R.O.C.

Conflicts of Interest: All authors declare no conflict of interest.

References

1. Gui, G.; Liu, M.; Tang, F.; Kato, N.; Adachi, F. 6G: Opening new horizons for integration of comfort, security, and intelligence. *IEEE Wirel. Commun.* **2020**, *27*, 126–132. [CrossRef]
2. Farasat, M.; Thalakotuna, D.N.; Hu, Z.; Yang, Y. A review on 5G sub-6 GHz base station antenna design challenges. *Electronics* **2021**, *10*, 2000. [CrossRef]

3. Tripathi, S.; Sabu, N.V.; Gupta, A.K.; Dhillon, H.S. Millimeter-Wave and Terahertz Spectrum for 6G Wireless. In *6G Mobile Wireless Networks*; Springer: Cham, Switzerland, 2021; pp. 82–121. [CrossRef]
4. Ikram, M.; Sultan, K.; Lateef, M.F.; Alqadami, A.S.M. A Road towards 6G Communication—A Review of 5G Antennas, Arrays, and Wearable Devices. *Electronics* **2022**, *11*, 169. [CrossRef]
5. Giordani, M.; Polese, M.; Mezzavilla, M.; Rangan, S.; Zorzi, M. Toward 6G networks: Use cases and technologies. *IEEE Commun. Mag.* **2020**, *58*, 55–61. [CrossRef]
6. Chowdhury, M.Z.; Shahjalal, M.; Hasan, M.K.; Jang, Y.M. The role of optical wireless communication technologies in 5G/6G and IoT solutions: Prospects, directions, and challenges. *Appl. Sci.* **2019**, *9*, 4367. [CrossRef]
7. Waterhouse, R.; Novack, D. Realizing 5G: Microwave photonics for 5G mobile wireless systems. *IEEE Microw. Mag.* **2015**, *16*, 84–92. [CrossRef]
8. Duan, B.Y. Evolution and innovation of antenna systems for beyond 5G and 6G. *Front. Inf. Technol. Electron. Eng.* **2020**, *21*, 1–3. [CrossRef]
9. Liaw, D.J.; Wang, K.L.; Huang, Y.C.; Lee, K.R.; Lai, J.Y.; Ha, C.S. Advanced Polyimide Materials: Syntheses, Physical Properties and Applications. *Prog. Polym. Sci.* **2012**, *37*, 907–974. [CrossRef]
10. Hao, Y.; Xiang, S.; Han, G.; Zhang, J.; Ma, X.; Zhu, C.; Guo, X.; Zhang, Y.; Han, Y.; Song, Z.; et al. Recent progress of integrated circuits and optoelectronic chips. *Sci. China Inf. Sci.* **2021**, *64*, 201401. [CrossRef]
11. Qian, C.; Fan, Z.-G.; Zheng, W.-W.; Bei, R.-X.; Zhu, T.-W.; Liu, S.-W.; Chi, Z.-G.; Aldred, M.P.; Chen, X.-D.; Zhang, Y.; et al. A Facile Strategy for Non-fluorinated Intrinsic Low-k and Low-loss Dielectric Polymers: Valid Exploitation of Secondary Relaxation Behaviors. *Chin. J. Polym. Sci.* **2020**, *38*, 213–219. [CrossRef]
12. Maier, G. Low dielectric constant polymers for microelectronics. *Prog. Polym. Sci.* **2001**, *26*, 3–65. [CrossRef]
13. Liu, Y.; Zhao, X.-Y.; Sun, Y.-G.; Li, W.-Z.; Zhang, X.-S.; Luan, J. Synthesis and applications of low dielectric polyimide. *Resour. Chem. Mater.* **2022**, in press. [CrossRef]
14. Bei, R.; Chen, W.; Zhang, Y.; Liu, S.; Chi, Z.; Xu, J. Progress of low dielectric constant polyimide films. *Insul. Mater.* **2016**, *49*, 1–11. [CrossRef]
15. Kourakata, Y.; Onodera, T.; Kasai, H.; Jinnai, H.; Oikawa, H. Ultra-low dielectric properties of porous polyimide thin films fabricated by using the two kinds of templates with different particle sizes. *Polymer* **2021**, *212*, 123115. [CrossRef]
16. Li, X.; Zhang, P.; Dong, J.; Gan, F.; Zhao, X.; Zhang, Q. Preparation of low-κ polyimide resin with outstanding stability of dielectric properties versus temperature by adding a reactive Cardo-containing diluent. *Compos. B Eng.* **2019**, *177*, 107401. [CrossRef]
17. Zhang, P.; Zhang, L.; Zhang, K.; Zhao, J.; Li, Y. Preparation of Polyimide Films with Ultra-Low Dielectric Constant by Phase Inversion. *Crystals* **2021**, *11*, 1383. [CrossRef]
18. Meador, M.; Mcmillon, E.; Anna, S.; Barrios, E.; Wilmoth, N.G.; Mueller, C.H.; Miranda, F.A. Dielectric and Other Properties of Polyimide Aerogels Containing Fluorinated Blocks. *ACS Appl. Mater. Interfaces* **2014**, *6*, 6062–6068. [CrossRef]
19. Wu, T.; Dong, J.; Gan, F.; Fang, Y.; Zhao, X.; Zhang, Q. Low dielectric constant and moisture-resistant polyimide aerogels containing trifluoromethyl pendent groups. *Appl. Surf. Sci.* **2018**, *440*, 595–605. [CrossRef]
20. Zhang, P.; Zhao, J.; Zhang, K.; Wu, Y.; Li, Y. Effect of co-solvent on the structure and dielectric properties of porous polyimide membranes. *J. Phys. D Appl. Phys.* **2018**, *51*, 215305. [CrossRef]
21. Chen, W.; Zhou, Z.; Yang, T.; Bei, R.; Zhang, Y.; Liu, S.; Chi, Z.; Chen, X.; Xu, J. Synthesis and properties of highly organosoluble and low dielectric constant polyimides containing non-polar bulky triphenyl methane moiety. *React. Funct. Polym.* **2016**, *108*, 71–77. [CrossRef]
22. Wang, Y.; Yang, Z.; Wang, H.; Li, E.; Yuan, Y. Investigation of PTFE-based ultra-low dielectric constant composite substrates with hollow silica ceramics. *J. Mater. Sci. Mater. Electron.* **2022**, *33*, 4550–4558. [CrossRef]
23. Moldoveanu, S.C.; David, V. Chapter 7—Mobile Phases and Their Properties. In *Essentials in Modern HPLC Separations*; Moldoveanu, S.C., David, V., Eds.; Elsevier: Amsterdam, The Netherlands, 2013. [CrossRef]
24. Hong, Z.; Dongyang, W.; Yong, F.; Hao, C.; Yusen, Y.; Jiaojiao, Y.; Liguo, J. Dielectric properties of polyimide/SiO2 hollow spheres composite films with ultralow dielectric constant. *Mater. Sci. Eng. B* **2016**, *203*, 13–18. [CrossRef]
25. Yang, P.; Zhao, D.; Margolese, D.I.; Chmelka, B.F.; Stucky, G.D. Generalized syntheses of large-pore mesoporous metal oxides with semicrystalline frameworks. *Nature* **1998**, *396*, 152–155. [CrossRef]
26. Kresge, C.T.; Leonowicz, M.E.; Roth, W.J.; Vartuli, J.C.; Beck, J.S. Ordered mesoporous molecular sieves synthesized by a liquid-crystal template mechanism. *Nature* **1992**, *359*, 710–712. [CrossRef]
27. Pang, J.-B.; Qiu, K.-Y.; Wei, Y. A novel non-surfactant pathway to hydrothermally stable mesoporous silica materials. *Microporous Mesoporous Mater.* **2000**, *40*, 299–304. [CrossRef]
28. Feng, Q.; Xu, J.; Dong, H.; Li, S.; Wei, Y. Synthesis of polystyrene–silica hybrid mesoporous materials via the non-surfactant-templated sol–gel process. *J. Mater. Chem.* **2000**, *10*, 2490–2494. [CrossRef]
29. Wei, Y.; Feng, Q.; Xu, J.; Dong, H.; Qiu, K.-Y.; Jansen, S.A.; Yin, R.; Ong, K.K. Polymethacrylate–Silica Hybrid Nanoporous Materials: A Bridge between Inorganic and Polymeric Molecular Sieves. *Adv. Mater.* **2000**, *12*, 1448–1450. [CrossRef]
30. Pang, J.; Qiu, K.; Wei, Y. Preparation of mesoporous silica materials with non-surfactant hydroxy-carboxylic acid compounds as templates via sol–gel process. *J. Non-Cryst. Solids* **2001**, *283*, 101–108. [CrossRef]
31. Hasegawa, M.; Hishiki, T. Poly(ester imide)s Possessing Low Coefficients of Thermal Expansion and Low Water Absorption (V). Effects of Ester-linked Diamines with Different Lengths and Substituents. *Polymers* **2020**, *12*, 859. [CrossRef]

32. Hegde, N.D.; Venkateswara Rao, A. Physical properties of methyltrimethoxysilane based elastic silica aerogels prepared by the two-stage sol–gel process. *J. Mater. Sci.* **2007**, *42*, 6965–6971. [CrossRef]
33. Mahadik, D.; Jung, H.-N.; Lee, Y.K.; Lee, K.-Y.; Park, H.-H. Elastic and Superhydrophobic Monolithic Methyltrimethoxysilane-based Silica Aerogels by Two-step Sol-gel Process. *J. Microelectron. Packag. Soc.* **2016**, *23*, 35–39. [CrossRef]
34. Tewari, P.; Rajagopalan, R.; Furman, E.; Lanagan, M.T. Control of interfaces on electrical properties of SiO2–Parylene-C laminar composite dielectrics. *J. Colloid Interface Sci.* **2009**, *332*, 65–73. [CrossRef] [PubMed]
35. Fahmy, A.; Mohamed, T.A.; Abu-Saied, M.; Helaly, H.; El-Dossoki, F. Structure/property relationship of polyvinyl alcohol/dimethoxydimethylsilane composite membrane: Experimental and theoretical studies. *Spectrochim. Acta Part A Mol. Biomol. Spectrosc.* **2020**, *228*, 117810. [CrossRef] [PubMed]
36. Gangan, A.; ElSabbagh, M.; Bedair, M.A.; Ahmed, H.M.; El-Sabbah, M.; El-Bahy, S.M.; Fahmy, A. Influence of pH values on the electrochemical performance of low carbon steel coated by plasma thin SiOxCy films. *Arab. J. Chem.* **2021**, *14*, 103391. [CrossRef]
37. Fahmy, A.; Abou-Saied, M.; Helaly, H.; El-Dessoki, F.; Mohamed, T.A. Novel PVA/Methoxytrimethylsilane elastic composite membranes: Preparation, characterization and DFT computation. *J. Mol. Struct.* **2021**, *1235*, 130173. [CrossRef]
38. Llusar, M.; Monrós, G.; Roux, C.; Pozzo, J.L.; Sanchez, C. One-pot synthesis of phenyl- and amine-functionalized silica fibers through the use of anthracenic and phenazinic organogelators. *J. Mater. Chem.* **2003**, *13*, 2505–2514. [CrossRef]
39. Wang, Y.Q.; Yang, C.M.; Zibrowius, B.; Spliethoff, B.; Lindén, M.; Schüth, F. Directing the formation of vinyl-functionalized silica to the hexagonal SBA-15 or large-pore Ia3d structure. *Chem. Mater.* **2003**, *15*, 5029–5035. [CrossRef]
40. Wang, X.; Lin, K.S.K.; Chan, A.J.C.C.; Cheng, S. Direct Synthesis and Catalytic Applications of Ordered Large Pore Aminopropyl-Functionalized SBA-15 Mesoporous Materials. *J. Phys. Chem. B* **2005**, *109*, 1763–1769. [CrossRef]
41. Krysztafkiewicz, A.; Jesionowski, T.; Binkowski, S. Precipitated silicas modified with 3-aminopropyltriethoxysilane. *Colloids Surf. A Physicochem. Eng. Asp.* **2000**, *173*, 73–84. [CrossRef]
42. Zheng, J.-Y.; Pang, J.-B.; Qiu, K.-Y.; Wei, Y. Synthesis of Mesoporous Silica Materials via Non-surfactant Templated Sol-Gel Route by Using Mixture of Organic Compounds as Template. *J. Sol-Gel Sci. Technol.* **2002**, *24*, 81–88. [CrossRef]
43. Okada, T.; Ando, S. Conformational characterization of imide compounds and polyimides using far-infrared spectroscopy and DFT calculations. *Polymer* **2016**, *86*, 83–90. [CrossRef]
44. Jaiboon, V.; Yoosuk, B.; Prasassarakich, P. Amine modified silica xerogel for H2S removal at low temperature. *Fuel Process. Technol.* **2014**, *128*, 276–282. [CrossRef]
45. Tung, H.-M.; Huang, J.-H.; Tsai, D.-G.; Ai, C.-F.; Yu, G.-P. Hardness and residual stress in nanocrystalline ZrN films: Effect of bias voltage and heat treatment. *Mater. Sci. Eng. A* **2009**, *500*, 104–108. [CrossRef]
46. Yoo, Y.J.; Ju, S.; Park, S.Y.; Kim, Y.J.; Bong, J.; Lim, T.; Kim, K.W.; Rhee, J.Y.; Lee, Y. Metamaterial Absorber for Electromagnetic Waves in Periodic Water Droplets. *Sci. Rep.* **2015**, *5*, 14018. [CrossRef] [PubMed]
47. *ASTM D882-18*; Standard Test Method for Tensile Properties of Thin Plastic Sheeting. ASTM International: West Conshohocken, PA, USA, 2018; p. 12.
48. Pearson, R.A.; Yee, A.F. Toughening mechanisms in elastomer-modified epoxies. *J. Mater. Sci.* **1989**, *24*, 2571–2580. [CrossRef]

Disclaimer/Publisher's Note: The statements, opinions and data contained in all publications are solely those of the individual author(s) and contributor(s) and not of MDPI and/or the editor(s). MDPI and/or the editor(s) disclaim responsibility for any injury to people or property resulting from any ideas, methods, instructions or products referred to in the content.

Article

Synthesis and DC Electrical Conductivity of Nanocomposites Based on Poly(1-vinyl-1,2,4-triazole) and Thermoelectric Tellurium Nanoparticles

Anna V. Zhmurova, Galina F. Prozorova, Svetlana A. Korzhova, Alexander S. Pozdnyakov *
and Marina V. Zvereva

A. E. Favorsky Irkutsk Institute of Chemistry, Siberian Branch of Russian Academy of Sciences, Favorsky 1, 664033 Irkutsk, Russia
* Correspondence: pozdnyakov@irioch.irk.ru; Tel.: +7-(3952)-426911

Citation: Zhmurova, A.V.; Prozorova, G.F.; Korzhova, S.A.; Pozdnyakov, A.S.; Zvereva, M.V. Synthesis and DC Electrical Conductivity of Nanocomposites Based on Poly(1-vinyl-1,2,4-triazole) and Thermoelectric Tellurium Nanoparticles. *Materials* 2023, 16, 4676. https://doi.org/10.3390/ma16134676

Academic Editors: Peijiang Liu and Liguo Xu

Received: 1 June 2023
Revised: 21 June 2023
Accepted: 23 June 2023
Published: 28 June 2023

Copyright: © 2023 by the authors. Licensee MDPI, Basel, Switzerland. This article is an open access article distributed under the terms and conditions of the Creative Commons Attribution (CC BY) license (https://creativecommons.org/licenses/by/4.0/).

Abstract: In this work, the structural characteristics and DC electrical conductivity of firstly synthesized organic–inorganic nanocomposites of thermoelectric Te^0 nanoparticles (1.4, 2.8, 4.3 wt%) and poly(1-vinyl-1,2,4-triazole) (PVT) were analyzed. The composites were characterized by high-resolution transmission electron microscopy, X-ray diffractometry, UV-Vis spectroscopy, and dynamic light scattering analysis. The study results showed that the nanocomposite nanoparticles distributed in the polymer matrix had a shape close to spherical and an average size of 4–18 nm. The average size of the nanoparticles was determined using the Brus model relation. The optical band gap applied in the model was determined on the basis of UV-Vis data by the Tauc method and the 10% absorption method. The values obtained varied between 2.9 and 5.1 nm. These values are in good agreement with the values of the nanoparticle size, which are typical for their fractions presented in the nanocomposite. The characteristic sizes of the nanoparticles in the fractions obtained from the Pesika size distribution data were 4.6, 4.9, and 5.0 nm for the nanocomposites with percentages of 1.4, 2.8, and 4.3%, respectively. The DC electrical conductivity of the nanocomposites was measured by a two-probe method in the temperature range of 25–80 °C. It was found that the formation of an inorganic nanophase in the PVT polymer as well as an increase in the average size of nanoparticles led to an increase in the DC conductivity over the entire temperature range. The results revealed that the DC electrical conductivity of nanocomposites with a Tellurium content of 2.8, 4.3 wt% at 80 °C becomes higher than the conventional boundary of 10^{-10} S/cm separating dielectrics and semiconductors.

Keywords: polymer nanocomposite; dielectric polymer; poly(1-vinyl-1,2,4-triazole); thermoelectric tellurium nanoparticles; DC electrical conductivity

1. Introduction

Electronic devices that consume electrical energy and inevitably produce parasitic heat are now widespread. The demand for systems that convert heat into electrical energy is therefore growing critically. It is known that the construction of the systems is possible in the form of a combination of thermoelectric coolers (TEC) and thermoelectric generators (TEG) [1]. The conversion efficiency of TEC and TEG is directly related to the thermoelectric figure of merit (ZT) of the materials implemented as thermoelectric cell legs. Currently, an active search for high-ZT thermoelectrics is underway. Particular attention is paid to the development of polymer-based thermoelectrics (PTs) due to their advantages over completely inorganic systems [2], such as low cost, environmental friendliness and diversity of synthesis methods, low thermal conductivity, relatively light weight, mechanical flexibility, etc. [3]. One important class of PTs are organo–inorganic nanocomposites that combine the thermoelectric properties of their inorganic nanophase (metal chalcogenides and oxides as well as elemental chalcogenes, in particular Te^0 [4–10]) with the performance and mechanical properties of polymer matrices. The PTs exhibit a relatively high Seebeck coefficient and low thermal conductivity. The use of insulating polymers as matrices for nanocomposites compared to conducting polymers allows the production of cheaper and more stable thermoelectrics with features important in practice (flexibility, extensibility, thermoplasticity).

The most widely applied polymers in PT creation are polyvinylidene fluoride [11], polyvinyl alcohol [12], polymethyl methacrylate [6], cellulose [5], and polyvinylpyrrolidone [13], etc. According to the literature, the direct incorporation of nanoparticles into the polymer matrix is achieved by combining the pre-synthesized nanoparticles and polymer in a solution with a subsequent drop casting, hot compaction [11,14], or vacuum filtration [11], mechanical pressing [11], or screen printing [15]. It should be noted that the above methods of PT creation are time consuming, often based on the application of special equipment providing sonication [16,17], high-pressure action [13], and high temperatures [10,12], as well as the use of organic toxic solvents and an extensive range of nanoparticle precursors (NaBr [5], Na_2TeO_3 [10], $NaBH_4$ [10], Na_2SeO_3 [11], TeO_2 [6,14], SeO_2 [13], NH_2OH [14], $CuSO_4$ [18]), which have toxic and environmentally unfriendly properties.

In this regard, the use of previously developed simple and environmentally friendly methods of producing chalcogen containing nanoparticles [19,20], particularly Te^0 ones from the commercially available bulk Te powder, as well as the application of the original polymer poly(1-vinyl-1,2,4-triazole) (PVT) in the form of a polymeric stabilizing matrix for the nanoparticle formation, seems highly promising for the synthesis of thermoelectric nanocomposites. The PVT exhibits a complex of practically important properties such as high hydrophilicity, solubility in polar organic solvents, ability for complexation and quaternization, chemical stability, biocompatibility, and thermal stability. In addition, triazole-containing polymers were previously experimentally found to behave as effective stabilizing matrices when forming metal-containing nanocomposites, exhibiting a synergy of unique polymer properties (solubility, biocompatibility, high coordination ability) and the optical, catalytic, and biological properties of metal nanoparticles [21–24]. At the same time, Te as a thermoelectric direct-gap semiconductor has been successfully used in the creation of polymeric organic–inorganic thermoelectrics with a p-type conductivity and both a conductive [25] and insulating matrix [10,26–28].

The paper presents the results of the polymer thermoelectrics synthesis of a Te nanophase and a PVT original polymer matrix with p-type conductivity and the study of the influence of nanocomposite structural features determined by the PVT: Te relation on their DC electrical conductivity.

2. Materials and Methods

2.1. Materials

NaOH (Reahim, Moscow, Russia), hydrazine hydrate (Reahim, Moscow, Russia), acetone (Reahim, Moscow, Russia), azobisisobutyronitrile (Sigma Aldrich, Burlington, VT, USA), dimethylformamide (Reahim, Moscow, Russia), and tellurium powder (Thermo Fisher, Kandel, Germany) were used without additional purification.

2.2. Methods

2.2.1. Synthesis of PVT

The synthesis of PVT was carried out by a radical polymerization method of 1-vinyl-1,2,4-triazole (VT) in the presence of an azobisisobutyronitrile (AIBN) initiator in a dimethylformamide (DMFA) medium under an argon atmosphere and a temperature of 60 °C during 6 h, according to the protocol detailed in the paper [24].

Briefly, VT (1.5 g, 16.0 mmol), AIBN (0.015 g, 0.09 mmol), and DMFA (1.0 g) were placed in a glass ampoule, flasked with argon, then sealed and incubated in the thermostat at 60 °C for 6 h. The resulting PVT was isolated and purified by a double resuspension in a mixture of ethanol and acetone (1:2), and then dried to a constant mass in a vacuum oven over phosphorus pentoxide at 50 °C. The average molecular weight of the polymer was 110 kDa and the weight average molecular weight was 211 kDa. The polydispersity coefficient was 1.92 (Figure 1). The PVT obtained in a 92% yield was a white powder, well soluble in water, DMF, and DMSO. The calculated % were C 50.52; H 5.26; and N 44.22 The found % were C 50.61; H 5.25; and N 44.14.

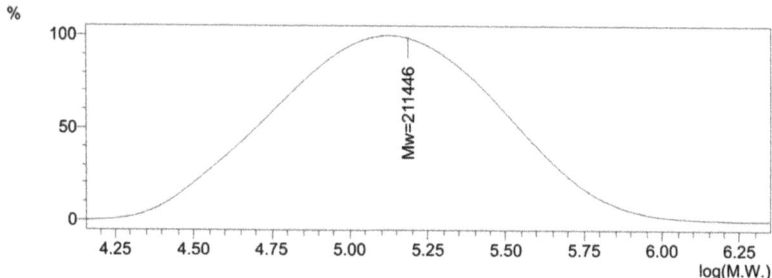

Figure 1. The molecular weight distribution of PVT.

Using the method of nuclear magnetic resonance ^1H and ^{13}C, the structure of the obtained polymer was confirmed (Figure 2). The ^1H NMR (DMSO-d_6, δ, ppm) was 8.08–7.41 (br m., 2H, triazole ring), 4.15–2.66 (br m., 1H, CH in the polymer backbone), 2.25–1.60 (br, 2H, CH_2 in the polymer backbone); the ^{13}C NMR (DMSO-d_6, δ, ppm) was 152.50–150.50, 145.00–142.5 (CH, triazole ring), 55.00–52.20 (CH in the polymer backbone), and 41.40–38.20 (probably the signals of carbon atoms of methylene groups CH_2 of the polymer backbone overlap with solvent signals).

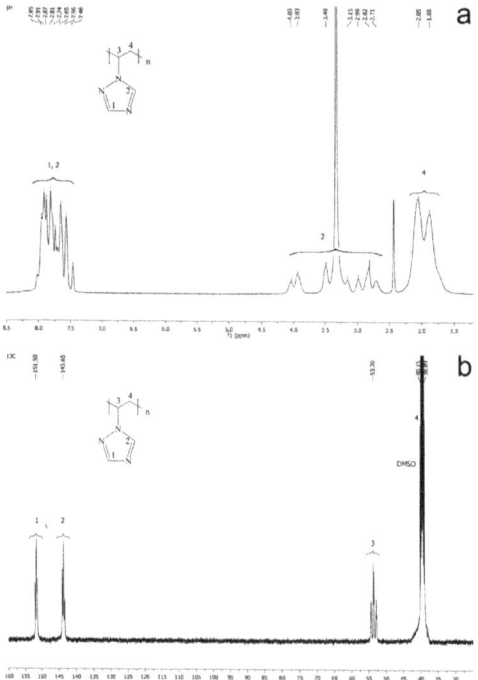

Figure 2. The ^1H (**a**) and ^{13}C NMR (**b**) spectra of PVT.

2.2.2. Synthesis of PVT–Te⁰NPs

To synthesize the PVT–Te^{0}NPs, the PVT (1 g) and distilled water (50 mL) were placed in the 100 mL flask. The reaction mixture was then stirred for 3 h at the temperature of 40 °C until the PVT was completely dissolved. The solution containing PVT was cooled to room temperature. The reaction mixture with tellurium anions (120–400 µL) that were pre-synthesized in accordance with the technique described in the articles [19,20] was added into a Te solution flask under constant stirring with a magnetic stirrer. The formation of the Te^0 nanoparticles was identified by a change in the color of the reaction mixture from colorless (native PVT solution) or red–purple (reaction mixture containing Te^{2-} anions) to gray. The synthesis duration was 25 min, after which the reaction mixture was precipitated in acetone cooled to 7 °C. The resulting fine precipitate was separated by centrifugation at 4000 rpm at 9 °C, washed three times with acetone, and dried in a vacuum at room temperature. For the PVT–Te^{0}NP (1.4%), the yield was 62%; calculated % were C 49.53; H 5.50; N 43.27; and Te 1.7. The found % were C 49.64; H 5.52; N 43.44; and Te 1.4. For the PVT–Te^{0}NPs (2.8%), the yield was 71%; the calculated % were C 48.05; H 5.36; N 43.19; and Te 3.4. The found % were C 48.95; H 5.56; N 42.71; Te 2.80. For the PVT–Te^{0}NPs (4.3%), the yield was 58%; the calculated % were C 48.00; H 5.56; N 41.04; and Te 5.4. The found % were C 48.30; H 5.67; N 41.73; and Te 4.3.

2.3. Equipment

2.3.1. Elemental Analysis (EA)

The elemental composition was determined by X-ray energy dispersive microanalysis using a Hitachi TM 3000 scanning electron microscope (Tokyo, Japan) with an SDD XFlash 430-4 X-ray detector and a Thermo Fisher Scientific Flash 2000 CHNS analyzer (Kandel, Germany).

2.3.2. X-ray Diffraction (XRD) Analysis

The XRD study was performed on a Bruker D8 ADVANCE diffractometer (Billerica, MA, USA) equipped with a Hebbel mirror, with Cu radiation in the locked coupled mode, with an exposure of 1 s for phase analysis and 3 s for the estimation of the cell parameter and the coherent scattering region size.

2.3.3. UV-Vis Spectroscopy

The UV-Vis spectra of 0.025% water solutions of PVT and the nanocomposites were recorded relative to distilled water in a 1 cm quartz cell on a Perkin Elmer LAMBDA 35 UV-Vis spectrophotometer (Waltham, MA, USA) in the wavelength range of 200–700 nm.

2.3.4. Dynamic Light Scattering (DLS)

The hydrodynamic radii (Rh) of pure PVT and PVT–Te^{0}NPs were determined by dynamic light scattering on a Photocor Compact-Z correlation spectrometer (Moscow, Russia) equipped by a 20 mV thermostabilized semiconductor laser (λ = 638 нм) at an angle of 90°. The autocorrelation function was analyzed by Dynals v.2 software (Tirat Carmel, Israel). The solutions for analysis were prepared by dissolving a 5 mg sample in 20 mL of distilled water at room temperature for 5 hours, pre-filtered through a 200 µm syringe filter. The time for each measurement was at least 200 s. The measurements were taken in triplicate and the mean was used.

2.3.5. Gel Permeation Chromatography

The molecular weight distribution of the PVT was measured using a gel permeation chromatograph Shimadzu LC-20 Prominence (Kyoto, Japan) with a differential refractive index detector, Shimadzu RID-20A. The chromatographic column was Agilent PolyPore 7.5 × 300 mm, PL1113-6500 (Santa Clara, CA, USA) with an appropriate pre-column. High purity N,N-dimethylformamide was used as a mobile phase (1 ml/min). The calibration was carried out using a series of polystyrene standards, consisting of samples with molecular weights from 162 to 6,570,000 g/mol.

2.3.6. Nuclear Magnetic Resonance

^{1}H and ^{13}C NMR spectra were recorded on a Bruker DPX-400 spectrometer (Billerica, MA, USA) at room temperature; ^{1}H, 400.13 MHz; ^{13}C, 100.62 MHz. The chemical shifts are given relative to the TMS.

2.3.7. High Resolution–Transmission Electron Microscopy (HR-TEM)

The transmission electron microphotographs were obtained with an FEI Tecnai G2 20F S-TWIN transmission electron microscope (Hillsboro, OR, USA) according to the procedure detailed in [29]. The nanoparticle size distribution was determined by the statistical treatment of the microphotographs using Gatan DigitalMicrograph v.3.5 software (Pleasanton, CA, USA) and Microsoft Office Excel (Redmond, WA, USA). The electronograms from the transmission electron microscope were processed and indicated with Process Diffraction v.8.7.1 (Budapest, Hungary) and CrysTBox v.1.1 software (Prague, Czech Republic) and the crystallographic database JCPDS-ICDD PDF-2.

2.3.8. Direct Current Electrical Conductivity Measurement

The DC conductivity was measured by a standard E6-13A teraohmmeter (Moscow, Russia) using a two-probe method within temperature range of 25–80 °C. A thermostat was used to maintain the temperature in the measuring cell. The powder samples were pressed in the form of pellets with a height of 0.2–0.6 mm and a radius of 1.5 mm. The DC conductivity was calculated using the following equation:

$$\sigma = \frac{d}{RA} \quad (1)$$

where A is the cross-section area (cm^2), d is thickness of the pellet (cm), and R is the resistance of the sample (Ω).

3. Results and Discussion

3.1. Synthesis of PVT–Te^{0}NPs

Water-soluble aggregation-stable PVT–Te^{0}NPs (1.4–4.3 wt% Te) were obtained by the oxidation of Te^{2-} anions to zero-valent tellurium (Te^0) in an aqueous PVT solution. The Te^{2-} anions were pre-generated by the reduction (activation) in commercial powdered tellurium with hydrazine hydrate in an alkaline medium (Figure 3), according to the method previously proposed in [19,20].

The process resulted in the powdered tellurium being completely dissolved to form highly reactive Te^{2-} anions in accordance with Equation (2):

$$2Te + 4NaOH + N_2H_4 \cdot H_2O = 2Na_2Te + N_2\uparrow + 5H_2O, \quad (2)$$

The weight content of tellurium in the nanocomposites was varied by changing the PVT:Te^{2-} ratio from 1:59 to 1:17. The conversion of Te^{2-} anions to the zero-valent state varied between 84% and 80%, decreasing with increasing PVT:Te^{2-} ratio. The Te^{2-} anions obtained have a limited time stability and high sensitivity to the conditions of synthesis because of their extremely high reactivity. Presumably, the formed sodium telluride is hydrolyzed in an aqueous medium to form H$_2$Te, which is further either removed from the reaction medium or oxidized by the oxygen present in the aqueous solution to Te^0 followed by the condensation of its atoms into nanoparticles and their subsequent stabilization. Probably the reason for the decrease in the Te^{2-} anion conversion to Te^0 is the part of H$_2$Te that is hydrolyzed and then volatilized from the reaction medium. The process can be identified by the presence of a radish smell associated with H$_2$Te. Under these conditions, we obtained Te⁰-containing PVT-based nanocomposites with yields of 58–71%, which are water soluble powders of grey color.

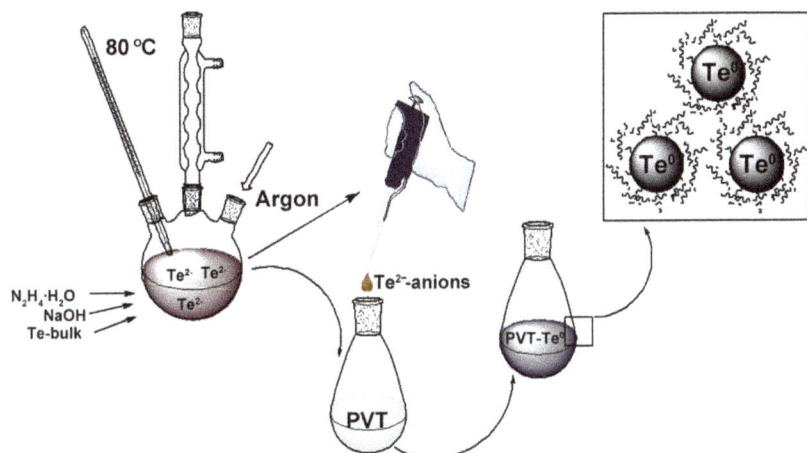

Figure 3. Scheme of PVT–Te⁰NPs synthesis.

3.2. XRD Analysis

According to X-ray diffraction analysis, the PVT–Te⁰NP (1.4 wt% Te) is X-ray amorphous, whereas the nanocomposites with a higher weight content have a two-phase amorphous-crystalline structure. Their diffractograms are characterized by an amorphous halo of the PVT phase as well as a number of reflexes corresponding to the crystalline phase of Te⁰ nanoparticles (Figure 4). We assume that the formation of Te⁰NPs in the hexagonal modification under the selected soft experimental conditions is probably due to the presence of the stabilizing matrix PVT in the reaction medium. In this case, the PVT molecules are capable of limiting the growth tendency of the material, which leads to the formation of hexagonal Te⁰ nanoparticles. Moreover, even in the case of the formation of a mixture of amorphous and crystalline tellurium under low-temperature conditions, the amorphous tellurium particles, due to their extremely low stability, are prone to re-dissolution during Ostwald ripening, and the released tellurium atoms can go to the completion of the crystal lattice of more stable hexagonal tellurium particles [30–32].

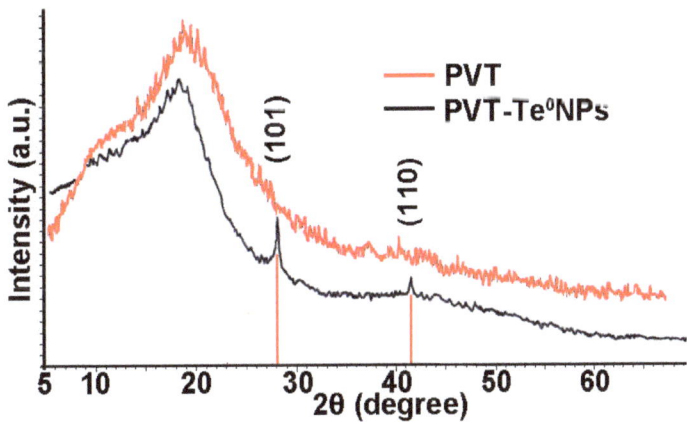

Figure 4. Typical XRD patterns for pure PVT and PVT–Te⁰NPs (4.3 wt% Te).

Thus, two broadened reflexes observed at angles of 27.55° and 40.45° in diffractograms for PVT–Te⁰NPs (4.3% Te) can be ascribed to the (101) and (110) planes of the Te^0 hexagonal lattice, respectively. The Te^0 nanocrystallite average size in the nanocomposite calculated by the Scherer formula is 40.7 nm.

3.3. HR-TEM Analysis

On the basis of HR-TEM data, it was found that PVT–Te⁰NPs are formed as high-contrast particles distributed in the PVT polymer matrix with a shape close to spherical. The nanoparticles have a pronounced tendency to partially agglomerate and aggregate into chains and dimers (Figure 5a). The size of the Te^0 nanoparticles formed in the PVT–Te⁰NPs (2.8 wt%) varies in the rather narrow range of 6–11 nm with a predominance (64%) of 8–9 nm particles and Te^0 nanoparticle average size of 8.5 nm (Figure 5b). The dark-field images reveal that the nanoparticles are clearly visualized as they contrast significantly with the surrounding matrix, confirming their crystalline structure (Figure 5c).

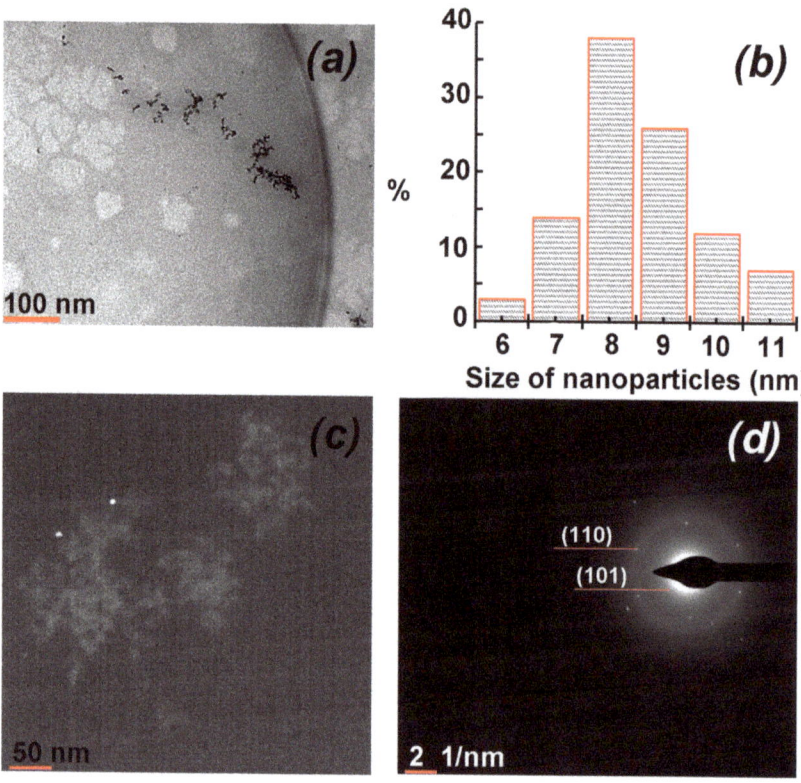

Figure 5. TEM image of PVT–Te⁰ nanocomposite (2.8 wt%) (**a**) in light field mode; (**b**) nanoparticle size distribution histogram of PVT–Te⁰ nanocomposite (2.8 wt%); (**c**) TEM image of PVT–Te⁰ nanocomposite (2.8 wt%) in dark field mode; (**d**) electronographic image of Te nanoparticles in PVT–Te⁰ nanocomposite (2.8 wt%).

The internal microstructure of the nanoparticles was also investigated by HR-TEM. The selected area electron diffraction (SAED) pattern of PVT–Te⁰NPs exhibits clear and discrete over-emission points, indicating a pronounced crystallinity of the nanoparticles (Figure 5d). The symmetrical rings with randomly distributed contrasting dots with no preferred orientation are observed in the nanocomposite (2.8 wt%) SAED pattern, indicating its polycrystalline nature. The two rings clearly visible in the SAED pattern were used as an initial data for the calculation of the interplanar distances of the nanoparticles. The distances were was found to be 3.2 Å and 2.2 Å for the first and second ring, respectively. These values are very close to the interplanar distance values of the elemental tellurium hexagonal lattice and correspond to its (101) and (110) crystallographic planes.

3.4. Dynamic Light Scattering Analysis

The study of aqueous solutions of PVT–Te^{0}NPs by the DLS method shows that the particle size distribution in terms of the scattering intensity is featured by bimodality. The colloids are characterized by two particle fractions (Figure 6).

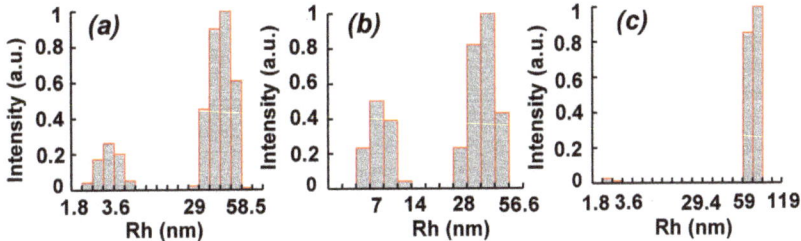

Figure 6. Distribution of the hydrodynamic radii (Rh) over the scattering intensity in PVT–Te^{0}NP nanocomposites with a Te content of (**a**) 1.4 wt%, (**b**) 2.8%, and (**c**) 4.3%.

Thus, the particle fractions with a hydrodynamic radius (Rh) of 3.6 nm and 45 nm were detected in an aqueous solution of PVT–Te^{0}NPs containing 1.4 wt% of Te (Figure 6a). Presumably, the first fast particle fraction is related to the presence of the individual PVT macromolecules in the nanocomposite water solution (Rh equals to 4.1 nm for PVT, whereas one is 3.1 nm for PVT–Te^{0}NPs). The second particle fraction with an average Rh of 45 nm probably originates from Te^0 nanoparticles formed in the PVT matrix or their agglomerates, which we detected by SEM. The increase in the Te weight content in the nanocomposite up to 4.3% is accompanied by a decrease in the fraction of the fast particle in its aqueous solution as well as an increase in the slow particle fraction. At the same time, there is an increase in the average Rh value of the slow particle fraction up to 75 nm, probably due to a growth of the average size of Te^0 nanoparticles themselves (Figure 6c). In addition, a significant narrowing of the particle dispersion of both slow and fast fractions of the PVT–Te^{0}NP solution with the highest Te content, as compared to the samples containing 1.4 and 2.8 wt% Te, should be noted. Thus, the minimum particle Rh range is observed to be 2.3–3.0 nm and 59–75 nm for the fast and slow particle fraction of PVT–Te^{0}NPs (4.3 wt%), respectively.

3.5. UV-Vis Spectroscopy

The study of optical absorption spectra in the range 200–700 nm of 0.025% aqueous solutions of PVT and PVT–Te^{0}NPs showed that the presence of an absorption band in the region of 276 nm (4.5 eV) is typical for all the samples (Figure 7). The 5 eV absorption band in the PVT spectrum isolated by deconvolution seems to be due to the availability of the C=N chromophore group in the polymer chains, which is characterized by a band at 4.7–5.4 eV in the absorption spectrum [33]. The deconvolution also reveals the presence of three bands in the region of 3.1, 4.5, and 5.6 eV in the absorption spectra of the nanocomposites under study. In contrast to the 4.5 eV absorption band, associated with the availability of PVT in nanocomposites and invariably present in the spectra of nanocomposites with different content of inorganic nanophase, the other two absorption bands undergo a blue shift with an increasing Te weight content. Thus, with a change in the inorganic nanophase content from 1.4 to 4.3% Te, a shift of the absorption bands from 5.8 to 4.5 and from 4.5 to 2.9 eV is observed (Figure 7).

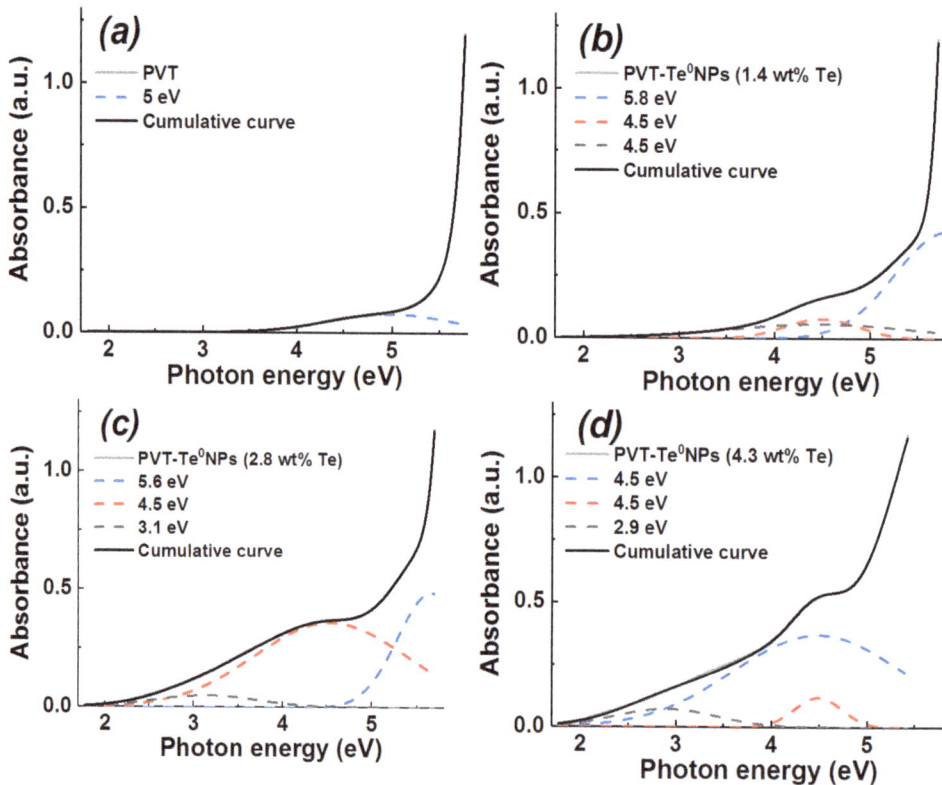

Figure 7. UV-Vis spectra and cumulative curves (solid lines) of 0.025% water solution of (**a**) pure PVT, and PVT–Te⁰NPs with a Te weight content of (**b**) 1.4%, (**c**) 2.8%, and (**d**) 4.3%. Photon energies corresponding to the peaks of the Gaussians are shown in the legends (dashed lines).

According to available data [34,35], the Te⁰ nanoparticles are characterized by the presence of an absorption band in the 270–300 nm region (4.6–4.1 eV). At the same time, the absorption spectrum of Te nanowires is featured by the availability of two bands in the region of 278 nm (4.5 eV) and 586 nm (2.1 eV) [36]. Hence, it can be concluded that the absorption bands in the region of 5.8–4.5 eV and 4.5–2.9 eV are directly related to the presence of Te⁰ nanoparticles in PVT–Te⁰NPs.

Based on the nanocomposite optical absorption spectra obtained, we determined their optical band gap, the average radius of nanoparticles, and the size particle distribution of the nanocomposite Te⁰ nanoparticles. The optical band gap was obtained in two ways: using the wavelength corresponding to 10% absorption in the measured optical absorption spectrum of the solutions studied [37] and by the Tauc method [38,39]. According to the Tauc method, the optical band gap of PVT–Te⁰NPs was determined by extrapolating (to the intersection with the abscissa axis) the linear sections of absorption spectra represented in Tauc coordinates (Figure 8a) using Equation (3):

$$\alpha h\nu = A\,(h\nu - E_g)^\gamma, \tag{3}$$

where α is the absorption coefficient defined by Beer–Lambert's law, $h\nu$ is the incident photon energy, A is a constant characterizing the degree of ordering of the material structure (for calculations we assumed that $A = 1$), E_g is the optical band gap, γ is an index describing transition process, and $\gamma = 1/2$ for direct allowed transitions. The resulting values are shown in Table 1. As can be seen from the table, the optical band gap decreases from 3.17 to 2.16 eV (Figure 8a) or from 2.79 to 2.05 eV with increasing inorganic nanophase content in the nanocomposites studied, depending on the method of Eg determination.

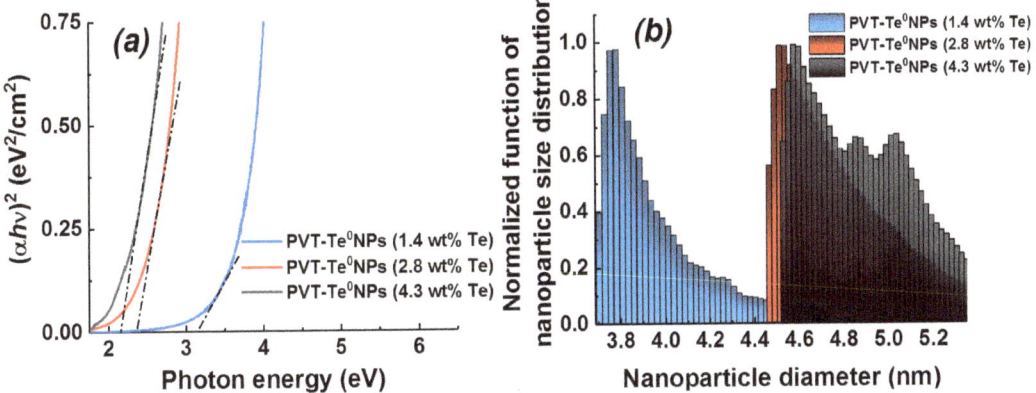

Figure 8. (a) Tauc plots for PVT–Te⁰NPs nanocomposites with different Te weight content, (b) dependence of the normalized nanoparticle size distribution function on nanoparticle diameter for PVT–Te⁰NPs with different Te weight content.

Table 1. Calculated values of the optical band gap, the mean nanoparticle diameter, and the "blue shift" for PVT–Te⁰ nanocomposites. Note: E_g^T: optical band gap calculated by Tauc method (eV); D^{BT}: mean nanoparticle diameter determined by the Brus equation using the value E_g^T (nm); BS^T: "blue" shift of E_g^T value from optical band gap of bulk Te (eV); E_g^{10}: the optical band gap calculated by the Tauc method (eV); D^{B10}: mean nanoparticle diameter determined by the Brus equation using the value E_g^T (nm); BS^{10}: "blue" shift of E_g^T value from the optical band gap of bulk Te (eV).

Nanocomposite	Tauc Method			10% Absorbance Method		
	E_g^T	D^{BT}	BS^T	E_g^{10}	D^{B10}	BS^{10}
PVT-Te⁰ (1.4 wt% Te)	3.17	2.9	2.84	2.79	4.2	2.46
PVT-Te⁰ (2.8 wt% Te)	2.36	3.0	2.03	2.11	5.0	1.78
PVT-Te⁰ (4.3 wt% Te)	2.16	3.2	1.83	2.05	5.1	1.72

As these values are larger than the band gap of bulk Te $E_g^{bulk} = 0.335$ [40] (Figure 8a), one can assume an appearance of the quantum confinement effect resulting in an Eg increase at the transition of material from a bulk to nanoscale state ("blue shift" of Eg). The effect has been well described in the literature [41]. The possibility of setting the band gap by controlling the size of semiconductor nanoparticles has previously been reported in studies of Se colloidal solutions [42], cobalt selenide nanostructure films [43], and ZnS nanoparticles stabilized with a PVA [44]. Taking into account the approximate spherical shape of the nanoparticles, it is possible to estimate the average size of the Te⁰ nanoparticles formed according to the Equation (4) proposed by L. Brus:

$$E_g = E_g^{bulk} + \frac{\hbar^2 \pi^2}{2r^2}\left(\frac{1}{m_e} + \frac{1}{m_h}\right) - \frac{1.8e^2}{4\pi\varepsilon\varepsilon_0 r} \quad (4)$$

where E_g is the nanoscale Te energy band gap, $E_g^{bulk} = 0.335$ эВ is the energy band gap of bulk Te, $\varepsilon = 23$ denotes the dielectric permittivity of the bulk Te, $m_h = m_h^* m_0 = 0.109 m_0$ is the effective mass of the hole in Te, $m_e = m_e^* m_0 = 0.05 m_0$ is the effective mass of electron in Te [40], m_0 represents the electron mass, $\hbar$ is the reduced Planck constant, r denotes the radius of the nanoparticle, ε_0 is the electric constant, and e represents the electron charge. The values of the nanoparticle average size (diameter) calculated without considering the third term in Equation (2) are shown in Table 1. The calculation results show that the average Te^0 nanoparticle size in the nanocomposite increases with a growing inorganic nanophase content. In this case, the value of the optical band gap, and therefore the "blue shift" is reduced. A similar trend was observed by Singh et al. in their study of the optical properties of Se quantum dots [42].

In order to determine the characteristic diameters of Te nanoparticles in the nanocomposites, the corresponding particle size distribution function curves were plotted according to the method proposed by Pesika [45]. This method makes it possible to determine the particle size distribution using the relation between the nanoparticle radius and the shift of the band gap formulated by Brus (Equation (4)):

$$N(r) = -\frac{1}{V}\left[\frac{dD}{dr}\right] = -\frac{1}{\frac{4\pi r^3}{3}}\left[\frac{dD}{d\lambda}\frac{d\lambda}{dr}\right]\bigg|_{\lambda=\frac{hc}{E_g(r)}} \quad (5)$$

where N(r) is the particle size distribution, D denotes the optical density obtained from the optical absorption spectra, r is the nanoparticle radius, V is the volume of a spherical nanoparticle, λ is a wavelength, E_g denotes the optical band gap of the nanoscale semiconductor, h is a Plank constant, and c is a light constant.

Figure 8b shows the calculated curves of the particle size distribution function of PVT–Te^0NPs as histograms. It was found that the PVT–Te^0NPs with inorganic nanophase content of 1.4 and 2.8% are characterized by log-normal size distributions with maximum values corresponding to 3.8 and 4.5 nm, respectively. The particle size distribution for the PVT–Te^0NP (4.3 wt%) has a more complex multimodal structure. The highest value of the distribution function is observed at 4.6 nm, while the peaks of the other two modes correspond to 4.9 and 5.0 nm. The resulting Te^0 nanoparticle diameters are of the same order of magnitude as their dimensional characteristics determined using HR-TEM and the Brus equation (Equation (4)).

3.6. DC Electrical Conductivity

The dependences of the specific DC electrical conductivity of PVT and PVT–Te^0NPs on temperature (in the temperature range of 25–80 °C) and the weight content of the inorganic nanophase were experimentally obtained to determine the presence of a particle dimensional effect on the electrical conductivity of the nanocomposites under study (Figure 9).

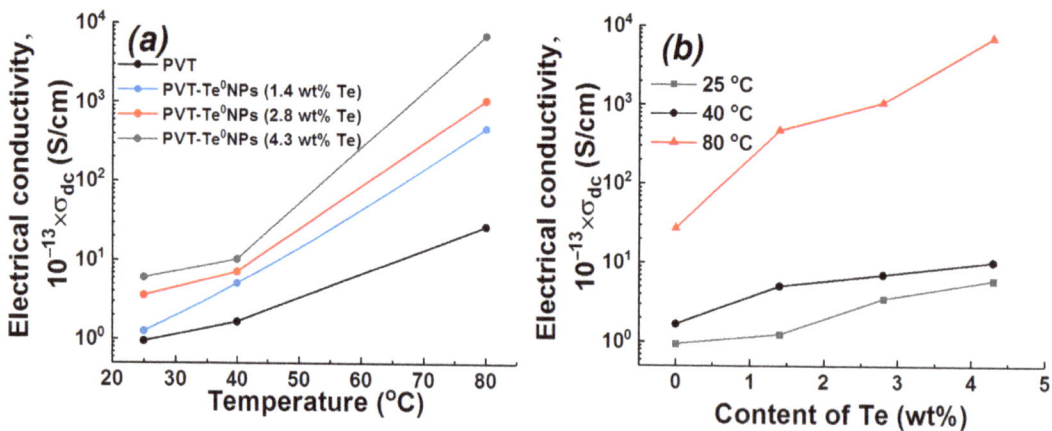

Figure 9. The dependence of electrical conductivity (**a**) on temperature and (**b**) on Te weight content for PVT and PVT–Te^0NPs.

The electrical conductivity values and the nature of its change with increasing temperature allow the initial PVT polymer to be classified as a dielectric (Figure 9a). The introduction of Te nanoparticles into the PVT matrix was found to result in an increase in the DC electrical conductivity of the nanocomposite. The electrical conductivity of the composite increases as the weight content of the inorganic nanophase rises (Figure 9b). At the same time, the electrical conductivity of the nanocomposites grows with increasing temperature (Figure 9a). It should be noted that the electrical conductivity of PVT–Te^{0}NPs with a Te content of 2.8 and 4.3% becomes greater than 10^{-10} S/cm at 80 °C, i.e., it overcomes the conventional dielectric-semiconductor boundary noted in the classification of materials by electrical conductivity value [46]. The behavior of the samples DC conductivity is common to dielectrics and semiconductors. The exponential growth of conductivity with an increasing temperature is characteristic of both semiconductor nanoparticles [47,48] and nanocomposites of dielectric polymers and semiconductor nanoparticles [49,50]. The variation of DC conductivity of PVA–Se nanocomposite with temperature showed the presence of nearly linear part and the part well described by the Vogel–Fulcher–Tammann relationship [49]. Sinha et al. attributed the conductivity behavior to the thermally activated hopping transport ions decoupled from the matrix and 3D variable range hopping of charge carriers. According to the literature data, the authors of paper [50] observed conductivity behavior most similar to the one we obtained. The introduction of TiO_2 nanoparticles into PVA–PEG–PVP matrix resulted in a grow of DC electrical conductivity. Meanwhile, the increase in weight content of inorganic nanophase from 2% to 8% also led to a rise in the nanocomposite DC conductivity. The authors have explained the behavior of PVA–PEG–PVP/TiO_2 nanocomposite conductivity by a reduction in their band gap and therefore thermally activated enhancement of charge carriers hopping motion between trapped sites.

The PVT–Te^{0}NPs nanocomposite is semiconductor nanoparticles non-uniformly distributed in the volume of the dielectric matrix. It can be represented as an organic–inorganic system of components, each exhibiting different physical properties (in particular, electrical conductivity) due to its microscopic characteristics. The macroscopic physical properties of a nanocomposite are due to its microstructural features [51]. Not only the matrix and nanofillers, but also the interphase, which is a system of formed interfaces at the nanoparticle-matrix interfaces, are considered to be components of a nanocomposite. Therefore, an increase in temperature does not uniformly affect the electrical conductivity of the nanocomposite components due to the different nature of the nanocomposite components. In a dielectric, for example, as the temperature rises, some of the electrons gain energy and participate in electrical conductivity. This happens more efficiently in a semiconductor. The transport of charge carriers in the interphase is enabled by an electron tunneling effect. This effect is particularly influential when the distance between the nanoparticles is significantly reduced. However, at high temperatures, the interface conductivity decreases due to an increase in the number of electron collisions. In addition, as the temperature rises, the thickness of the interface layer may decrease, resulting in a reduction in electrical conductivity.

In the case of PVT–Te^{0}NPs, the increase in their electrical conductivity, in comparison with the matrix, can be explained by the appearance of the charge-carrier transport across the crystalline semiconductor and interface conduction due to the electron tunneling effect. As the weight content of the inorganic nanophase increases, the nanoparticles become larger, i.e., the regions of successful charge carrier transport through the semiconductor increase with a simultaneous decrease in the interphase effect. This leads to an increase in the degree of ordering in the composite system, which translates into the increased electrical conductivity. In addition, the observed increase in DC electrical conductivity may be related to the possible formation and increase in the number of local conductive channels created when the average interparticle distance decreases [52].

4. Conclusions

Thus, we have first obtained the nanocomposites of an original poly(1-vinyl-1,2,4-triazole) polymer matrix and Te^0 thermoelectric nanoparticles with varying amounts of inorganic phase in the range of 1.4–4.3%. The synthesized nanocomposites have been characterized using a set of complementary modern methods (HR-TEM, XRD, DLS, UV-Vis spectroscopy). Nanocomposites have been found to form as nanoparticles dispersed in a polymeric matrix and sized between 4 and 18 nm. Using the optical spectroscopy data as well as computational values calculated by the Brus formula and the Pesika method, the mean diameters and characteristic nanoparticle diameters of the nanoparticle fractions present in the composite, in the range of 4.2–5.1 and 3.8–5.0 nm, were determined for the nanocomposites containing 1.4%, 2.8%, and 4.3%. Based on the temperature dependence of the DC electrical conductivity in the range 25–80 °C, it was found that the introduction of the inorganic nanophase in the PVT dielectric polymer as well as an increase in the average size of Te^0 nanoparticles leads to an increase in electrical conductivity over the entire temperature range. Herewith, the DC electrical conductivity values of PVT–Te^0NPs with a Te weight content of 2.8% and 4.3% become higher than 10^{-10} S/cm at 80 °C, i.e., they exceed the conventional boundary of dielectric–semiconductor in the classification of materials by electrical conductivity value. The results obtained suggest that varying the Te weight content in the synthesis process of PVT-based nanocomposites enables one to obtain nanocomposites with a required value of electrical conductivity. Taking into account the rather high Seebeck coefficient of Te^0 nanoparticles and probably the low thermal conductivity provided by dielectric polymer PVT matrix, this could result in the directional design of thermoelectric nanocomposites with a high thermoelectric figure of merit.

Author Contributions: Conceptualization, A.V.Z. and M.V.Z.; methodology, A.V.Z., M.V.Z. and G.F.P.; software, A.V.Z. and M.V.Z.; validation, A.V.Z., M.V.Z. and G.F.P.; formal analysis, A.V.Z., M.V.Z., G.F.P. and A.S.P.; investigation, A.V.Z., M.V.Z., G.F.P. and S.A.K.; resources, A.V.Z., M.V.Z. and S.A.K.; data curation, M.V.Z.; writing—original draft preparation, M.V.Z., A.V.Z. and G.F.P.; writing—review and editing, M.V.Z., A.V.Z., A.S.P. and G.F.P.; visualization, M.V.Z. and A.V.Z.; supervision, M.V.Z.; project administration, M.V.Z., A.V.Z. and A.S.P. All authors have read and agreed to the published version of the manuscript.

Funding: This research was funded by the Ministry of Science and Higher Education of the Russian Federation.

Institutional Review Board Statement: Not applicable.

Informed Consent Statement: Not applicable.

Data Availability Statement: The data presented in this study are available from the corresponding author upon request.

Acknowledgments: The authors wish to thank the Isotopegeochemical research center for Collective Use (A.P. Vinogradov Institute of Geochemistry of the Siberian Branch of the Russian Academy of Sciences) and the Baikal Analytical Centre (Irkutsk Institute of Chemistry of the Siberian Branch of the Russian Academy of Sciences) for providing the equipment. The high-resolution transmission electron microscopy was performed using equipment from the Baikal Nanotechnology Center, Irkutsk National Research Technical University.

Conflicts of Interest: The authors declare no conflict of interest.

References

1. Kwan, T.H.; Wu, X.; Yao, Q. Complete implementation of the combined TEG-TEC temperature control and energy harvesting system. *Control Eng. Pract.* **2020**, *95*, 104224. [CrossRef]
2. Pan, S.; Wang, T.; Jin, K.; Cai, X. Understanding and designing metal matrix nanocomposites with high electrical conductivity: A review. *J. Mater. Sci.* **2022**, *57*, 6487–6523. [CrossRef]
3. Li, J.; Huckleby, A.B.; Zhang, M. Polymer-based thermoelectric materials: A review of power factor improving strategies. *J. Mater.* **2022**, *8*, 204–220. [CrossRef]
4. Zhou, C.; Dun, C.; Ge, B.; Wang, K.; Shi, Z.; Liu, G.; Carroll, D.L.; Qiao, G. Highly robust and flexible n-type thermoelectric film based on Ag_2Te nanoshuttle/polyvinylidene fluoride hybrids. *Nanoscale* **2018**, *10*, 14830–14834. [CrossRef] [PubMed]
5. Zhao, X.; Han, W.; Jiang, Y.; Zhao, C.; Ji, X.; Kong, F.; Xu, W.; Zhang, X. A honeycomb-like paper-based thermoelectric generator based on a Bi_2Te_3/bacterial cellulose nanofiber coating. *Nanoscale* **2019**, *11*, 17725–17735. [CrossRef] [PubMed]

6. Kim, S.; Ryu, S.H.; Kwon, Y.-T.; Lim, H.-R.; Park, K.-R.; Song, Y.; Choa, Y.-H. Synthesis and thermoelectric characterization of high density Ag$_2$Te nanowire / PMMA nanocomposites. *Mater. Chem. Phys.* **2017**, *190*, 187–193. [CrossRef]
7. Andzane, J.; Buks, K.; Bitenieks, J.; Bugovecka, L.; Kons, A.; Merijs-Meri, R.; Svirksts, J.; Zicans, J.; Erts, D. p-Type PVA/MWCNT-Sb$_2$Te$_3$ Composites for Application in Different Types of Flexible Thermoelectric Generators in Combination with n-Type PVA/MWCNT-Bi$_2$Se$_3$ Composites. *Polymers* **2022**, *14*, 5130. [CrossRef]
8. Morad, M.; Fadlallah, M.M.; Hassan, M.A.; Sheha, E. Evaluation of the effect of V$_2$O$_5$ on the electrical and thermoelectric properties of poly (vinyl alcohol)/graphene nanoplatelets nanocomposite. *Mater. Res. Express* **2016**, *3*, 035015. [CrossRef]
9. Wang, J.; Du, Y.; Qin, J.; Wang, L.; Meng, Q.; Li, Z.; Shen, S.Z. Flexible Thermoelectric Reduced Graphene Oxide/Ag$_2$S/Methyl Cellulose Composite Film Prepared by Screen Printing Process. *Polymers* **2022**, *14*, 5437. [CrossRef]
10. Dun, C.; Hewitt, C.A.; Huang, H.; Montgomery, D.S.; Xu, J.; Carroll, D.L. Flexible thermoelectric fabrics based on self-assembled tellurium nanorods with a large power factor. *Phys. Chem. Chem. Phys.* **2015**, *17*, 8591–8595. [CrossRef]
11. Pammi, S.V.N.; Jella, V.; Choi, J.S.; Yoon, S.G. Enhanced thermoelectric properties of flexible Cu$_{2-x}$Se (x ≥ 0.25) NW/polyvinylidene fluoride composite films fabricated via simple mechanical pressing. *J. Mater. Chem. C* **2017**, *5*, 763–769. [CrossRef]
12. Oopathump, C.; Boonthuma, D.; Smith, S.M. Effect of Poly (Vinyl Alcohol) on thermoelectric properties of sodium cobalt oxide. In Key Engineering Materials. *Key Eng. Mater.* **2019**, *798*, 304–309. [CrossRef]
13. Jiang, C.; Wei, P.; Ding, Y.; Cai, K.; Tong, L.; Gao, Q.; Lu, Y.; Zhao, W.; Chen, S. Ultrahigh performance polyvinylpyrrolidone/Ag$_2$Se composite thermoelectric film for flexible energy harvesting. *Nano Energy* **2021**, *80*, 105488. [CrossRef]
14. Kim, S.; Lee, Y.-I.; Ryu, S.H.; Hwang, T.-Y.; Song, Y.; Seo, S.; Yoo, B.; Lim, J.-H.; Cho, H.-B.; Myung, N.V.; et al. Synthesis and thermoelectric characterization of bulk-type tellurium nanowire/polymer nanocomposites. *J. Mater. Sci.* **2017**, *52*, 12724–12733. [CrossRef]
15. Li, X.; Du, Y.; Meng, Q. Flexible ball-milled Bi$_{0.4}$Sb$_{1.6}$Te$_3$/methyl cellulose thermoelectric films fabricated by screen-printing method. *Funct. Mater. Lett.* **2021**, *14*, 2151034. [CrossRef]
16. Dun, C.; Hewitt, C.A.; Huang, H.; Hu, J.; Montgomery, D.S.; Nie, W.; Jiang, Q.; Carroll, D.L. Layered Bi$_2$Se$_3$ nanoplate/polyvinylidene fluoride composite based n-type thermoelectric fabrics. *ACS Appl. Mater. Inter.* **2015**, *7*, 7054–7059. [CrossRef]
17. Bitenieks, J.; Buks, K.; Merijs-Meri, R.; Andzane, J.; Ivanova, T.; Bugovecka, L.; Voikiva, V.; Zicans, J.; Erts, D. Flexible N-Type Thermoelectric Composites Based on Non-Conductive Polymer with Innovative Bi$_2$Se$_3$-CNT Hybrid Nanostructured Filler. *Polymers* **2021**, *13*, 4264. [CrossRef]
18. Qin, J.; Du, Y.; Meng, Q.; Ke, Q. Flexible thermoelectric Cu–Se nanowire/methyl cellulose composite films prepared via screen printing technology. *Compos. Commun.* **2023**, *38*, 101467. [CrossRef]
19. Lesnichaya, M.V.; Zhmurova, A.V.; Sapozhnikov, A.N. Synthesis and Characterization of Water-Soluble Arabinogalactan-Stabilized Bismuth Telluride Nanoparticles. *Russ. J. Gen. Chem.* **2021**, *91*, 1379–1386. [CrossRef]
20. Zvereva, M.V.; Zhmurova, A.V. Synthesis, structure, and spectral properties of ZnTe-containing nanocomposites based on arabinogalactan. *Russ. J. Gen. Chem.* **2022**, *92*, 1995–2004. [CrossRef]
21. Pozdnyakov, A.; Emel'yanov, A.; Kuznetsova, N.; Ermakova, T.; Bolgova, Y.; Trofimova, O.; Albanov, A.; Borodina, T.; Smirnov, V.; Prozorova, G. A Polymer Nanocomposite with CuNP Stabilized by 1-Vinyl-1,2,4-Triazole and Acrylonitrile Copolymer. *Synlett* **2015**, *27*, 900–904. [CrossRef]
22. Zezin, A.; Danelyan, G.; Emel'yanov, A.; Zharikov, A.; Prozorova, G.; Zezina, E.; Korzhova, S.; Fadeeva, T.; Abramchuk, S.; Shmakova, N.; et al. Synthesis of Antibacterial Polymer Metal Hybrids in Irradiated Poly-1-vinyl-1,2,4-triazole Complexes with Silver Ions: PH Tuning of Nanoparticle Sizes. *Appl. Organomet. Chem.* **2022**, *36*, e6581. [CrossRef]
23. Pozdnyakov, A.; Emel'yanov, A.; Ivanova, A.; Kuznetsova, N.; Semenova, T.; Bolgova, Y.; Korzhova, S.; Trofimova, O.; Fadeeva, T.; Prozorova, G. Strong Antimicrobial Activity of Highly Stable Nanocomposite Containing AgNPs Based on Water-Soluble Triazole-Sulfonate Copolymer. *Pharmaceutics* **2022**, *14*, 206. [CrossRef] [PubMed]
24. Prozorova, G.F.; Pozdnyakov, A.S. Synthesis, Properties, and Biological Activity of Poly(1-vinyl-1,2,4-triazole) and Silver Nanocomposites Based on It. *Polym. Sci. Ser. C* **2022**, *64*, 62–72. [CrossRef]
25. Li, C.; Jiang, F.; Liu, C.; Wang, W.; Li, X.; Wang, T.; Xu, J. A simple thermoelectric device based on inorganic/organic composite thin film for energy harvesting. *Chem. Eng. J.* **2017**, *320*, 201–210. [CrossRef]
26. Lin, S.; Li, W.; Chen, Z.; Shen, J.; Ge, B.; Pei, Y. Tellurium as a high-performance elemental thermoelectric. *Nat. Commun.* **2016**, *7*, 10287. [CrossRef]
27. Blackburn, J.L.; Ferguson, A.J.; Cho, C.; Grunlan, J.C. Thermoelectric materials: Carbon-nanotube-based thermoelectric materials and devices. *Adv. Mater.* **2018**, *30*, 1870072. [CrossRef]
28. Fan, Z.; Zhang, Y.; Pan, L.; Ouyang, J.; Zhang, Q. Recent developments in flexible thermoelectrics: From materials to devices. *Renew. Sustain. Energy Rev.* **2021**, *137*, 110448. [CrossRef]
29. Lesnichaya, M.V.; Tsivileva, O.M. Arabinogalactan-Stabilized Selenium Sulfide Nanoparticles and Their Fungistatic Activity Against Phytophthora cactorum. *J. Clust. Sci.* **2022**, 1–12. [CrossRef]
30. Hu, H.; Zeng, Y.; Gao, S.; Wang, R.; Zhao, J.; You, K.; Song, Y.; Xiao, Q.; Cao, R.; Li, J.; et al. Fast solution method to prepare hexagonal tellurium nanosheets for optoelectronic and ultrafast photonic applications. *J. Mater. Chem. C* **2021**, *9*, 508–516. [CrossRef]
31. Liu, C.; Wang, R.; Zhang, Y. Tellurium Nanotubes and Chemical Analogues from Preparation to Applications: A Minor Review. *Nanomaterials* **2022**, *12*, 2151. [CrossRef]

32. Lu, Q.; Gao, F.; Komarneni, S. Biomolecule-Assisted Reduction in the Synthesis of Single-Crystalline Tellurium Nanowires. *Adv. Mater.* **2004**, *16*, 1629–1632. [CrossRef]
33. Ioffe, B.V.; Kostikov, R.R.; Razin, V.V. *Physical Methods for Determining the Structure of Organic Molecules*; Leningrad University Press: Leningrad, USSR, 1976.
34. Rosales-Conrado, N.; Gómez-Gómez, B.; Matías-Soler, J.; Pérez-Corona, M.T.; Madrid-Albarrán, Y. Comparative study of tea varieties for green synthesis of tellurium-based nanoparticles. *Microchem. J.* **2021**, *169*, 106511. [CrossRef]
35. Zambonino, M.C.; Quizhpe, E.M.; Jaramillo, F.E.; Rahman, A.; Vispo, N.S.; Jeffryes, C.; Dahoumane, S.A. Green synthesis of selenium and tellurium nanoparticles: Current trends, biological properties and biomedical applications. *Int. J. Mol. Sci.* **2021**, *22*, 989. [CrossRef] [PubMed]
36. Qian, H.-S.; Yu, S.-H.; Gong, J.-Y.; Luo, L.-B.; Fei, L.-F. High-quality luminescent tellurium nanowires of several nanometers in diameter and high aspect ratio synthesized by a poly (vinyl pyrrolidone)-assisted hydrothermal process. *Langmuir* **2006**, *22*, 3830–3835. [CrossRef]
37. Wallace, A.M.; Curiac, C.; Delcamp, J.H.; Fortenberry, R.C. Accurate determination of the onset wavelength (λ_{onset}) in optical spectroscopy. *J. Quant. Spectrosc. Radiat. Transf.* **2021**, *265*, 107544. [CrossRef]
38. Tauc, J.; Grigorovici, R.; Vancu, A. Optical properties and electronic structure of amorphous germanium. *Phys. Stat. Sol.* **1966**, *15*, 627–637. [CrossRef]
39. Davis, E.A.; Mott, N.F. Conduction in non-crystalline systems V. Conductivity, optical absorption and photoconductivity in amorphous semiconductors. *Philos. Mag.* **1970**, *22*, 0903–0922. [CrossRef]
40. Madelung, O. (Ed.) Elements. In *Semiconductors: Other than Group IV Elements and III–V Compounds*; Springer: Berlin, Germany, 1992; pp. 4–7. [CrossRef]
41. De Mello Donega, C. *Nanoparticles: Workhorses of Nanoscience*; Springer: Berlin, Germany, 2014.
42. Singh, S.C.; Mishra, S.K.; Srivastava, R.K.; Gopal, R. Optical properties of Selenium quantum dots produced with laser irradiation of water suspended Sc nanoparticles. *J. Phys. Chem. C* **2010**, *114*, 17374–17384. [CrossRef]
43. Ghobadi, N.; Khazaie, F. Fundamental role of the pH on the nanoparticle size and optical band gap in cobalt selenide nanostructure films. *Opt. Quant. Electron.* **2016**, *48*, 165. [CrossRef]
44. Borah, J.P.; Barman, J.; Sarma, K.C. Structural and optical properties of ZnS nanoparticles. *Chalcogenide Lett.* **2008**, *5*, 201–208.
45. Pesika, N.S.; Stebe, K.J.; Searson, P.C. Relationship between absorbance spectra and particle size distributions for quantum-sized nanocrystals. *J. Phys. Chem. B* **2003**, *107*, 10412–10415. [CrossRef]
46. Prokhorov, A.M. (Ed.) *Physical Encyclopedic Dictionary*; Soviet Encyclopedia: Moscow, USSR, 1983.
47. Ansari, M.M.N.; Khan, S. Structural, electrical and optical properties of sol-gel synthesized cobalt substituted MnFe2O4 nanoparticles. *Physica B* **2017**, *520*, 21–27. [CrossRef]
48. Mehraj, S.; Ansari, M.S.; Al-Ghamdi, A.A. Annealing dependent oxygen vacancies in SnO2 nanoparticles: Structural, electrical and their ferromagnetic behavior. *Mater. Chem. Phys.* **2016**, *171*, 109–118. [CrossRef]
49. Sinha, S.; Chatterjee, S.K.; Ghosh, J.; Meikap, A.K. Electrical transport properties of polyvinyl alcohol–selenium nanocomposite films at and above room temperature. *J. Mater. Sci.* **2015**, *50*, 1632–1645. [CrossRef]
50. Agool, I.R.; Kadhim, K.J.; Hashim, A. Preparation of (polyvinyl alcohol–polyethylene glycol–polyvinyl pyrrolidinone–titanium oxide nanoparticles) nanocomposites: Electrical properties for energy storage and release. *Int. J. Plast. Technol.* **2016**, *20*, 121–127. [CrossRef]
51. Xia, X.; Weng, G.J.; Zhang, J.; Li, Y. The effect of temperature and graphene concentration on the electrical conductivity and dielectric permittivity of graphene–polymer nanocomposites. *Acta Mech.* **2020**, *231*, 1305–1320. [CrossRef]
52. Du, B.X.; Han, C.; Li, J.; Li, Z. Temperature-dependent DC conductivity and space charge distribution of XLPE/GO nanocomposites for HVDC cable insulation. *IEEE Trans. Dielectr. Electr. Insul.* **2020**, *27*, 418–426. [CrossRef]

Disclaimer/Publisher's Note: The statements, opinions and data contained in all publications are solely those of the individual author(s) and contributor(s) and not of MDPI and/or the editor(s). MDPI and/or the editor(s) disclaim responsibility for any injury to people or property resulting from any ideas, methods, instructions or products referred to in the content.

Article

Puncture Resistance and UV aging of Nanoparticle-Loaded Waterborne Polyurethane-Coated Polyester Textiles

Domenico Acierno [1,*], Lucia Graziosi [1] and Antonella Patti [2]

1. Regional Center of Competence New Technologies for Productive Activities Scarl, Via Nuova Agnano 11, 80125 Naples, Italy; lucia.graziosi93@gmail.com
2. Department of Civil Engineering and Architecture (DICAr), University of Catania, Viale Andrea Doria 6, 95125 Catania, Italy; antonella.patti@unict.it
* Correspondence: acierno@crdctecnologie.it

Abstract: The goal of this research was to investigate the effect of different types of nanoparticles on the UV weathering resistance of polyurethane (PU) treatment in polyester-based fabrics. In this regard, zinc oxide nanoparticles (ZnO), hydrophilic silica nanoparticles (SiO_2 (200)), hydrophobic silica nanoparticles (SiO_2 (R812)), and carbon nanotubes (CNT) were mixed into a waterborne polyurethane dispersion and impregnated into textile samples. The puncturing resistance of the developed specimens was examined before and after UV-accelerated aging. The changes in chemical structure and surface appearance in nanoparticle-containing systems and after UV treatments were documented using microscopic pictures and infrared spectroscopy (in attenuated total reflectance mode). Polyurethane impregnation significantly enhanced the puncturing strength of the neat fabric and reduced the textile's ability to be deformed. However, after UV aging, mechanical performance was reduced both in the neat and PU-impregnated specimens. After UV treatment, the average puncture strength of all nanoparticle-containing systems was always greater than that of aged fabrics impregnated with PU alone. In all cases, infrared spectroscopy revealed some slight differences in the absorbance intensity of characteristic peaks for polyurethane polymer in specimens before and after UV rays, which could be related to probable degradation effects.

Keywords: nanoparticles; waterborne polyurethane; UV aging; mechanical resistance; polyester fabrics

1. Introduction

In 2019, synthetic fibers accounted for more than 60% of global consumption of natural and man-made fibers, with polyester accounting for 53%, cotton accounting for 24%, cellulosic accounting for 6%, and wool accounting for 1%.

Polyester fibers have a specific gravity of 1.22–1.38 g/cm^3 [1], are wrinkle resistant [2], and have a moisture regain of roughly 0.4%. Mechanical properties are frequently dependent on fiber drawing; as the polymer chain alignment increases, so does the tensile strength and modulus of polyester fibers [3]. PET fibers are widely used in woven and knitted garments and upholstery, including pants, skirts, stockings, carpets, and blankets, as well as industrial threads and ropes, protective textiles (work wear, conveyor belt, and jet engines (abradable seal) [3].

Coating or impregnation are two common methods, generally considered in the textile industry, for improving the final performances of treated fabrics and providing a proper finishing [4]. In recent years, many researchers investigated the puncture resistance behavior of textile materials, and it was discovered that the perforating properties of textiles were affected by coating technologies [5]. The body protection for law enforcement, security, and military personnel has been the primary focus of investigations on the stab and puncture resistance of fabrics [6]. Firouzi et al. investigated nylon-based coating to improve the penetration resistance of ultra-high molecular weight polyethylene (UHMWPE) against spikes or blades [7]. The puncture performance of woven high modulus polypropylene (HMPP) fabric impregnated with dispersions constituted of fumed silica nanoparticles, carbon nanotubes (CNTs), and polyethylene glycol (PEG) was investigated in [8].

Polyurethane is widely employed in advanced coating technology because it improves the quality, appearance, and durability of coated substrates [9]. A polyurethane and pre-vulcanized natural rubber latex coating was tested on polyester textiles to obtain high performance in water vapor permeability, water pressure resistance, and stretchability [10]. Polyurethane formulations have traditionally been manufactured with hazardous organic solvents such as toluene, xylene, and formaldehyde. Its manufacturing process is responsible for volatile organic compounds emissions, relative ozone formation potential, and relative carcinogenic risk [11,12]. However, green materials have become popular in the coating industry as a result of growing environmental concerns and the depletion of petroleum supplies [13]. As an alternative to solvent-based polyurethane dispersions, water-based polyurethane dispersions (WPU) with minimal volatile organic compounds (VOCs) and no co-solvents were developed [14]. The addition of nanoparticles to WPU has been demonstrated useful to improve the overall performance of WPU-based products and attain the same requirements as their solvent-based equivalents [11,15].

Recently, nanomaterials have been used to modify the characteristics of organic and inorganic matrices and improve their durability against UV weathering. Polyetheretherketone (PEEK) was loaded with sub-micrometric titanium dioxide particles to increase the UV resistance of the basic polymer [16]. Gorrasi and Sorrentino examined the influence of carbon nanotubes on the photo-oxidation of composites made of polylactic acid [17]. Cheraghian and Wistuba investigated the effect of fumed silica nanoparticles on the mechanical and chemical properties of bitumen after UV aging [18]. The addition of nano-zinc oxide particles was discovered to minimize the photo-degradation of the polyurethane film and shield it from the detrimental impacts of UV light [19].

In this investigation, various types of nanoparticles (zinc oxide nanoparticles, hydrophilic and hydrophobic silica nanoparticles, and carbon nanotubes) were combined into aqueous polyurethane dispersions and impregnated into polyester-based fabrics. The mechanical performance of produced specimens was evaluated mainly in terms of puncture strength. The measurement of fabric puncture strength has been regarded as a practical and useful approach to analyzing the quality of textiles used in daily life, particularly in the industry production of bags, luggage, and suitcases. Puncture resistance might represent a frequent stress condition to which materials may be subjected during common uses, potentially damaging and compromising integrity and functionality. To verify treatment durability, mechanical properties were also evaluated following an accelerated aging process. Environmental variables such as UV radiation, heat from the sun, and moisture can cause polymer-based materials to degrade by affecting mechanical properties and limiting product life span. Understanding the effects of UV radiation on the mechanical performance of textiles can be useful in predicting the durability of final products. Infrared spectroscopy and images of sample surfaces were collected on specimens before and after UV exposure as primary characterization tools to analyze the chemical and visible changes caused by potential breakdown of the applied PU polymer.

2. Materials and Methods

2.1. Materials

In this study, a technical fabric (FAB) with a woven structure constructed entirely of polyester fibers (100%) and with an area density of roughly 300 g/m^2 was explored. The aliphatic polyester-based polyurethane waterborne dispersion (IDROCAP 983 PF, anionic nature, dry residue at 130 °C 1 h = 35%, pH = 7.8) was kindly supplied by ICAP-Sira Chemicals and Polymers s.p.a., Parabiago-Milan, Italy. Hydrophilic fumed nano-silica (here identified as SiO_2 (200)) (cod. AEROSIL 200, an average particle size of 12 nm, a specific surface area of 175–225 m^2/g) and hydrophobic organosilane modified silica produced by treating fumed SiO_2 with hexamethyldisilazane (HMDS) (here identified as

SiO$_2$ (R812)) (cod. AEROSIL R812, an average particle size of 7 nm, a specific surface area of 260 ± 30 m^2/g) were acquired by Evonik Resource Efficiency GmbH (Germany). Carbon nanotubes (CNT) (cod. NC 3150, length < 1 µm, diameter: 9.5 nm and purity > 95%) were supplied by Nanocyl S.A. (Sambreville—Belgium). Zinc oxide (ZnO, average particle size of 100 nm) was purchased from Sigma Aldrich Co. LLC., Milan, Italy.

2.2. Sample Preparation

A square piece of fabric (20 cm in size) was soaked in 13 g of aqueous polyurethane (Fab-PU). The wet specimens were dried in a climatic chamber (mod. 250E, Angelantoni Industrie spa, Perugia, Italy) at 25 °C and 50% humidity. Other impregnating dispersions were produced by combining the components, namely nanoparticles (hydrophilic and hydrophobic SiO$_2$, ZnO, CNT) and aqueous polyurethane (13 g), for 15 min with magnetic stirring at 800 rpm. The nanoparticle weight contents of 1% and 4% were related to the nominal solid polyurethane component in the dispersion (see details in Table 1). A dry time of about 4 days was determined through preliminary testing when the sample weight stabilized with time.

Table 1. Developed formulations to impregnate a single piece of fabric (20 × 20 cm^2).

	WPU	ZnO	SiO$_2$ (200)	SiO$_2$ (R812)	CNT
FAB-PU	13 g	/	/	/	/
FAB-PU + 1% ZnO	13 g	1% in wt. (* = 0.04 g)	/	/	/
FAB-PU + 4% ZnO	13 g	4% in wt. (* = 0.18 g)	/	/	/
FAB-PU + 1% SiO$_2$ (200)	13 g	/	1% in wt. (* = 0.04 g)	/	/
FAB-PU + 4% SiO$_2$ (200)	13 g	/	4% in wt. (* = 0.18 g)	/	/
FAB-PU + 1% SiO$_2$ (R812)	13 g	/	/	1% in wt. (* = 0.04 g)	/
FAB-PU + 4% SiO$_2$ (R812)	13 g	/	/	4% in wt. (* = 0.18 g)	/
FAB-PU + 1% CNT	13 g	/	/	/	1% in wt. (* = 0.04 g)
FAB-PU + 4% CNT	13 g	/	/	/	4% in wt. (* = 0.18 g)

(*) with respect to PU solid content in waterborne polyurethane dispersion (4.55 g of solid PU).

UV-accelerated aging was carried out for 5 days at 35 °C in an oven equipped with a low-pressure mercury lamp (90% irradiation at 254 nm, 10% irradiation at 185 nm).

2.3. Characterization Techniques

The puncture resistance of the treated fabrics was evaluated on a dynamometer (cod. TENSOMETER 2020) produced by Alpha Technologies INSTRON (Norwood, MA, USA). The specimens were held between two annular flanges with internal diameters of 14 cm, and firmly secured with four screws. The upper mobile grip was equipped with a rounded tip spike (3 mm in diameter) and traveled down perpendicular to the sample at a speed of 50 mm/min. Tensile 2020 was the management software. The maximum force sustained by the specimen before breaking corresponds to the highest point achieved on the load–displacement curve. This value was considered the puncture resistance. Measurements were repeated 9 times for each sample. Two different specimens were prepared and tested for each formulation. The maximum change in material length before breaking caused by the piercing probe's perpendicular pulling force was identified as displacement. Figure 1 depicts the locations of the punching points on the sample surface during testing.

An infrared spectrometer (model Spectrum 65 FT IR) from Perkin Elmer (Waltham, MA, USA) was used to examine aged specimens in attenuated total reflectance (ATR) mode. During the test, infrared radiation was transmitted via a diamond crystal upon which the sample was placed. Initially, the background spectrum was collected using an empty and clean crystal. A wavenumber range of 650–4000 cm^{-1}, a resolution of 4 cm^{-1}, and a scan number of 16 were set during the experiment.

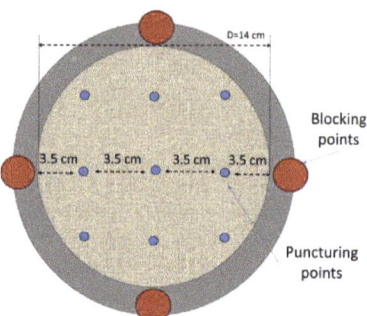

Figure 1. Schematic of puncturing positions on the sample surface.

The effect of impregnating dispersions on the appearance of the textile surface was examined through the auxilium of a Wi-Fi digital microscope 1000× with manual focal length (resolution 1280 × 1024).

3. Results

3.1. Visual Aspects of Sample Surface

The effect of the treatment on the final aesthetics of the fabric was evaluated in relation to the color of the base surface. In this instance, a beige/white fabric was used. The surface appearance of the textile surface of each prepared specimen was reported in Figure 2.

The textile impregnation with aqueous polyurethane (Figure 2b) did not appear to significantly alter the surface properties of basic fabric (Figure 2a). A very thin layer seemed to coat the fibers on the surface, making them slightly brighter.

When the hydrophilic silica and zinc oxide nanoparticles were added into dispersion (Figure 2c,f, respectively), only at the microscopic level an aesthetic change in the visible characteristics of the material can be discerned, becoming less brilliant and whiter. The presence of nanoparticles in the soaking solution had a significant impact when the hydrophobic nanosilica or carbon nanotubes were embodied in the aqueous medium, as seen in Figure 2d,e, respectively. In the first situation, the fabric surface was colored with a brighter white, whereas in the second case, the fabric surface darkened.

Touching and stroking the treated fabric sample resulted in no nanoparticle detachment or dispersion in the surroundings. Polyurethane was found to be necessary in order to adhere the nanoparticles to the textile structure and prevent their leakage into the environment during use.

Although the effectiveness of keeping the physical and mechanical properties of the polymer is critical, there is frequently another issue that can cause concerns regarding the product quality: "Yellowing or Pinking". This color-change issue could severely limit the final product approval. Such discoloration was mainly attributed to a side reaction of the polymer itself during the manufacturing process or to by-products of the stabilization processes of antioxidants and stabilizers [20]. By comparing the color of aged samples to the initial ones at the end of the UV irradiation time, a yellowing phenomenon of the textile surface in specimens incorporating all the different types of nanoparticles (ZnO and hydrophilic and hydrophobic silica) could be observed (Figure 3).

Thus, the presence of particles appeared ineffective in preventing the yellowing of treated textiles. Due to the darker surface in CNT-based specimens, this effect was not evident.

Figure 2. Images of sample surfaces for (**a**) FAB; (**b**) FAB-PU; (**c**) FAB-PU + 4% SiO$_2$ (200); (**d**) FAB-PU + 4% SiO$_2$ (R812); (**e**) FAB-PU + 4% CNT; (**f**) FAB-PU + 4% ZnO.

Typically, several research findings highlighted the benefits of incorporating nanoparticles into a polymer matrix to minimize UV-induced photo-oxidation, which was thought to be the primary cause of polymer yellowing. For example, micrometric zinc oxide particles were found to improve the stability against UV and hydrophobicity without affecting the final mechanical performance of polyester-coated specimens [21]. The optimal ZnO content for impregnated fabrics was set at 3–5 wt.%. The photoprotective effect was attested through UV reflectance measurements via a spectrometer [21]. Zinc oxide nanoparticles were synthesized and applied on cotton and wool fabrics for UV shielding in [22]. Assessments of the treatment's efficiency were based on UV-Vis spectrophotometry and the computation of the ultraviolet protection factor (UPF). Grigoriadou et al. [23] studied the UV stability of high-density polyethylene (HDPE) with several types of nanoparticles (montmorillonite, silica, and carbon nanotubes). Nanocomposites were aged using UV irradiation at 280 nm at a constant temperature (25 °C) for up to 200 h. The results showed a first increase in tensile characteristics after 100 h of irradiation, followed by a decrease, with an effect more evident in the presence of nanoparticles. FTIR measurements revealed a significant degradation of HDPE polymer, especially with the addition of silica and montmorillonite. The authors concluded that such particles enhanced HDPE photo-oxidation.

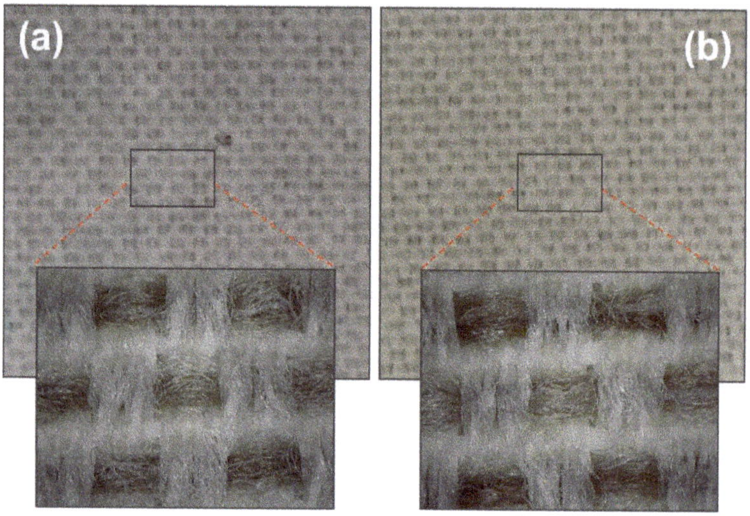

Figure 3. Impregnated textiles with WPUD containing 4% in wt. of hydrophilic nano-SiO$_2$ before (**a**) and after (**b**) UV treatment.

3.2. Puncture Performance

Table 2 summarizes the puncture load and displacement for impregnated fabrics containing zinc oxide, hydrophilic silica, hydrophobic silica, and carbon nanotubes before and after UV aging. For comparison, data from basic fabric and impregnated fabric have been reported. Results were displayed in terms of mean (MN), median (MD), maximum (MAX) and minimum (MIN) values, and standard deviations (STD). Figure 4 summarizes the average puncture load for each developed system before and after UV.

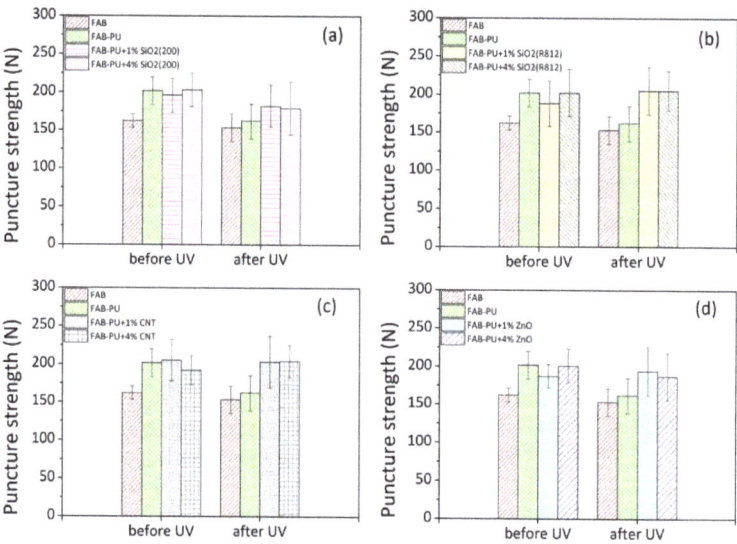

Figure 4. Average puncture strength of PU-impregnated specimens containing (**a**) hydrophilic SiO$_2$; (**b**) hydrophobic SiO$_2$; (**c**) CNT; (**d**) ZnO before and after UV aging.

Table 2. Statistical analysis of puncture data (load and displacement) for PU-impregnated specimens: mean (MN), median (MD), maximum (MAX) and minimum (MIN) values, and standard deviation (STD).

	Load					Displacement				
Before UV	MN	MD	MIN	MAX	STD	MN	MD	MIN	MAX	STD
FAB	162	157	152	182	9	28	26	21	51	7
FAB-PU	202	201	170	244	28	26	26	15	35	20
FAB-PU + 1% ZnO	187	188	158	214	16	20	18	15	26	4
FAB-PU + 4% ZnO	201	197	161	241	22	21	21	14	35	5
FAB-PU + 1% SiO$_2$ (200)	196	195	153	239	22	22	21	12	33	6
FAB-PU + 4% SiO$_2$ (200)	203	205	153	237	22	21	19	13	33	6
FAB-PU + 1% SiO$_2$ (R812)	188	190	127	246	30	19	17	11	44	8
FAB-PU + 4% SiO$_2$ (R812)	202	211	162	250	31	19	18	12	27	5
FAB-PU + 1% CNT	205	207	161	244	27	18	18	12	27	3
FAB-PU + 4% CNT	203	204	144	240	34	19	19	15	33	4
After UV										
FAB	153	149	134	194	18	25	23	16	44	7
FAB-PU	162	156	134	201	23	20	18	15	33	5
FAB-PU + 1% ZnO	194	197	144	261	32	20	19	14	31	5
FAB-PU + 4% ZnO	187	181	139	258	31	19	18	13	29	4
FAB-PU + 1% SiO$_2$ (200)	182	184	137	246	28	18	17	10	29	5
FAB-PU + 4% SiO$_2$ (200)	179	180	127	237	35	19	17	10	38	7
FAB-PU + 1% SiO$_2$ (R812)	189	190	130	239	31	18	17	13	31	5
FAB-PU + 4% SiO$_2$ (R812)	205	205	163	245	26	17	16	11	27	4
FAB-PU + 1% CNT	192	200	159	212	19	16	17	11	23	3
FAB-PU + 4% CNT	204	205	162	235	21	16	15	12	23	4

There were very small effects of UV radiation on the mechanical performance of the basic material (FAB specimens). The original polyester had an average puncturing load of 162 N, which was 153 N after UV treatment. On the contrary, a strong decrease in puncture strength was seen for impregnated textiles (FAB-PU samples). In this case, the average load started at ~202 N and arrived at ~162 N. The negative impact of UV weathering on the benefits of polyurethane treatment of the fabric surface was attributed to the potential destruction of molecular bonds in the PU polymer applied to the textile surfaces (as further confirmed through ATR measurements). The presence of nanoparticles in polyurethane dispersion, regardless of filler type, did not significantly increase mechanical resistance against the piercing probe. However, the reinforcing effect of nanoparticles was observed following UV aging. The puncturing strength of aged specimens containing nanoparticles was superior to that of aged specimens treated only with PU. The data was statistically examined with analysis of variance at one factor (One-way ANOVA) comparing values from treated samples with polyurethane loaded with nanoparticles to ones from samples treated solely with polyurethane before and after UV (Figure 5), p-value approach was used to interpret the relevance of differences. When p is less than 0.05 ($p < 0.05$), there is enough evidence to determine that the effect is significant.

Initially (before UV), systems containing nanoparticles behave similarly to systems impregnated with polyurethane alone ($p > 0.05$). After UV, in all cases, p-values were much lower than 0.05. This underlined the strong effect of nanoparticles on the puncture strength of treated textiles, particularly in the case of carbon nanotubes and hydrophobic silica nanoparticles.

The average displacement of the beginning fabric, which was roughly 28 mm, was slightly reduced by treating the textile with the aqueous polymer solution. When nanoparticles were added to the impregnating dispersion, especially carbon nanotubes, the capacity of the textile to be deformed during the puncturing test was reduced even further by stiffening the textile structure. After UV exposure, the treated textiles became more and more rigid and less deformable.

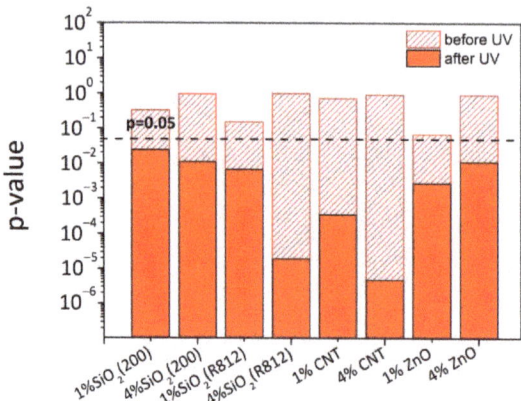

Figure 5. *p*-values from the analysis of variance of the puncture strength of PUD-impregnated specimens in comparison to FAB-PU systems before and after UV exposure.

These findings were consistent with previous literature studies on the same topics. Polyurethane served as a protective support in basic textiles, preserving and strengthening the fibers and threads. Common coating and laminating processes were demonstrated to be effective in enhancing the stiffness of treated fabrics because the applied polymer on textile structures operated by binding together weft and warp threads. Mechanical strength increased as a result of more filaments sharing the mechanical load. The previous research activities demonstrated a variety of advantages to using polyurethane in the textile weave of synthetic fabrics [24–26]. The impregnation treatment had no influence on the appearance of virgin textile material despite an increase in rigidity and weight. PU-treated samples outperformed untreated samples in terms of breaking tensile load, abrasion resistance, water repellency, and waterproofness. [25]. The effect of aqueous polyurethane dispersion containing hydrophilic or hydrophobic silica nanoparticles on the quasi-static perforating features of polypropylene-based textiles was attested in [24]. As piercing probes, a spherical spike and a pointed blade were used in quasi-static perforation experiments. The most significant results, i.e., the greatest rise in blade strength and piercing strength compared to the neat material, were obtained through the incorporation of the two additives, nano SiO_2 and crosslinker, into the polyurethane. The beneficial effect of nano SiO_2 on the mechanical properties of produced textiles was attributed to the efficient inhibition of fibers from sliding easily during the puncture testing by determining more energy spent in friction.

3.3. Infrared Spectroscopic Measurements

FT-IR spectroscopy, in attenuated total reflectance mode, was utilized to identify and qualitatively analyze the chemical changes caused by ultraviolet irradiation in polyurethane-based systems.

The ATR-FTIR spectra of basic and treated polyester-based fabrics (before and after UV weathering) were analyzed in Figure 6.

From the left to the right, different absorption bands were recognized for the basic PET textile [27] (Figure 6a): (i) at 2920 cm^{-1} and 2850 cm^{-1} assigned to asymmetrical and symmetrical stretching of methylene groups (CH_2); (ii) at 1712 cm^{-1} due to the stretching of carboxylic ester group (C=O); (iii) at 1455 cm^{-1} assigned to C–H deformation; (iv) at 1245 cm^{-1} caused by asymmetric C-C-O stretching in the aromatic ring; (v) at 1095 cm^{-1} due to the ester C–O–C stretching. The absorption peak at 1407 cm^{-1}, corresponding to the aromatic ring, was the characteristic peak of PET [28].

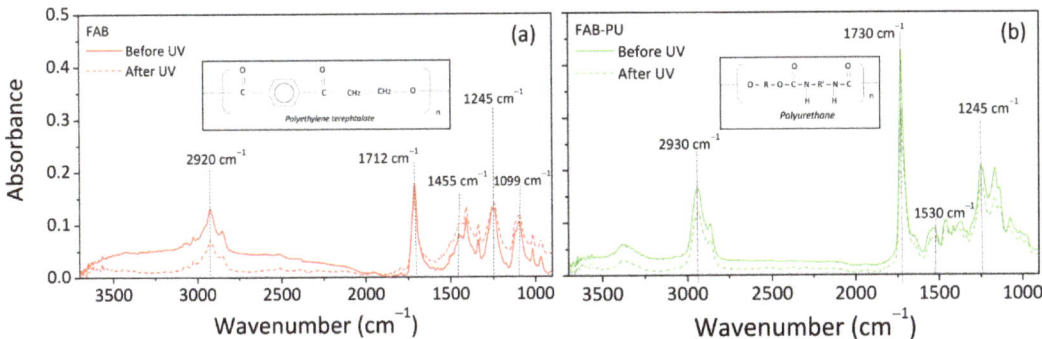

Figure 6. ATR spectra before and after UV of FAB (**a**) and FAB-PU (**b**) specimens.

A board band at 3450 cm^{-1} was found in basic textiles before UV aging, usually associated with the stretching of hydroxyl groups. In this case, this band could be due to the possible moisture absorption on the sample surface, potentially removed during the UV treatment in the oven at 35 °C. Furthermore, the polyester structure includes ester, alcohol, anhydride, aromatic rings, and heterocyclic aromatic rings. Alcohol could react with anhydride to form ester groups. This indicates that the polyester could still contain residual reactants such as alcohol and anhydride [28]. The peaks at 1712 and 1095 cm^{-1} were considered a sign of polyester breaking under certain conditions [28]. Except for the region between 3500 and 2500 cm^{-1}, there was no significant alteration in the base textiles downstream of the UV aging compared to the original scenario.

On the other side, the main characteristic bands of polyurethane polymer were distinguished in treated polyester-based textiles (Figure 6b): N-H stretching vibrations in the range of 3500–3300 cm^{-1}, N-H bending and C-N stretching at 1530 cm^{-1}, C-H asymmetric and symmetric stretching at 2930 cm^{-1} and 2855 cm^{-1}, respectively, C=O stretching at 1730 cm^{-1}, and C-O vibration at 1245 cm^{-1} [29,30]. The distinctions in spectra of basic PET and samples treated with PU confirmed the efficiency of the treatment on the sample's surface.

A comparison of ATR spectra of samples before and after UV aging was reported in Figure 7 for systems containing zinc oxide (a), hydrophilic silica (b), and hydrophobic silica (c). This investigation was not performed in the case of systems containing carbon nanotubes since this filler is a black substance that absorbed all of the radiation, resulting in a signal with a lot of noise. Li et al. [31] reported infrared spectroscopic measurements on basic, ammino-treated, or acid-treated multi-walled carbon nanotubes. In the instance of unfunctionalized carbon nanotubes, the signal was just noisy, with no evidence of specific peaks or absorption bands.

In all the cases, the peak intensity of the aforesaid bands (typical of PU polymers), intended as the absorbance value at the maximum point, was slightly decreased in specimens after UV exposure. This result was considered a sign of polyurethane chemical degradation following the UV treatment. The loss in absorbance (L_λ) was calculated as the ratio of the recorded intensities at the specific wavenumbers (λ) of 2950 cm^{-1}, 1750 cm^{-1}, 1530 cm^{-1}, and 1250 cm^{-1} before UV ($A^\lambda_{beforeUV}$) and after UV ($A^\lambda_{afterUV}$) (Table 3) according to Equation (1):

$$L_\lambda : \frac{A^\lambda_{afterUV}}{A^\lambda_{beforeUV}} \qquad (1)$$

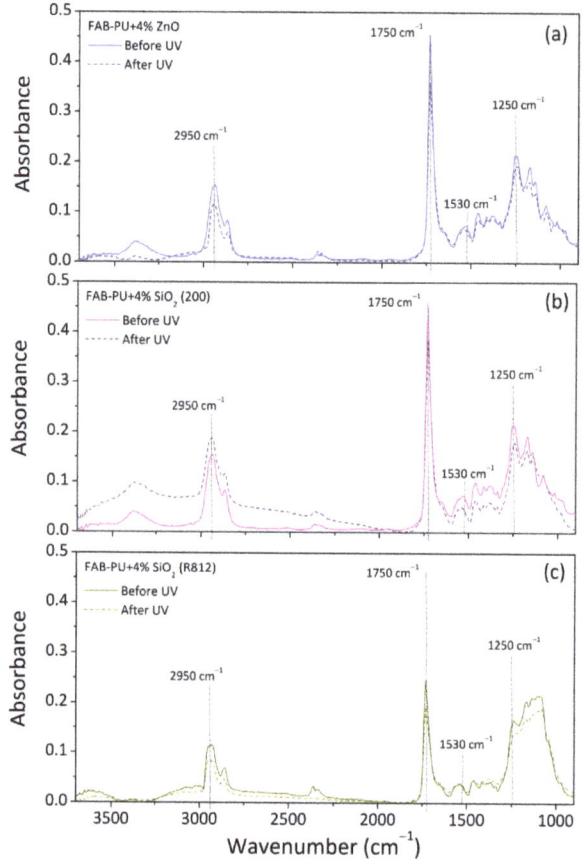

Figure 7. ATR spectra before and after UV of (**a**) FAB-PU + 4% ZnO; (**b**) FAB-PU + 4% SiO$_2$ (200); (**c**) FAB-PU + 4% SiO$_2$ (R812).

Table 3. Loss in absorbance (L$_\lambda$: L$_{2930}$, L$_{1730}$, L$_{1530}$, L$_{1245}$) at the respective wavenumber (λ: 2950 cm^{-1}, 1750 cm^{-1}, 1530 cm^{-1}, 1250 cm^{-1}) for each specimen.

	L$_{2930}$	L$_{1730}$	L$_{1530}$	L$_{1245}$
FAB-PU	0.73	0.72	0.80	0.84
FAB-PU + 4% ZnO	0.75	0.85	0.67	0.81
FAB-PU + 4% SiO$_2$ (200)	1.25	0.85	0.66	0.80
FAB-PU + 4% SiO$_2$ (R812)	0.77	0.83	0.89	0.87

Depending on whether the PU polymer is aliphatic or aromatic, photo-oxidation could take different courses. In the case of aliphatic PU (as in this case study), firstly, a mechanism of random chain scission could occur (Norrish Type I reaction) by leading to the formation of free radicals (Figure 8). These radical species could abstract hydrogen from CH$_2$ groups and generate polymer peroxy radical (PO$_2$•) and polymer hydroperoxide (POOH). The POOH could then be converted into polymer oxy (PO•) radicals if the scission reaction occurs, or hydroxyl group (POH) if the abstraction hydrogen reaction occurs. The degradation pathway could end with the interaction of different radicals with each other via the crosslinking reaction [32].

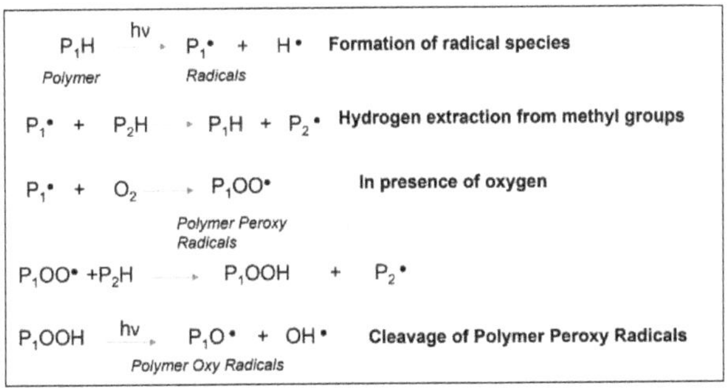

$$P_1H \xrightarrow{h\nu} P_1\bullet + H\bullet \quad \text{Formation of radical species}$$
Polymer   Radicals

$$P_1\bullet + P_2H \rightarrow P_1H + P_2\bullet \quad \text{Hydrogen extraction from methyl groups}$$

$$P_1\bullet + O_2 \rightarrow P_1OO\bullet \quad \text{In presence of oxygen}$$
Polymer Peroxy Radicals

$$P_1OO\bullet + P_2H \rightarrow P_1OOH + P_2\bullet$$

$$P_1OOH \xrightarrow{h\nu} P_1O\bullet + OH\bullet \quad \text{Cleavage of Polymer Peroxy Radicals}$$
Polymer Oxy Radicals

Figure 8. Possible reactions occurred during the UV-induced degradation of PU polymer.

As concerning the nanoparticles, zinc oxide displayed an absorption signal at wavenumbers less than 500 cm^{-1}, not reached in the present case [33]. Hydrophilic silica (Aerosil 200) displayed two characteristic peaks in the range of 1050–780 cm^{-1} attributed to the asymmetric–symmetric vibration of the Si-O-Si bond (siloxane groups) [34]. Hydrophobic silica (Aerosil R812) displayed a recognizable peak in areas of C-H stretching bands (2850–3000 cm^{-1}), considered correlated to the alkyl groups of organo-silane modifiers on the surface of silica nanoparticles, and the main peak at 1100 cm^{-1} corresponding to the symmetric Si-O-Si stretching [35].

Following UV irradiation, the intensity of absorption bands of hydroxyl radicals (OH$\bullet$) in the area of 3600–3200 cm^{-1} rose only for FAB-PU + 4% SiO$_2$ specimens. This could be due to the formation of hydroxyl radicals during the cleavage of polymer peroxy radicals (as indicated in the last reaction reported in Figure 8), which were next linked to silica, given the hydrophilic nature of nanoparticles.

4. Conclusions

This work investigated the mechanical characteristics, in terms of puncture strength, of impregnated polyester-based textiles with aqueous polyurethane solutions containing different types of nanoparticles, such as zinc oxide, functionalized silica (both hydrophilic and hydrophobic), and carbon nanotubes. All systems were evaluated using a puncturing test, infrared spectroscopy, and microscopic examination before and after UV-accelerated aging.

The results revealed a change in the appearance of treated specimens containing nanoparticles, particularly hydrophobic silica and carbon nanotubes. Such nanoparticles remained more distributed on the sample surface without spreading onto the textile structure. Furthermore, in the case of CNT, the textile surface retained its black hue. The mechanical resistance of PU-impregnated specimens was raised compared to basic fabric. A negligible effect on the average puncture strength of PU-treated textiles was verified when nanoparticles were introduced to the textile treatment. On the other side, the stiffness of neat fabric became higher when the polyurethane was applied to the textile surface.

Following UV aging, the yellowing effect was also visible in nanoparticle-containing specimens. A loss of mechanical performance was verified for basic fabric and PU-impregnated specimens. When nanoparticles were present on the fabric surface, the average puncture strength of treated specimens remained greater than the average value obtained in fabric treated with polyurethane alone (without nanoparticles). This was attributed to an increase in energy spent in friction during the passage of the puncturing probe in the textile surface covered by nanoparticles. In such systems, infrared spectroscopy revealed a weak drop in absorbance intensity for polyurethane absorption bands, intended as potential polymer deterioration during UV weathering.

Finally, the addition of nanoparticles to impregnated fabrics did not significantly inhibit polymer deterioration after UV treatment, as evidenced by yellowing in treated specimens. However, the nanoparticles seemed to positively contribute as reinforcement of

PU polymer into the textile structure to overall improve the mechanical strength, especially after UV exposure.

Author Contributions: Conceptualization, D.A.; data curation, L.G. and A.P.; writing—original draft preparation, D.A.; writing—review and editing, A.P. All authors have read and agreed to the published version of the manuscript.

Funding: This research received no external funding.

Institutional Review Board Statement: Not applicable.

Informed Consent Statement: Not applicable.

Data Availability Statement: The data presented in this study are available on request from the corresponding author.

Conflicts of Interest: The authors declare no conflict of interest.

References

1. Ravi, R.S.S.; Risfanali, S.R.; Raja, R.R. Wear Resistance and Water Absorption Study of Sic Reinforced Polyester Composite. *Mater. Today Proc.* **2018**, *5*, 14567–14572.
2. Looney, F.S.; Handy, C.T. The Effects of Construction on the Wear Wrinkling of Dacron®/Wool Suitings. *Text. Res. J.* **1968**, *38*, 989–998. [CrossRef]
3. Ahmad, S.; Ullah, T.; Ziauddin. Fibers for Technical Textiles. In *Topics in Mining, Metallurgy and Materials Engineering*; Ahmad, S., Rasheed, A., Nawab, Y., Eds.; Springer Science and Business Media Deutschland GmbH: Berlin/Heidelberg, Germany, 2020; pp. 21–47.
4. Singha, K. A Review on Coating & Lamination in Textiles: Processes and Applications. *Am. J. Polym. Sci.* **2012**, *2*, 39–49. [CrossRef]
5. Xu, J.; Zhang, Y.; Wang, Y.; Song, J.; Zhao, Y.; Zhang, L. Quasi-static puncture resistance behaviors of architectural coated fabric. *Compos. Struct.* **2021**, *273*, 114307. [CrossRef]
6. Kanesalingam, S.; Nayak, R.; Wang, L.; Padhye, R.; Arnold, L. Stab and puncture resistance of silica-coated Kevlar–wool and Kevlar–wool–nylon fabrics in quasistatic conditions. *Text. Res. J.* **2019**, *89*, 2219–2235. [CrossRef]
7. Firouzi, D.; Foucher, D.A.; Bougherara, H. Nylon-coated ultra high molecular weight polyethylene fabric for enhanced penetration resistance. *J. Appl. Polym. Sci.* **2014**, *131*, 40350. [CrossRef]
8. Hasanzadeh, M.; Mottaghitalab, V.; Babaei, H.; Rezaei, M. The influence of carbon nanotubes on quasi-static puncture resistance and yarn pull-out behavior of shear-thickening fluids (STFs) impregnated woven fabrics. *Compos. Part A Appl. Sci. Manuf.* **2016**, *88*, 263–271. [CrossRef]
9. Somarathna, H.M.C.C.; Raman, S.N.; Mohotti, D.; Mutalib, A.A.; Badri, K.H. The use of polyurethane for structural and infrastructural engineering applications: A state-of-the-art review. *Constr. Build. Mater.* **2018**, *190*, 995–1014. [CrossRef]
10. Pongsathit, S.; Chen, S.Y.; Rwei, S.P.; Pattamaprom, C. Eco-friendly high-performance coating for polyester fabric. *J. Appl. Polym. Sci.* **2019**, *136*, 48002. [CrossRef]
11. Patti, A.; Acierno, D. Structure-property relationships of waterborne polyurethane (WPU) in aqueous formulations. *J. Vinyl Addit. Technol.* **2023**, *29*, 589–606. [CrossRef]
12. Yang, H.-H.; Gupta, S.K.; Dhital, N.B. Emission factor, relative ozone formation potential and relative carcinogenic risk assessment of VOCs emitted from manufacturing industries. *Sustain. Environ. Res.* **2020**, *20*, 28–43. [CrossRef]
13. Honarkar, H. Waterborne polyurethanes: A review. *J. Dispers. Sci. Technol.* **2018**, *39*, 507–516. [CrossRef]
14. Chattopadhyay, D.K.; Raju, K.V.S.N. Structural engineering of polyurethane coatings for high performance applications. *Prog. Polym. Sci.* **2007**, *32*, 352–418. [CrossRef]
15. Patti, A.; Acierno, D. Waterborne polyurethane in textiles treatment. In *Waterborne Polyurethanes (WBPUs): Production, Chemistry and Applications*; Mohammadi, A., Ed.; Nova Science Publishers, Inc.: Hauppauge, NY, USA, 2023; pp. 251–272.
16. Bragaglia, M.; Cherubini, V.; Nanni, F. PEEK -TiO_2 composites with enhanced UV resistance. *Compos. Sci. Technol.* **2020**, *199*, 108365. [CrossRef]
17. Gorrasi, G.; Sorrentino, A. Photo-oxidative stabilization of carbon nanotubes on polylactic acid. *Polym. Degrad. Stab.* **2013**, *98*, 963–971. [CrossRef]
18. Cheraghian, G.; Wistuba, M.P. Effect of Fumed Silica Nanoparticles on Ultraviolet Aging Resistance of Bitumen. *Nanomater.* **2021**, *11*, 454. [CrossRef]
19. Rashvand, M.; Ranjbar, Z.; Rastegar, S. Nano zinc oxide as a UV-stabilizer for aromatic polyurethane coatings. *Prog. Org. Coatings* **2011**, *71*, 362–368. [CrossRef]
20. Allen, N.S.; Edge, M.; Hussain, S. Perspectives on yellowing in the degradation of polymer materials: Inter-relationship of structure, mechanisms and modes of stabilisation. *Polym. Degrad. Stab.* **2022**, *201*, 109977. [CrossRef]
21. Broasca, G.; Borcia, G.; Dumitrascu, N.; Vrinceanu, N. Characterization of ZnO coated polyester fabrics for UV protection. *Appl. Surf. Sci.* **2013**, *282*, 974–980. [CrossRef]

22. Becheri, A.; Dürr, M.; Lo Nostro, P.; Baglioni, P. Synthesis and characterization of zinc oxide nanoparticles: Application to textiles as UV-absorbers. *J. Nanoparticle Res.* **2008**, *10*, 679–689. [CrossRef]
23. Grigoriadou, I.; Paraskevopoulos, K.M.; Chrissafis, K.; Pavlidou, E.; Stamkopoulos, T.G.; Bikiaris, D. Effect of different nanoparticles on HDPE UV stability. *Polym. Degrad. Stab.* **2011**, *96*, 151–163. [CrossRef]
24. Patti, A.; Acierno, D. The effect of silica/polyurethane waterborne dispersion on the perforating features of impregnated polypropylene-based fabric. *Text. Res. J.* **2020**, *90*, 1201–1211. [CrossRef]
25. Patti, A.; Costa, F.; Perrotti, M.; Barbarino, D.; Acierno, D. Polyurethane Impregnation for Improving the Mechanical and the Water Resistance of Polypropylene-Based Textiles. *Materials* **2021**, *14*, 1951. [CrossRef] [PubMed]
26. Patti, A.; Acierno, D. The Puncture and Water Resistance of Polyurethane- Impregnated Fabrics after UV Weathering. *Polymers* **2019**, *12*, 15. [CrossRef]
27. Aljoumaa, K.; Abboudi, M. Physical ageing of polyethylene terephthalate under natural sunlight: Correlation study between crystallinity and mechanical properties. *Appl. Phys. A* **2015**, *122*, 6. [CrossRef]
28. Li, L.; Frey, M.; Browning, K.J. Biodegradability Study on Cotton and Polyester Fabrics. *J. Eng. Fiber. Fabr.* **2010**, *5*, 42–53. [CrossRef]
29. Lando, G.A.; Marconatto, L.; Kessler, F.; Lopes, W.; Schrank, A.; Vainstein, M.H.; Weibel, D.E. UV-Surface Treatment of Fungal Resistant Polyether Polyurethane Film-Induced Growth of Entomopathogenic Fungi. *Int. J. Mol. Sci.* **2017**, *18*, 1536. [CrossRef]
30. Sung, L.P.; Jasmin, J.; Cu, X.; Nguyen, T.; Martins, J.W. Use of laser scanning confocal microscopy for characterizing changes in film thickness and local surface morphology of UV-exposed polymer coatings. *J. Coatings Technol. Res.* **2004**, *1*, 267–276. [CrossRef]
31. Li, L.; Wang, J.; Liu, W.; Wang, R.; Yang, F.; Hao, L.; Zheng, T.; Jiao, W.; Jiang, L. Remarkable improvement in interfacial shear strength of carbon fiber/epoxy composite by large-scare sizing with epoxy sizing agent containing amine-treated MWCNTs. *Polym. Compos.* **2018**, *39*, 2734–2742. [CrossRef]
32. Xie, F.; Zhang, T.; Bryant, P.; Kurusingal, V.; Colwell, J.M.; Laycock, B. Degradation and stabilization of polyurethane elastomers. *Prog. Polym. Sci.* **2019**, *90*, 211–268. [CrossRef]
33. Nagaraju, G.; Udayabhanu; Shivaraj; Prashanth, S.A.; Shastri, M.; Yathish, K.V.; Anupama, C.; Rangappa, D. Electrochemical heavy metal detection, photocatalytic, photoluminescence, biodiesel production and antibacterial activities of Ag–ZnO nanomaterial. *Mater. Res. Bull.* **2017**, *94*, 54–63. [CrossRef]
34. Cohen, S.; Chejanovsky, I.; Suckeveriene, R.Y. Grafting of Poly(Ethylene Imine) to Silica Nanoparticles for Odor Removal from Recycled Materials. *Nanomaterials* **2022**, *12*, 2237. [CrossRef] [PubMed]
35. Dolatzadeh, F.; Moradian, S.; Jalili, M.M. Influence of various surface treated silica nanoparticles on the electrochemical properties of SiO_2/polyurethane nanocoatings. *Corros. Sci.* **2011**, *53*, 4248–4257. [CrossRef]

Disclaimer/Publisher's Note: The statements, opinions and data contained in all publications are solely those of the individual author(s) and contributor(s) and not of MDPI and/or the editor(s). MDPI and/or the editor(s) disclaim responsibility for any injury to people or property resulting from any ideas, methods, instructions or products referred to in the content.

Article

Lipid Corona Formation on Micro- and Nanoplastic Particles Modulates Uptake and Toxicity in A549 Cells

Anna Daniela Dorsch [1], Walison Augusto da Silva Brito [1,2], Mihaela Delcea [3], Kristian Wende [1,*] and Sander Bekeschus [1,4,*]

[1] ZIK *plasmatis*, Leibniz Institute for Plasma Science and Technology (INP), Felix-Hausdorff-Str. 2, 17489 Greifswald, Germany
[2] Department of General Pathology, State University of Londrina, Rodovia Celso Garcia Cid, Londrina 86057-970, Brazil
[3] Institute of Biochemistry, University of Greifswald, Felix-Hausdorff-Str. 4, 17487 Greifswald, Germany
[4] Clinic and Policlinic for Dermatology and Venerology, Rostock University Medical Center, Strempelstr. 13, 18057 Rostock, Germany
* Correspondence: kristian.wende@inp-greifswald.de (K.W.); sander.bekeschus@inp-greifswald.de (S.B.)

Citation: Dorsch, A.D.; da Silva Brito, W.A.; Delcea, M.; Wende, K.; Bekeschus, S. Lipid Corona Formation on Micro- and Nanoplastic Particles Modulates Uptake and Toxicity in A549 Cells. *Materials* 2023, *16*, 5082. https://doi.org/10.3390/ma16145082

Academic Editors: Liguo Xu and Peijiang Liu

Received: 23 June 2023
Revised: 15 July 2023
Accepted: 17 July 2023
Published: 19 July 2023

Copyright: © 2023 by the authors. Licensee MDPI, Basel, Switzerland. This article is an open access article distributed under the terms and conditions of the Creative Commons Attribution (CC BY) license (https://creativecommons.org/licenses/by/4.0/).

Abstract: Plastic waste is a global issue leaving no continents unaffected. In the environment, ultraviolet radiation and shear forces in water and land contribute to generating micro- and nanoplastic particles (MNPP), which organisms can easily take up. Plastic particles enter the human food chain, and the accumulation of particles within the human body is expected. Crossing epithelial barriers and cellular uptake of MNPP involves the interaction of plastic particles with lipids. To this end, we generated unilamellar vesicles from POPC (1-palmitoyl-2-oleoyl-glycero-3-phosphocholine) and POPS (1-palmitoyl-2-oleoyl-sn-glycero-3-phospho-L-serine) and incubated them with pristine, carboxylated, or aminated polystyrene spheres (about 1 μm in diameter) to generate lipid coronas around the particles. Lipid coronas enhanced the average particle sizes and partially changed the MNPP zeta potential and polydispersity. In addition, lipid coronas led to significantly enhanced uptake of MNPP particles but not their cytotoxicity, as determined by flow cytometry. Finally, adding proteins to lipid corona nanoparticles further modified MNPP uptake by reducing the uptake kinetics, especially in pristine and carboxylated plastic samples. In conclusion, our study demonstrates for the first time the impact of different types of lipids on differently charged MNPP particles and the biological consequences of such modifications to better understand the potential hazards of plastic exposure.

Keywords: A549 cells; dynamic light scattering; microplastic; surface plasmon resonance; unilamellar vesicles

1. Introduction

Plastic waste pollution in the environment has increased drastically in the past decades [1,2]. A plentitude of plastic types can be found in soils or water bodies [3,4]. Due to environmental impacts, such as ultraviolet (UV) radiation and shearing forces, the waste breaks down into micro- and nanoplastic particles (MNPP) [5]. As their numbers increase, a plentitude of particles is found in different areas of the world, even in isolated mountain regions and deep-sea sediments [6,7]. These increasing numbers of MNPP, especially in water and air, threaten (aquatic) ecosystems and even the human food chain [8,9]. There are two prevailing perspectives regarding the categorization of the term nanoplastic. Some favor the classification of particles smaller than 1000 nm, while others suggest the use of this term only for particles smaller than 100 nm [10–12]. An upper limit of 100 nm is commonly used by general consensus [13]. As in the environment, the size of plastic particles is likely a continuum. Therefore, it seems feasible to include both terms (microplastic and nanoplastic) in a single acronym (MNPP). The plastic particles used in this study were mostly about 1 μm.

In addition to the known formation of protein coronas around ingested nanoparticles [14,15], interactions of environmentally relevant nanoparticles with lipids [16–18] and modulation of their cell entry pathways can also be expected. Lipid corona formation might also result in (local) lipid oxidation, further increasing the physiologic impact on cells and mammal organisms [19,20].

In this study, two highly amphiphilic glycerophospholipids (GP), POPC (1-palmitoyl-2-oleoyl-sn-glycero-3-phosphocholine) and POPS (1-palmitoyl-2-oleoyl-sn-glycero-3-phospho-L-serine), as well as a combination of both, were investigated. The investigated glycerophospholipids have a similar tail structure, while their head group structure was different [21]. With these different head group structures, the lipids' charges also differ. POPC is a neutrally charged lipid [22], while POPS is known to be negatively charged [23]. Glycerophospholipids are the main structural lipid components of eukaryotic cell membranes, with phosphatidylcholine representing more than 50% of all cell membrane phospholipids [21]. In line with this, POPC and POPS are described as lipids of the extracellular membrane across all tissue types according to the human metabolome database. POPC is especially found in the placenta, while POPS can be found in the brain, heart, kidney, and liver tissue. Among other lipids, POPC is also an important factor in lung health, as it is a component of pulmonary surfactants crucial for lung health and compliance [24], while POPS is an important factor in apoptosis and enzyme activation [25].

This study aims to understand the formation of lipid coronas with three different lipid compositions and three different MNPP surface charges. In addition, the lipid corona plastic nanoparticles were tested in A549 human lung cells to investigate the impact of these particles on cells. In addition to dermal and enteral uptake, breathing and transmission via the lung tissue is one of the possible routes suggested for MNPP for entering the human body. Even though they are of malignant origin, human A549 lung cancer cells were chosen as a model cell line in this study to address this aspect.

2. Materials and Methods

2.1. Liposome Preparation by Extrusion

Unilamellar vesicle liposomes were prepared using an extrusion method with modifications from to original protocol [26,27]. Briefly, the liposomes were prepared from 1-palmitoyl-2-oleoyl-glycero-3-phosphocholine (16:0-18:1 PC (POPC); Avanti Polar Lipids, Alabaster, AL, USA) and 1-palmitoyl-2-oleoyl-sn-glycero-3-phospho-L-serine (16:0-18:1 PS (POPS); Avant Polar Lipids, Alabaster, AL, USA) at 1 mM (final concentration). The lipids were used as mono-lipids or combined with POPC:POPS (75 μM:25 μM) to form liposomes. Therefore, the respective amount of lipid was diluted in a glass tube using 2 mL chloroform (Carl Roth, Karlsruhe, Germany) prior to chloroform evaporation in an N_2-evaporator (TurboVap; Biotage, Uppsala, Sweden) to form a lipid film. The lipid film was rehydrated in an inorganic solvent (i.e., Dulbecco's phosphate-buffered saline (DPBS); BioWest, Nuaillé, France) and heated to 40 °C in a water bath. After dissolving the lipid film in the inorganic solvent, the solution was either frozen in liquid nitrogen for 1 min or completely frozen, depending on the volume of the inorganic solvent. Afterward, the frozen solution was reheated to 40 °C, and this freeze–thaw process was repeated 5 times. To form unilamellar vesicles of a uniform size, the solution was extruded 13 times through a *Nuclepore* membrane filter with a pore size of 0.1 μm (Whatman products; Cytiva, Marlborough, MA, USA) using a mini extruder (Avanti Polar Lipids, Alabaster, AL, USA) on a 40 °C heating plate. The extruded unilamellar vesicles were stored at 4 °C for a maximum of 3 days.

2.2. Interaction of Plastic Particles and Unilamellar Vesicles

To investigate the effect of the interaction between differently composed unilamellar vesicles and plastic particles, polystyrene (PS) particles (Polysciences Europe, Hirschberg,

Germany) with an average size of 1 μm without further modification (PS) or with different surface modifications, such as carboxylation (PS-COOH) or amination (PS-NH$_2$), were incubated with unilamellar vesicles. The particles were diluted to 1.25 mg/mL in DPBS and incubated in equal volumes with unilamellar vesicles (1 mM of POPC, POPS, or POPC:POPS at 75 μM:25 μM) for 1 h at 37 °C in a ThermoMixer (Eppendorf, Hamburg, Germany) with an agitation of 300 rpm (Figure 1a). After incubation, suspensions were centrifuged (4 °C) for 45 min at 10,000× g, the supernatant was discarded to remove unbound liposomes, and the remaining pellet was resuspended in DPBS.

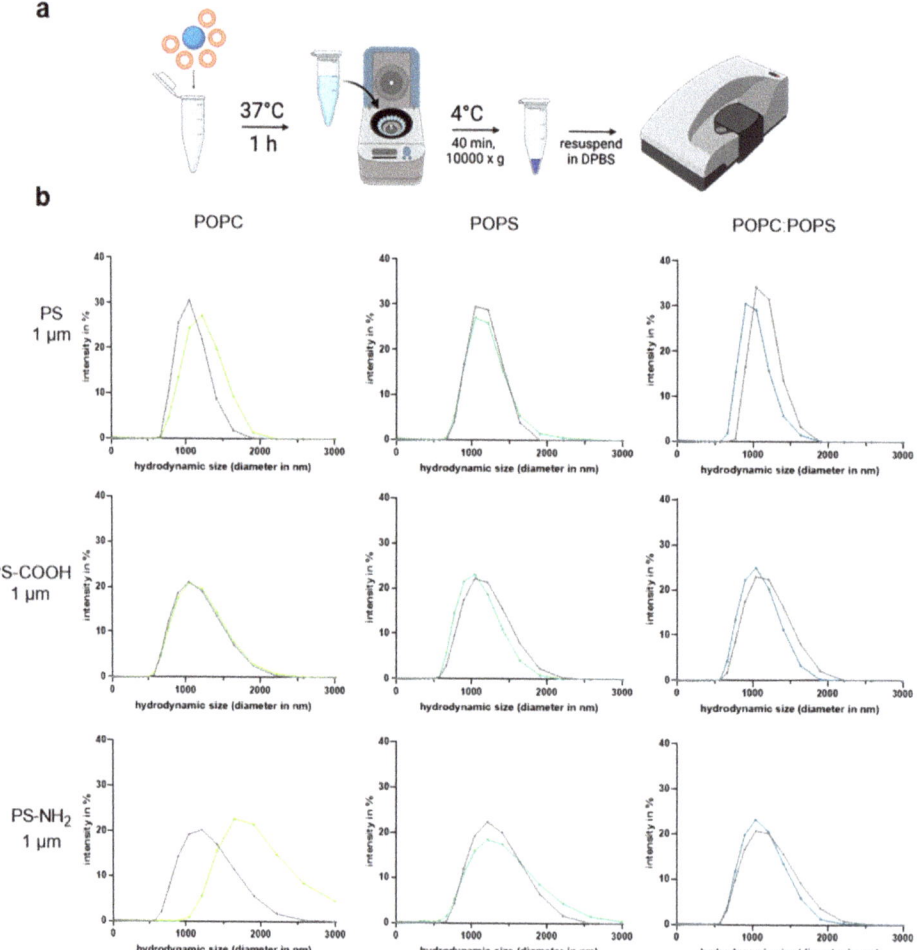

Figure 1. Dynamic light scattering. (a) interaction protocol of polystyrene MNPP and unilamellar vesicles; (b) hydrodynamic size distribution (nm) and intensity weighted (%). Data are presented as the mean of three independent experiments. The plastic particles alone measured via DLS are shown as grey curves, while green (POPC), turquoise (POPS), and blue (POPC:POPS) curves show lipids incubated with respective plastic particles. POPC = 1-palmitoyl-2-oleoyl-glycero-3-phosphocholine; POPS = 1-palmitoyl-2-oleoyl-sn-glycero-3-phospho-L-serine; PS = polystyrene; PS-COOH = carboxylated polystyrene; PS-NH2 = aminated polystyrene. Statistical analysis is from Table 1.

Table 1. Dynamic light scattering quantification. Polydispersity index (PDI) and Z-average (in nm) measurements of polystyrene MNPP and unilamellar vesicles. Data are presented as mean ± SD of three independent experiments. DPBS = Dulbecco's phosphate-buffered saline; POPC = 1-palmitoyl-2-oleoyl-glycero-3-phosphocholine; POPS = 1-palmitoyl-2-oleoyl-sn-glycero-3-phospho-L-serine; PS = polystyrene; PS-COOH = carboxylated polystyrene; PS-NH_2 = aminated polystyrene. Significant differences from the DPBS control groups are indicated in bold (* $p < 0.05$; ** $p < 0.01$; *** $p < 0.001$; n.s. = non-significant).

		PS	PS-COOH	PS-NH_2
DPBS	PDI	0.42 ± 0.05	0.21 ± 0.05	0.19 ± 0.04
POPC	PDI	0.41 ± 0.15 n.s.	0.20 ± 0.10 n.s.	0.21 ± 0.10 n.s.
POPS	PDI	0.42 ± 0.06 n.s.	0.26 ± 0.05 **	0.28 ± 0.10 *
POPC:POPS	PDI	0.55 ± 0.12 **	0.27 ± 0.05 *	0.24 ± 0.03 **
DPBS	Z-average	1753 ± 224.10	1193 ± 45.08	1247 ± 38.26
POPC	Z-average	1878 ± 108.70 n.s.	1193 ± 57.60 n.s.	2365 ± 552.7 ***
POPS	Z-average	1785 ± 131.90 n.s.	1195 ± 24.67 n.s.	1414 ± 329.9 n.s.
POPC:POPS	Z-average	1937 ± 193.70 *	1226 ± 35.87 n.s.	1188 ± 16.00 **

2.3. Dynamic Light Scattering and Zeta Potential

All samples were subjected to dynamic light scattering (DLS). For this, the resuspended particle pellets were diluted in DPBS and measured in DTS1070 cuvettes (Malvern Panalytical, Kassel, Germany) using a Zetasizer Ultra device (Malvern Panalytical, Kassel, Germany). Measurements were performed in technical triplicates both for the hydrodynamic size as well as for the zeta potential. Hydrodynamic size measurements were performed at 25 °C with an equilibration time of 60 s. Zeta potential measurements were performed at 25 °C with an equilibration time of 120 s for a total of 25 runs. The DLS data were exported from the manufacturer's software ZSXplorer (Malvern Panalytical, Kassel, Germany).

2.4. Surface Plasmon Resonance (SPR)

An L1 sensor chip (Cytiva, Marlborough, MA, USA) was coated with unilamellar vesicles consisting of either POPC (1 mM) or a combination of POPC:POPS (75 µM:25 µM). The coating of the sensor chip surfaces was performed in a manual run, starting with three injections of 20 mM 3-[(3-Cholamidopropyl)-dimethylammonio]-1-propansulfonat (CHAPS; Carl Roth, Karlsruhe, Germany) at a flow rate of 10 µL/min for 60 s. After changing the flow rate to 5 µL/min, 1 mM of unilamellar vesicles was injected for 16 min. The unbound unilamellar vesicles were removed by a 60 s injection of 10 mM NaOH Carl Roth, Karlsruhe, Germany) at a 30 µL/min flow rate. To prove that the sensor chip surface was fully coated with the unilamellar vesicles, 0.125 mg/mL bovine serum albumin (BSA; Carl Roth, Karlsruhe, Germany) was injected for 60 s at a flow rate of 10 µL/min. The change in relative units (RU) was investigated after injecting pristine PS into the lipid-coated sensor chip surface. For this, the 1 µm plastic particles were diluted to a concentration of 0.625 mg/mL in DPBS and injected at a flow rate of 10 µL/min for 90 s to allow the interaction with the lipid-coated sensor chip surface. After the experiments, the lipid coating was removed from the sensor chip surface with three 90 s injections of 20 mM CHAPS at a 10 µL/min flow rate.

2.5. In Vitro Experiments

Unilamellar vesicles (1 mM) were incubated in equal volumes with three different 1 µm free-labeled plastic particles (PS, PS-COOH, and PS-NH_2) for 1 h at 37 °C with horizontal agitation (300 rpm). Vesicle mixtures or control solution (DPBS) was then applied to A549 cells (ATTC. CCL-185), which were seeded at 1×10^4 cells in 24-well flat-bottom plates (Thermo Fischer, Dreieich, Germany) for attachment 24 h before treatment in Dulbecco's Modified Eagle Medium (DMEM; Pan-Biotech, Aidenbach, Germany) with high glucose (4.5 g/L), supplemented with or without 10% fetal calf serum (FCS; Sigma-Aldrich, Taufkirchen, Germany). A549 cells were incubated with 10 µg/mL (final concentration) of the pre-incubated particles for 3 h to 24 h at 37 °C, 5% CO_2, and 95% humidity. After incubation, the cells were subjected to microscopy or detached using Accutase (BioLegend, Amsterdam, The Netherlands) for flow cytometry.

2.6. Flow Cytometry

The effects of MNPP on A549 cells were investigated using flow cytometry (CytoFLEX S; Beckman-Coulter, Krefeld, Germany) 24 h after exposure. The cells were harvested, washed in PBS, and resuspended in 96-well plates with PBS. At least 50,000 cells were acquired per sample. Forward and side-scatter signals were analyzed, and their intensities were quantified using Kaluza Analysis 2.2 software (Beckman-Coulter, Krefeld, Germany). Gating strategies in forward scatter plots were used to determine the percentage of dead cells.

2.7. Microscopy Analysis

After exposure of A549 cells to lipid-coated and control MNPP for 24 h, the cells were washed with PBS to remove the remaining particles. Nine fields of view per well were captured using a 20× air objective (NA 0.4; Zeiss, Jena, Germany) and a high-content imaging system (Operetta CLS; PerkinElmer, Hamburg, Germany). After flatfield and brightfield corrections, cells were segmented using digital phase contrast. Algorithm-based quantitative image analysis was performed using *Harmony* 4.9 software (PerkinElmer, Hamburg, Germany) to analyze cell roundness and cell area (μm^2).

2.8. Statistical Analysis

Prism 9.5.1 (GraphPad Software, San Diego, CA, USA) was used for graphs and statistical analyses. If not stated otherwise, data are displayed as mean ± standard deviation (SD). The data sets were tested for normal distribution using the D'Agostino and Pearson tests. For normally distributed data sets, one-way analysis of variance (ANOVA) was used to compare more than two groups. Two groups were tested using a *t*-test. The levels of significance were indicated as follows: $\alpha = 0.05$ (*), $\alpha = 0.01$ (**), and $\alpha = 0.001$ (***).

3. Results

3.1. Type of Lipid Corona Modulated MNPP Properties

The interaction of differently surface-modified polystyrene MNPP and a unilamellar vesicle model composed of different lipids was investigated using DLS. Changes in hydrodynamic size, as well as zeta potential, were investigated after the incubation of the components (Figure 1b). After the incubation of 1 µm pristine PS particles with a combination of the investigated lipids, POPC:POPS (75 µM:25 µM), a significant change in the polydispersity index (PDI) was measured (Table 1). Several changes in size and PDI were observed, depending on the type of lipid incubated. For the PS-COOH particles, POPS and POPC:POPS significantly increased the PDI. The incubation of PS-NH2 particles with POPC and POPS unilamellar vesicles showed changes in the investigated Z-average as well as in the corresponding PDIs.

Subsequent analyses of the suspensions' zeta potential were performed to investigate the surface charge effects after the exposure of the PS plastic particles to different unilamellar vesicles (Figure 2). Here, the incubation of PS and PS-COOH particles with POPC:POPS liposomes showed a significant increase in the measured zeta potential, whereas the incubation with POPC showed a significant decrease and increase in zeta potential in PS-COOH and PS-NH_2 MNPP, respectively. Next, the interaction of the PS particles with a lipid-coated gold sensor chip was investigated using surface plasmon resonance (Figure 3). The data indicated a better interaction between the gold sensor chip, which was only coated using POPC, and a change in the relative response units (RU) of around 500 RU after injection of the PS particles. Coating the gold sensor surface with a lipid mixture of

POPC:POPS resulted in a response of only 50 RU after the incubation with PS particles, indicating a non-sufficient interaction between the particles and the coated sensor surface. Altogether, these results indicated that the type of lipid corona is decisive in altering MNPP surface characteristics. At the same time, the initial plastic surface charge is also relevant regarding potential modifications by lipid coronas.

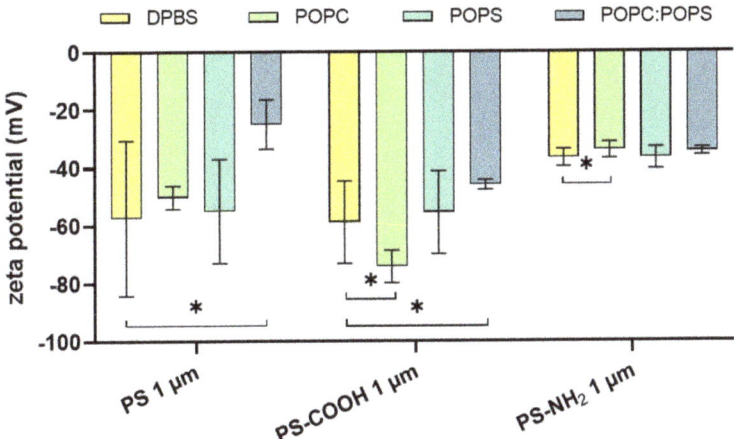

Figure 2. Zeta potential. Changes in zeta potential in millivolt (mV) after incubation of pristine (left), carboxylated (middle), and aminated (right) 1 µm plastic particles with the respective unilamellar vesicles. Data are presented as mean ± SD of three independent experiments. Statistical analysis was performed using one-way analysis of variances (ANOVA) with Dunnett's post hoc test against DPBS (without liposome incubation), and are indicated as follows: * $p < 0.05$; DPBS = Dulbecco's phosphate-buffered saline; POPC = 1-palmitoyl-2-oleoyl-glycero-3-phosphocholine; POPS = 1-palmitoyl-2-oleoyl-sn-glycero-3-phospho-L-serine; PS = polystyrene; PS-COOH = carboxylated polystyrene; PS-NH$_2$ = aminated polystyrene.

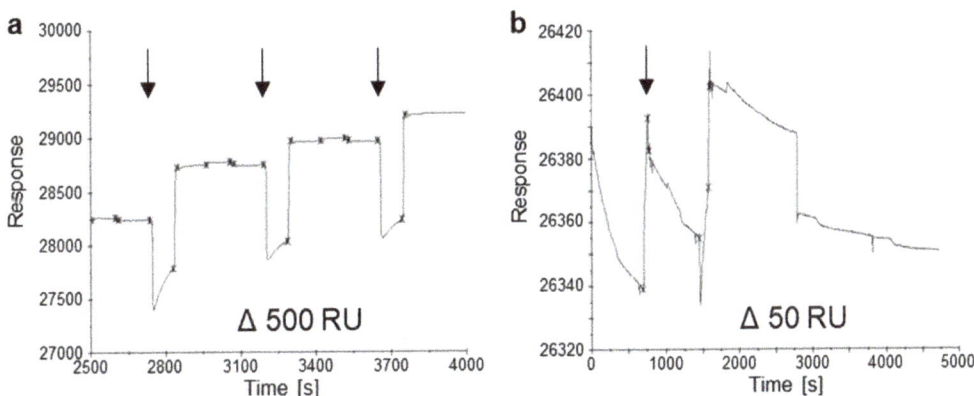

Figure 3. Surface plasmon resonance responses. (**a**) Time (s) dependent responses after injection of PS particles into a POPC-coated sensor surface; (**b**) time-dependent responses after the injection of PS particles to a sensor surface coated with POPC:POPS (75 µM:25 µM). RU = response units. Arrows mark the injection timing. Arrows indicate starting of regeneration phases of the sensor.

3.2. Lipid Coronas Affected Nanoparticle Uptake in Cells

After investigating the effects of the lipid coronas on MNPP surface characteristics, we investigated the biological consequences of the lipid coronas. A549 cells were incubated under nine different MNPP conditions and investigated using flow cytometry 24 h later (Figure 4a). Cytochalasin B (CytoB) was used as a positive control to inhibit actin filament polymerization and cell growth to increase granule formation in cells, leading to increased side-scatter signals (Figure 4b). The side-scatter intensity of 1 µm PS and PS-COOH particles was significantly affected by the presence of POPC and POPS unilamellar vesicles (Figure 4c). Interestingly, no effect was observed with POPC:POPS liposomes.

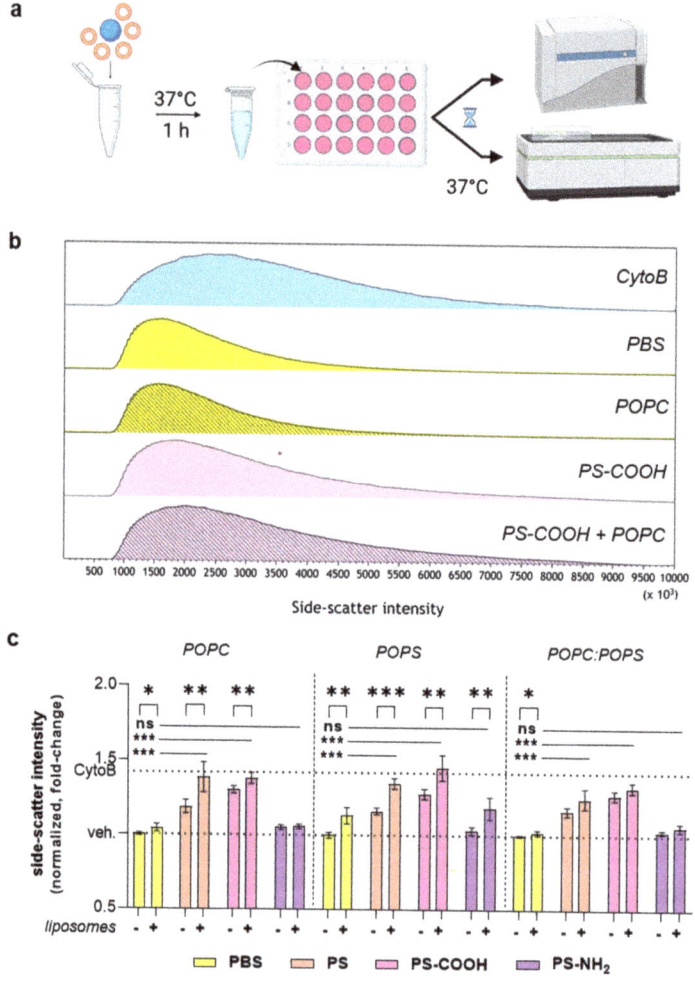

Figure 4. Cellular plastic particle uptake using flow cytometry. (a) Experimental setup of the pre-incubation of particles with different unilamellar vesicles and subsequent exposure to A549 cells; (b) representative side-scatter histograms of cells using flow cytometry; (c) side-scatter intensity quantification (%) in conditions with (+) or without (-) liposomes. Data are presented as mean ± SD of three independent experiments. Statistical analysis was performed using one-way analysis of variances (ANOVA) with Dunnett's post hoc test against DPBS (with liposome incubation) or t-test, comparing the results with and without liposomes, which are indicated as follows: * $p < 0.05$; ** $p < 0.01$; *** $p < 0.001$;. n.s. = non-significant; DPBS = Dulbecco's phosphate-buffered saline; POPC = 1-palmitoyl-2-oleoyl-glycero-3-phosphocholine; POPS = 1-palmitoyl-2-oleoyl-sn-glycero-3-phospho-L-serine; PS = polystyrene; PS-COOH = carboxylated polystyrene; PS-NH$_2$ = aminated polystyrene.

Next, we inferred whether MNPP lipid coronas affected the cell morphology (Figure A1). Using algorithm-driven quantitative image analysis, we found no effect of the lipid coronas of MNPP on average cell areas (Figure 5a) and cell roundness (Figure 5b). In contrast, significant differences were observed concerning cell death, which increased when A549 cells were cultured with lipid corona-modified PS-NH$_2$ MNPP under all conditions (Figure 5c). However, the PS and PS-COOH mixtures with POPC and POPS showed decreased cell death compared to their control counterparts. In addition, for PS-COOH incubated with POPC:POPS, decreased cytotoxicity was observed, whereas for PS with POPC:POPS, cytotoxicity remained unaltered.

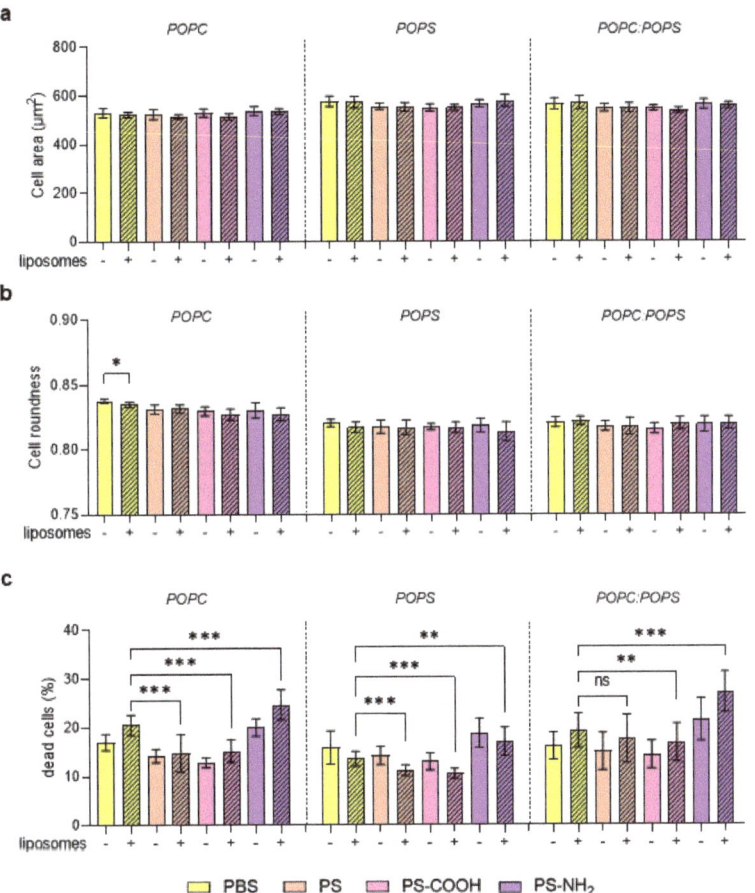

Figure 5. **Cell morphology and viability analysis.** (a) Cell areas (μm^2), (b) cell roundness (a.u.), and (c) percentage of dead cells in A549 cells incubated with different pre-incubated particles. Data are presented as mean ± SD of three independent experiments. Statistical analysis was performed using ANOVA, comparing results with and without liposomes, which are indicated as follows: * $p < 0.05$; ** $p < 0.01$; *** $p < 0.001$; n.s. = non-significant; DPBS = Dulbecco's phosphate-buffered saline; POPC = 1-palmitoyl-2-oleoyl-glycero-3-phosphocholine; POPS = 1-palmitoyl-2-oleoyl-sn-glycero-3-phospho-L-serine; PS = polystyrene; PS-COOH = carboxylated polystyrene; PS-NH$_2$ = aminated polystyrene.

In addition to lipids, proteins are ubiquitous biomolecules present in the body. To this end, we investigated the effect of surplus proteins on the interaction of particles with A549 cells. In this experiment, fluorescently labeled PS particles (PS-NH$_2$ 55 nm; PS 190 nm; PS 1040 nm) were pre-incubated with 1 mM POPC unilamellar vesicles for 1 h at 37 °C prior to the exposure of A549 cells to these pre-coated particles for different time points. Under one condition, the cells were washed with PBS, and the cell medium was replaced with a cell culture medium not supplemented with 10% FCS, whereas under the other experimental conditions, the cells were washed with PBS and supplemented with a cell culture medium containing 10% FCS.

MNPP uptake was determined by fluorescence using flow cytometry. The presence of FCS led to a significant decrease in the MNPP uptake, independent of their size (Figure 6). The largest effect, however, was observed for the 190 nm particles. Pristine particles (190 nm and 1040 nm) showed increased uptake over time, while positively surface-charged particles (NH$_2$) were taken up immediately and were already at a maximum of 30 min post-addition. Altogether, the type of lipid corona significantly affected MNPP uptake into cells, which was the case for the total uptake and its kinetics. Additionally, the particle toxicity was differently affected by the different conditions applied.

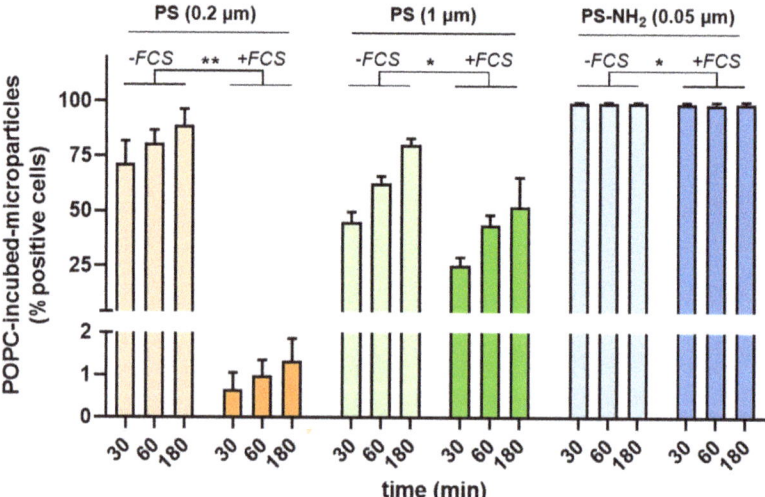

Figure 6. Contribution of FCS to A549 cell particle uptake. Measurement of the uptake of plastic particles (pre-incubated with POPC) in A549 cell using flow cytometry and quantified for small (0.2 μm and 1.0 μm) and aminated (55 nm) particles in the presence and absence of FCS quantified at three different time points post-exposure (30 min, 60 min, and 180 min). Data are presented as mean ± SD of three independent experiments. Statistical analysis was performed using multiple ratio-paired t-tests, comparing results with and without FCS, which are indicated as follows: * $p < 0.05$; ** $p < 0.01$; POPC = 1-palmitoyl-2-oleoyl-glycero-3-phosphocholine; PS = polystyrene; PS-NH$_2$ = aminated polystyrene.

4. Discussion

Environmental pollution with micro- and nanoplastic particles (MNPP) has evolved over the last decades, infesting the human food chain [8,9,28]. With this rising problem of environmental pollution, the contact of mammalian and human bodies with MNPP is inevitable. Although the interaction of proteins with model plastic particles is studied well [29–31], little is known about the interaction of lipids with model polystyrene particles [32]. A corona forms around the particle when plastic particles contact biomolecules like proteins, lipids, and sugars. In the case of well-studied protein coronas, a soft and hard corona is formed around the particles, determining the plastic particles' fate in organisms [33,34]. Our study investigated the interaction of a unilamellar vesicle model with polystyrene particles modified with different surface chemistries. The model lipids chosen for this study are cell membrane-relevant POPC and POPS [21,35].

Different studies describe the interactions between nanomaterials and lipids in different bodily fluids. Hellstrand and colleagues investigated the interactions of human plasma lipids with nanomaterials and showed that whole lipoprotein complexes are prone to interact with nanomaterials [36]. Our study also suggests an interaction of lipids with polystyrene material, depending on the different surface chemistries. Especially after incubation with suspensions containing a high amount of neutral lipids like POPC, the hydrodynamic size of the positively charged aminated PS particles (PS-NH$_2$) increased.

The interaction and the formation of lipid coronas of physiological lipids in mouse serum, like triglycerol and cholesterol, were studied by Lima et al. using PS-COOH particles of different sizes [33]. The study observed that cholesterol binding depends largely on the size of the investigated particles and less on the surface area, whereas triglyceride binding seems to be affected by the surface area of the investigated particles [33]. In contrast, our carboxylated particles did not show increased hydrodynamic sizes independent of the lipid corona type. The reason might be that the structure of the investigated lipids differs from the structures of cholesterol and triglycerides investigated in other studies.

Cholesterol inherits a unique structure with a hydrocarbon tail, a sterol core out of four hydrocarbon rings, and a hydroxyl group, where the hydrocarbon tail and the ring system are non-polar and consequently do not mix with water [37]. Bloodstream transport is, therefore, only possible if water-insoluble lipids like cholesterols are attached to proteins in the bloodstream, the so-called apolipoproteins [38]. Possibly, the attachment of lipids like cholesterol to the lipoproteins is the driving force in the interaction of the lipids with PS particles.

Once particles enter body fluids, corona formation is described as being driven by the reduction of the high surface energy of the particles and is, therefore, a highly dynamic process [33]. Over time, the biocorona composition changes by replacing biomolecules attached early to the particle surface with other biomolecules with a higher affinity to the polymer surface until an equilibrium is reached, which is described as the Vroman effect [39]. In these studies, proteins in the investigated solution, like blood serum, seem to be the driving force in corona formation, during which lipid attachment is more likely an additional effect. Our experiments only investigated solutions containing unilamellar vesicles without a protein admixture in the suspension. The vesicles used in this study were ~100 nm, with hydrophilic head groups located outside the vesicles, whereas the hydrophobic tail structures were centered in a lipidic bilayer. With this stable unilamellar vesicle system, the attachment of the lipids to the particle surface seems to be inhibited.

In the case of positively charged PS-NH$_2$ particles, it seems that the driving force of interaction is the charge difference between the particles and the unilamellar vesicles. Theoretically, the unilamellar vesicles composed of POPC should be neutral in charge, but in zeta potential measurements, they show a negative zeta potential. This might be an effect of the suspension medium, in which DPBS, a buffer containing negatively charged phosphates, was used. Potentially, this medium affected the overall charge of the unilamellar vesicles in this suspension and enforced the interaction of the positively charged particles with the unilamellar vesicles containing POPC.

The effect of the buffer medium might also explain why suspensions of only negatively charged unilamellar vesicles, like POPS, did not increase the hydrodynamic size of any of the tested MNPP. Repulsion effects between the unilamellar vesicles and the buffer medium should also be considered, i.e., the particles may not come close enough to the unilamellar vesicles for any interaction to occur. Conversely, the disruption of lipid bilayers is faster and induces a greater leakage with positively charged particles than with negatively charged particles [40]. This effect seems to be due to the surface modification and is independent of the particles' core material, so the amination of the polystyrene particles can be the reason for their interaction behavior [40,41].

With the POPC:POPS liposome solutions, the PDI and zeta potential changed significantly. PDI changes indicate aggregates formed in the investigated suspension after incubating the liposome suspension with the particles [42]. The significant change in the PDI after the incubation of PS particles with POPC:POPS liposomes, and under some conditions, significant changes in the hydrodynamic size average, indicate that lipid–lipid aggregates might have formed and that an interaction between the particles and the POPC:POPS liposomes is unlikely. This fact can also be assumed since, in the surface plasmon resonance experiments, no interaction between the PS particles and the POPC:POPS-covered sensor chip surface could be measured (Figure 3b). In contrast, the highly significant change in the hydrodynamic size after the incubation of PS-NH$_2$ with the POPC liposome suspension without changing the PDI indicates an attachment of the lipids to the investigated particle without forming aggregates. Incubation of POPC:POPS liposomes with PS-NH$_2$ showed a significant difference in the average hydrodynamic size. The data also suggested the formation of lipid–lipid aggregates in the PDI.

Zeta potential change is an important measurement to characterize the suspension's stability and investigate the surface functionality of particle dispersions [43]. The resulting zeta potential occurs due to the interaction of ions solubilized in the measuring buffer with the functional groups on the particles' surface. Suspensions with zeta potentials ranging between −30 mV and +30 mV are considered stable due to the sufficient repulsive force in the system to avoid aggregate formation in the solution [44]. The change in zeta potential after the incubation of the unilamellar vesicles consisting of POPC:POPS with PS particles indicates that the solution stabilizes in the presence of POPC:POPS liposomes. In a suspension system considered stable, repulsion occurs between the components, so an interaction seems very unlikely.

The significant decrease in zeta potential after adding POPC liposomes to PS-COOH particles suggests that the suspension is no longer stable and is more prone to form aggregates. Yet, the formation of aggregates was not observed in the analysis of hydrodynamic sizes. We assume that due to repulsion effects, no aggregates and no interaction of the particles and the added liposomes were formed due to the same surface charges of the components. While the interaction of PS particles with POPC liposomes showed changes in neither the hydrodynamic size nor the PDI, an interaction of these components was observed in SPR analysis. This might be due to a larger lipid interaction surface and, therefore, more attraction of the particles to the lipid-coated sensor surface than to the liposomes.

A549 cells responded differently to MNPP, with and without lipid corona. Cytotoxicity was significantly altered, as previously shown by others [45,46] and ourselves [47], and PS-NH$_2$ showed more pronounced effects. Interestingly, we presume that particle size may be a distinguishing factor in aminated particle cytotoxicity. Smaller particles, such as 55 nm, are readily cytotoxic, as previously demonstrated [47], while larger particles, such as the 1 µm PS-NH$_2$ particles used in our study, were cytotoxic, but not of the same magnitude. This was potentially due to the limited uptake of positively charged particles at that size, as suggested by the side-scatter quantification in our study (Figure 4c). Interestingly, POPC lipid addition to these particles did not affect the uptake rates either. This is notable since this condition showed the most significant size change effect in our study (Figure 1b). This can be explained by the size of the pre-incubated particles, which are too large to interact with the cells [48]. Moreover, POPC incubation seemed to mask the uptake-increasing effects of POPS on PS-NH$_2$, as incubation with POPC:POPS PS-NH$_2$ did not differ from particles without adding lipids.

Generally, pre-incubated PS and PS-COOH particles seemed to interact more in our A549 cell model. This is concerning because environmental particles that were ingested or inhaled also had a kind of "pre-coating" with different biomolecules before or while entering the human food chain [49]. This corona formation around the environmentally relevant particles can also alter the interaction behavior of the particles within cells and organisms. The particles can also transport chemicals to mammalian organisms, such as ethane (DDT), polychlorinated biphenyls (PCB), and so-called endocrine disruptors, potentially affecting the hormone systems [50].

Our study could not provide direct evidence of lipid corona formation on MNPP. Nevertheless, DLS experiments have shown increasing polystyrene particle sizes following incubation with lipids, suggesting the latter's coating on the plastic particles. Alternatively, future studies could perform mass spectrometry analysis to provide unambiguous evidence of lipid attachment to the MNPP. The biological relevance of our study is also potentially hampered by the fact that MNPP modification by, e.g., protein or lipid coronas, would occur already before entering the cells. This could be due to, for example, the mucus that covers the epithelium. In addition, major MNPP modifications may also be formed only after uptake into the circulatory system. Future in vivo trials may reveal the presence and relevance of MNPP modifications via the respiratory route.

5. Conclusions

The interaction between lipids and plastic particles was studied using a liposome model, and polystyrene (PS) particles were subjected to DLS, zeta potential measurements, and SPR analysis. The type of lipid and plastic particle surface charge also affected the uptake and toxicity of human A549 lung cells differently. In future studies, polymethyl methacrylate (PMMA), polyethylene terephthalate (PET), and polypropylene (PP) particles in the sub-micron range should be studied, and for PS model particles, both dynamics and lipid oxidation should be investigated via high-resolution mass spectrometry. These studies are needed to prove the dependencies of cell interactions on lipid charge, nanoparticle material (PMMA/PVC), and surface modification.

Author Contributions: Conceptualization, K.W. and S.B.; methodology, A.D.D., W.A.d.S.B. and S.B.; software, A.D.D.; validation, A.D.D. and W.A.d.S.B.; formal analysis, A.D.D. and W.A.d.S.B.; investigation, A.D.D. and W.A.d.S.B.; resources, K.W. and S.B.; data curation, A.D.D., W.A.d.S.B. and S.B.; writing—original draft preparation, A.D.D. and S.B.; writing—review and editing, W.A.d.S.B., M.D. and K.W.; visualization, A.D.D., W.A.d.S.B. and S.B.; supervision, M.D. and K.W.; project administration, K.W.; funding acquisition, M.D., K.W. and S.B. All authors have read and agreed to the published version of the manuscript.

Funding: This work was funded by the Meta-ZIK project PlasMark sponsored by the German Federal Ministry of Education and Research (BMBF), grant number 03Z22D511. The funding source had no role in the design of this study or its execution, analyses, interpretation of the data, or decision to publish results.

Institutional Review Board Statement: Not applicable.

Informed Consent Statement: Not applicable.

Data Availability Statement: The underlying data of this study are available from the corresponding author upon reasonable request.

Acknowledgments: The authors acknowledge the technical support provided by Stephanie Betancourt, Felix Niessner, and Henry Skowski (all INP, Greifswald, Germany). The figure design was supported by a commercial license of biorender.com.

Conflicts of Interest: The authors have no conflict of interest to declare.

Appendix A

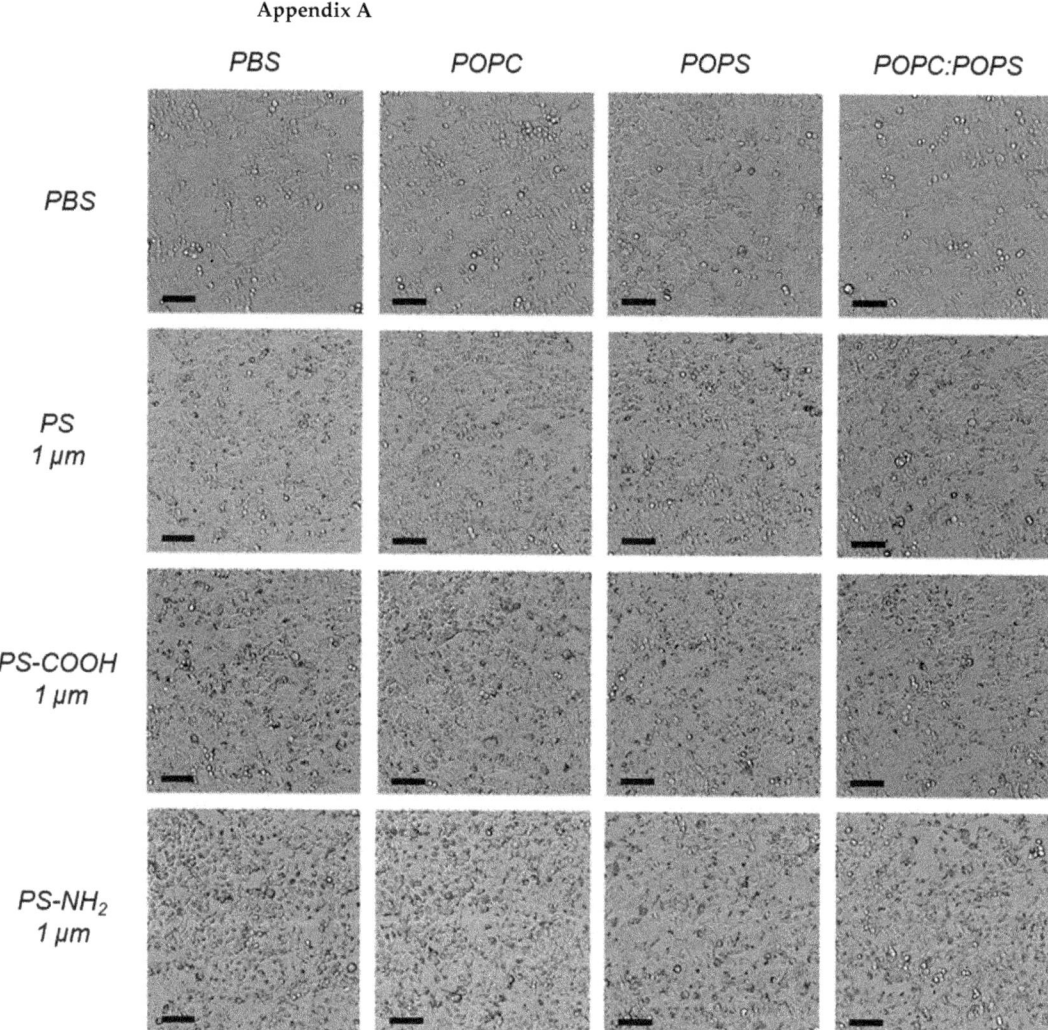

Figure A1. Representative microscopic A549 cell images. Images were taken after the exposure of A549 cells to different types of PS particles pre-incubated with or without different types of liposomes. Especially PS-COOH and PS-NH$_2$ MNPP incubation with A549 cells increased the appearance of intracellular structures with higher density (dark spots) indicative of particle uptake. This phenomenon can be quantified as an increased side-scatter signal in flow cytometry. The overall cellular size and shape were not affected by MNPP incubation.

References

1. Geyer, R.; Jambeck, J.R.; Law, K.L. Production, use, and fate of all plastics ever made. *Sci. Adv.* **2017**, *3*, e1700782. [CrossRef]
2. Ostle, C.; Thompson, R.C.; Broughton, D.; Gregory, L.; Wootton, M.; Johns, D.G. The rise in ocean plastics evidenced from a 60-year time series. *Nat. Commun.* **2019**, *10*, 1622. [CrossRef] [PubMed]
3. Alimi, O.S.; Farner Budarz, J.; Hernandez, L.M.; Tufenkji, N. Microplastics and Nanoplastics in Aquatic Environments: Aggregation, Deposition, and Enhanced Contaminant Transport. *Environ. Sci. Technol.* **2018**, *52*, 1704–1724. [CrossRef] [PubMed]
4. Wanner, P. Plastic in agricultural soils—A global risk for groundwater systems and drinking water supplies?—A review. *Chemosphere* **2021**, *264*, 128453. [CrossRef]

5. Paul, M.B.; Stock, V.; Cara-Carmona, J.; Lisicki, E.; Shopova, S.; Fessard, V.; Braeuning, A.; Sieg, H.; Bohmert, L. Micro- and nanoplastics—Current state of knowledge with the focus on oral uptake and toxicity. *Nanoscale Adv.* **2020**, *2*, 4350–4367. [CrossRef]
6. Napper, I.E.; Davies, B.F.R.; Clifford, H.; Elvin, S.; Koldewey, H.J.; Mayewski, P.A.; Miner, K.R.; Potocki, M.; Elmore, A.C.; Gajurel, A.P.; et al. Reaching New Heights in Plastic Pollution—Preliminary Findings of Microplastics on Mount Everest. *One Earth* **2020**, *3*, 621–630. [CrossRef]
7. Barrett, J.; Chase, Z.; Zhang, J.; Holl, M.M.B.; Willis, K.; Williams, A.; Hardesty, B.D.; Wilcox, C. Microplastic Pollution in Deep-Sea Sediments from the Great Australian Bight. *Front. Mar. Sci.* **2020**, *7*, 576170. [CrossRef]
8. Rhodes, C.J. Solving the plastic problem: From cradle to grave, to reincarnation. *Sci. Prog.* **2019**, *102*, 218–248. [CrossRef]
9. Mamun, A.A.; Prasetya, T.A.E.; Dewi, I.R.; Ahmad, M. Microplastics in human food chains: Food becoming a threat to health safety. *Sci. Total Environ.* **2023**, *858*, 159834. [CrossRef] [PubMed]
10. Du, T.; Yu, X.; Shao, S.; Li, T.; Xu, S.; Wu, L. Aging of Nanoplastics Significantly Affects Protein Corona Composition Thus Enhancing Macrophage Uptake. *Environ. Sci. Technol.* **2023**, *57*, 3206–3217. [CrossRef]
11. Qu, M.; Miao, L.; Chen, H.; Zhang, X.; Wang, Y. SKN-1/Nrf2-dependent regulation of mitochondrial homeostasis modulates transgenerational toxicity induced by nanoplastics with different surface charges in *Caenorhabditis elegans*. *J. Hazard. Mater.* **2023**, *457*, 131840. [CrossRef] [PubMed]
12. Gigault, J.; Halle, A.T.; Baudrimont, M.; Pascal, P.Y.; Gauffre, F.; Phi, T.L.; El Hadri, H.; Grassl, B.; Reynaud, S. Current opinion: What is a nanoplastic? *Environ. Pollut.* **2018**, *235*, 1030–1034. [CrossRef] [PubMed]
13. Hartmann, N.B.; Huffer, T.; Thompson, R.C.; Hassellov, M.; Verschoor, A.; Daugaard, A.E.; Rist, S.; Karlsson, T.; Brennholt, N.; Cole, M.; et al. Are We Speaking the Same Language? Recommendations for a Definition and Categorization Framework for Plastic Debris. *Environ. Sci. Technol.* **2019**, *53*, 1039–1047. [CrossRef]
14. Ducoli, S.; Federici, S.; Nicsanu, R.; Zendrini, A.; Marchesi, C.; Paolini, L.; Radeghieri, A.; Bergese, P.; Depero, L.E. A different protein corona cloaks "true-to-life" nanoplastics with respect to synthetic polystyrene nanobeads. *Environ. Sci.-Nano* **2022**, *9*, 1414–1426. [CrossRef]
15. Kihara, S.; van der Heijden, N.J.; Seal, C.K.; Mata, J.P.; Whitten, A.E.; Koper, I.; McGillivray, D.J. Soft and Hard Interactions between Polystyrene Nanoplastics and Human Serum Albumin Protein Corona. *Bioconjug. Chem.* **2019**, *30*, 1067–1076. [CrossRef]
16. Cao, J.Y.; Yang, Q.; Jiang, J.; Dalu, T.; Kadushkin, A.; Singh, J.; Fakhrullin, R.; Wang, F.J.; Cai, X.M.; Li, R.B. Coronas of micro/nano plastics: A key determinant in their risk assessments. *Part. Fibre Toxicol.* **2022**, *19*, 55. [CrossRef]
17. Maity, A.; De, S.K.; Bagchi, D.; Lee, H.; Chakraborty, A. Mechanistic Pathway of Lipid Phase-Dependent Lipid Corona Formation on Phenylalanine-Functionalized Gold Nanoparticles: A Combined Experimental and Molecular Dynamics Simulation Study. *J. Phys. Chem. B* **2022**, *126*, 2241–2255. [CrossRef]
18. Kurepa, J.; Shull, T.E.; Smalle, J.A. Metabolomic analyses of the bio-corona formed on TiO(2) nanoparticles incubated with plant leaf tissues. *J. Nanobiotechnol.* **2020**, *18*, 28. [CrossRef]
19. Yang, H.; Zhou, M.; Li, H.; Wei, T.; Tang, C.; Zhou, Y.; Long, X. Effects of Low-level Lipid Peroxidation on the Permeability of Nitroaromatic Molecules across a Membrane: A Computational Study. *ACS Omega* **2020**, *5*, 4798–4806. [CrossRef]
20. Gaschler, M.M.; Stockwell, B.R. Lipid peroxidation in cell death. *Biochem. Biophys. Res. Commun.* **2017**, *482*, 419–425. [CrossRef]
21. Sonnino, S.; Chiricozzi, E.; Grassi, S.; Mauri, L.; Prioni, S.; Prinetti, A. Gangliosides in Membrane Organization. *Prog. Mol. Biol. Transl. Sci.* **2018**, *156*, 83–120. [CrossRef]
22. Dickey, A.; Faller, R. Examining the contributions of lipid shape and headgroup charge on bilayer behavior. *Biophys. J.* **2008**, *95*, 2636–2646. [CrossRef] [PubMed]
23. Jurkiewicz, P.; Cwiklik, L.; Vojtiskova, A.; Jungwirth, P.; Hof, M. Structure, dynamics, and hydration of POPC/POPS bilayers suspended in NaCl, KCl, and CsCl solutions. *Biochim. Biophys. Acta* **2012**, *1818*, 609–616. [CrossRef]
24. Geng, Y.; Cao, Y.; Li, Y.; Zhao, Q.; Liu, D.; Fan, G.; Tian, S. A Deeper Insight into the Interfacial Behavior and Structural Properties of Mixed DPPC/POPC Monolayers: Implications for Respiratory Health. *Membranes* **2022**, *13*, 33. [CrossRef]
25. Pan, J.; Cheng, X.; Monticelli, L.; Heberle, F.A.; Kucerka, N.; Tieleman, D.P.; Katsaras, J. The molecular structure of a phosphatidylserine bilayer determined by scattering and molecular dynamics simulations. *Soft Matter* **2014**, *10*, 3716–3725. [CrossRef]
26. Bangham, A.D.; Horne, R.W. Negative Staining of Phospholipids and Their Structural Modification by Surface-Active Agents as Observed in the Electron Microscope. *J. Mol. Biol.* **1964**, *8*, 660–668. [CrossRef]
27. Zhang, H. Thin-Film Hydration Followed by Extrusion Method for Liposome Preparation. In *Liposomes: Methods and Protocols*; D'Souza, G.G.M., Ed.; Springer New York: New York, NY, USA, 2017; pp. 17–22. [CrossRef]
28. Rhodes, C.J. Plastic pollution and potential solutions. *Sci. Prog.* **2018**, *101*, 207–260. [CrossRef] [PubMed]
29. Jasinski, J.; Wilde, M.V.; Voelkl, M.; Jerome, V.; Frohlich, T.; Freitag, R.; Scheibel, T. Tailor-Made Protein Corona Formation on Polystyrene Microparticles and its Effect on Epithelial Cell Uptake. *ACS Appl. Mater. Interfaces* **2022**, *14*, 47277–47287. [CrossRef] [PubMed]
30. Abdelkhaliq, A.; van der Zande, M.; Punt, A.; Helsdingen, R.; Boeren, S.; Vervoort, J.J.M.; Rietjens, I.; Bouwmeester, H. Impact of nanoparticle surface functionalization on the protein corona and cellular adhesion, uptake and transport. *J. Nanobiotechnol.* **2018**, *16*, 70. [CrossRef] [PubMed]
31. Winzen, S.; Schoettler, S.; Baier, G.; Rosenauer, C.; Mailaender, V.; Landfester, K.; Mohr, K. Complementary analysis of the hard and soft protein corona: Sample preparation critically effects corona composition. *Nanoscale* **2015**, *7*, 2992–3001. [CrossRef]

32. Mahmoudi, M.; Landry, M.P.; Moore, A.; Coreas, R. The protein corona from nanomedicine to environmental science. *Nat. Rev. Mater.* **2023**, *8*, 422–438. [CrossRef] [PubMed]
33. Lima, T.; Bernfur, K.; Vilanova, M.; Cedervall, T. Understanding the Lipid and Protein Corona Formation on Different Sized Polymeric Nanoparticles. *Sci. Rep.* **2020**, *10*, 1129. [CrossRef]
34. Xiao, Q.; Zoulikha, M.; Qiu, M.; Teng, C.; Lin, C.; Li, X.; Sallam, M.A.; Xu, Q.; He, W. The effects of protein corona on in vivo fate of nanocarriers. *Adv. Drug Deliv. Rev.* **2022**, *186*, 114356. [CrossRef]
35. Singh, M.; Kumar, V.; Sikka, K.; Thakur, R.; Harioudh, M.K.; Mishra, D.P.; Ghosh, J.K.; Siddiqi, M.I. Computational Design of Biologically Active Anticancer Peptides and Their Interactions with Heterogeneous POPC/POPS Lipid Membranes. *J. Chem. Inf. Model.* **2020**, *60*, 332–341. [CrossRef]
36. Hellstrand, E.; Lynch, I.; Andersson, A.; Drakenberg, T.; Dahlback, B.; Dawson, K.A.; Linse, S.; Cedervall, T. Complete high-density lipoproteins in nanoparticle corona. *FEBS J.* **2009**, *276*, 3372–3381. [CrossRef] [PubMed]
37. Cerqueira, N.M.; Oliveira, E.F.; Gesto, D.S.; Santos-Martins, D.; Moreira, C.; Moorthy, H.N.; Ramos, M.J.; Fernandes, P.A. Cholesterol Biosynthesis: A Mechanistic Overview. *Biochemistry* **2016**, *55*, 5483–5506. [CrossRef]
38. Ohkawa, R.; Low, H.; Mukhamedova, N.; Fu, Y.; Lai, S.J.; Sasaoka, M.; Hara, A.; Yamazaki, A.; Kameda, T.; Horiuchi, Y.; et al. Cholesterol transport between red blood cells and lipoproteins contributes to cholesterol metabolism in blood. *J. Lipid Res.* **2020**, *61*, 1577–1588. [CrossRef]
39. Jung, S.Y.; Lim, S.M.; Albertorio, F.; Kim, G.; Gurau, M.C.; Yang, R.D.; Holden, M.A.; Cremer, P.S. The Vroman effect: A molecular level description of fibrinogen displacement. *J. Am. Chem. Soc.* **2003**, *125*, 12782–12786. [CrossRef] [PubMed]
40. Moghadam, B.Y.; Hou, W.-C.; Corredor, C.; Westerhoff, P.; Posner, J.D. Role of Nanoparticle Surface Functionality in the Disruption of Model Cell Membranes. *Langmuir* **2012**, *28*, 16318–16326. [CrossRef]
41. Saptarshi, S.R.; Duschl, A.; Lopata, A.L. Interaction of nanoparticles with proteins: Relation to bio-reactivity of the nanoparticle. *J. Nanobiotechnol.* **2013**, *11*, 1–12. [CrossRef]
42. Thilak Mudalige, H.Q.; Haute, D.V.; Ansar, S.M.; Paredes, A.; Ingle, T. Characterization of Nanomaterials: Tools and Challenges. *Nanomater. Food Appl.* **2019**, 313–353. [CrossRef]
43. Midekessa, G.; Godakumara, K.; Ord, J.; Viil, J.; Lattekivi, F.; Dissanayake, K.; Kopanchuk, S.; Rinken, A.; Andronowska, A.; Bhattacharjee, S.; et al. Zeta Potential of Extracellular Vesicles: Toward Understanding the Attributes that Determine Colloidal Stability. *ACS Omega* **2020**, *5*, 16701–16710. [CrossRef] [PubMed]
44. Samimi, S.; Maghsoudnia, N.; Eftekhari, R.B.; Dorkoosh, F. Lipid-Based Nanoparticles for Drug Delivery Systems. In *Characterization and Biology of Nanomaterials for Drug Delivery*; Elsevier: Amsterdam, The Netherlands, 2019; pp. 47–76. [CrossRef]
45. Busch, M.; Bredeck, G.; Kampfer, A.A.M.; Schins, R.P.F. Investigations of acute effects of polystyrene and polyvinyl chloride micro- and nanoplastics in an advanced in vitro triple culture model of the healthy and inflamed intestine. *Environ. Res.* **2020**, *193*, 110536. [CrossRef] [PubMed]
46. He, Y.; Li, J.; Chen, J.; Miao, X.; Li, G.; He, Q.; Xu, H.; Li, H.; Wei, Y. Cytotoxic effects of polystyrene nanoplastics with different surface functionalization on human HepG2 cells. *Sci. Total Environ.* **2020**, *723*, 138180. [CrossRef]
47. da Silva Brito, W.A.; Singer, D.; Miebach, L.; Saadati, F.; Wende, K.; Schmidt, A.; Bekeschus, S. Comprehensive in vitro polymer type, concentration, and size correlation analysis to microplastic toxicity and inflammation. *Sci. Total Environ.* **2022**, *854*, 158731. [CrossRef] [PubMed]
48. Gaspar, T.R.; Chi, R.J.; Parrow, M.W.; Ringwood, A.H. Cellular Bioreactivity of Micro- and Nano-Plastic Particles in Oysters. *Front. Mar. Sci.* **2018**, *5*, 345. [CrossRef]
49. Wheeler, K.E.; Chetwynd, A.J.; Fahy, K.M.; Hong, B.S.; Tochihuitl, J.A.; Foster, L.A.; Lynch, I. Environmental dimensions of the protein corona. *Nat. Nanotechnol.* **2021**, *16*, 617–629. [CrossRef]
50. Darbre, P.D. Chemical components of plastics as endocrine disruptors: Overview and commentary. *Birth Defects Res.* **2020**, *112*, 1300–1307. [CrossRef]

Disclaimer/Publisher's Note: The statements, opinions and data contained in all publications are solely those of the individual author(s) and contributor(s) and not of MDPI and/or the editor(s). MDPI and/or the editor(s) disclaim responsibility for any injury to people or property resulting from any ideas, methods, instructions or products referred to in the content.

Article

Suppressing Viscous Fingering in Porous Media with Wetting Gradient

Xiongsheng Wang [1,2], Cuicui Yin [2,*], Juan Wang [2], Kaihong Zheng [2], Zhengrong Zhang [1], Zhuo Tian [2] and Yongnan Xiong [2]

[1] School of Materials and Energy, Guangdong University of Technology, Guangzhou 510006, China
[2] Guangdong Provincial Key Laboratory of Metal Toughening Technology and Application, National Engineering Research Center of Powder Metallurgy of Titanium & Rare Metals, Institute of New Materials, Guangdong Academy of Sciences, Guangzhou 510651, China
* Correspondence: yincuicui@gdinm.com; Tel.: +86-020-87716039

Abstract: The viscous fingering phenomenon often occurs when a low-viscosity fluid displaces a high-viscosity fluid in a homogeneous porous media, which is an undesirable displacement process in many engineering applications. The influence of wetting gradient on this process has been studied over a wide range of capillary numbers (7.5×10^{-6} to 1.8×10^{-4}), viscosity ratios (0.0025 to 0.04), and porosities (0.48 to 0.68), employing the lattice Boltzmann method. Our results demonstrate that the flow front stability can be improved by the gradual increase in wettability of the porous media. When the capillary number is less than 3.5×10^{-5}, the viscous fingering can be successfully suppressed and the transition from unstable to stable displacement can be achieved by the wetting gradient. Moreover, under the conditions of high viscosity ratio ($M > 0.01$) and large porosity ($\Phi > 0.58$), wetting gradient improves the stability of the flow front more significantly.

Keywords: wetting gradient; suppressed viscous fingering; porous media; multiphase flow; immiscible fluid displacement; lattice Boltzmann simulation

Citation: Wang, X.; Yin, C.; Wang, J.; Zheng, K.; Zhang, Z.; Tian, Z.; Xiong, Y. Suppressing Viscous Fingering in Porous Media with Wetting Gradient. *Materials* 2023, *16*, 2601. https://doi.org/10.3390/ma16072601

Academic Editor: Bernard Dominique

Received: 30 January 2023
Revised: 10 March 2023
Accepted: 21 March 2023
Published: 24 March 2023

Copyright: © 2023 by the authors. Licensee MDPI, Basel, Switzerland. This article is an open access article distributed under the terms and conditions of the Creative Commons Attribution (CC BY) license (https:// creativecommons.org/licenses/by/ 4.0/).

1. Introduction

The displacement of immiscible two-phase fluids in porous media is an important subject in natural and engineering fields, including oil exploration [1], carbon dioxide storage [2], fuel cells [3,4], batteries [5], water purification [6], general electrochemical energy storage [7,8] etc. For example, in oil exploration, the preferential flow of displacement fluids will bypass larger areas of oil, trapping some of the oil in the reservoir, thus reducing oil recovery efficiency [9]. In terms of carbon dioxide storage, supercritical carbon dioxide is injected into a deep saline aquifer. The viscous stability of the primary drainage process is of major interest for the injection of carbon dioxide in saline aquifers, since it determines the spread of the carbon dioxide plume in the target aquifer and consequently the initial utilization of the pore space for carbon dioxide storage [10]. It is of great significance for understanding the flow laws of immiscible fluids in porous media.

In order to better study the flow law of fluid in porous media, it is necessary to quantitatively analyze the results. There are several parameters for quantitative analysis, including the displacement efficiency, saturation, interface length and fractal dimension. The displacement efficiency refers to the percentage of the volume of the displaced fluid flowing out and the initial filling volume [11]. The saturation refers to the percentage of the displacement fluid volume in the total pore volume [12]. The saturation can be the percentage at different times, which is different from displacement efficiency. The length of the interface is defined as the length of the interface between the displacing fluid and the displaced fluid. It is often employed to characterize the instability of the fluid front. The longer the interface length, the more unstable the flow along the front of the fluid. The fractal dimension is described as a measure of the space-filling capacity of a pattern, which is used as an indicator to represent the complex geometric form, to compare the changes of details in the pattern, and to reflect the effectiveness of complex shapes occupying space. It is a measure of the irregularity of complex shapes. The more complex the fractal dimension is, the more complex the shape is [13].

The flow of fluids in porous media can be divided into three types: stable displacement, capillary fingering and viscous fingering [13]. There exist two dimensionless parameters by which to characterize the flow of fluids in porous media, including the viscosity ratio $M = \mu_1/\mu_2$ and the capillary number $Ca = \mu_1 U/\sigma$; here, μ_1 is the viscosity of the displacing fluid. μ_2 is the viscosity of the displaced fluid, U is the rate at which the displaced fluid is injected, and σ is the surface tension. The phase diagrams of M and Ca have been summarized [14], which is of great significance for further understanding the flow laws of fluids in porous media. Zhang [15] and Zheng [16] have also summarized the relevant phase diagrams when they study the flow of fluids in porous media. The phase diagrams summarized by these researchers are not identical in the boundary ranges of different flow patterns, which may be due to the wettability, gravity and surface roughness [17].

Zhao et al. [18] studied the influence of wetting angle between fluid and porous media, and found that the influence of wettability is not monotonic. Jung et al. [19] studied the influence of wettability by combining experiments and simulations. Their research shows that wettability has a great impact on the displacement process of immiscible fluid. When the wetting angle between the displacement fluid and the porous medium is less than 80°, the saturation of the final displacement fluid is higher because the interface of the fluid front moves smoothly and the displacement fluid is not trapped. Even when studying the influence of wettability, the law of influence on wettability is not completely consistent. On the one hand, the pore channels in porous media are uneven, especially in the displacement problem in porous media in the engineering field, where the pores are very irregular. Even in the displacement problem in homogeneous porous media, the nonuniformity of the flow front interface will also have a great impact [20]. Golmohammadi et al. [21] used experimental methods to study the comprehensive effect of gravity and wettability on fluid flow in two-dimensional porous media, and found that gravity plays an important role in improving the filling effect. Hu et al. [22] studied the influence of roughness on the flow pattern in rock fractures and quantized the energy dissipated in the process of fluid invasion dominated by capillary force. Moreover, scholars have also made some progress in improving the stability of the fluid front flow in porous media. Rabbani et al. [11] proposed a method of suppressing viscous fingering by designing the gradual and monotonic variation of pore sizes along the front path. Lu et al. [23] investigated the influence of pore size gradient and pore-scale disorder on the displacement process when a non-wetting fluid displaces a wetting fluid, and found that a sufficiently large gradient can completely suppress capillary fingering.

With the development of science and technology, the investigation of immiscible fluid flow in porous media is not limited to experiments, and many scholars have also tried to use numerical simulation methods to carry out corresponding research. The lattice Boltzmann method (LBM) is a mesoscopic numerical simulation method between macro and micro. Compared with macro methods, it is more convenient for dealing with a complex boundary and, at the same time, it overcomes the limitations of the micro calculation method on size. It has the advantages of the automatic capture of a two-phase interface without manual processing or flexible boundary processing, and is suitable for parallel computing [24]. The lattice Boltzmann method has been widely used for simulating multiphase flow [25], micro-nano scale flow [26], turbulence [27], flow-induced vibrations [28], heat and mass transfer [29], and porous media flow problems [30,31]. Liu et al. [30] simulated the immiscible flow of wetting fluid and non-wetting fluid in two porous medias. Their simulation results confirmed that three different displacement modes are related to capillary number, viscosity ratio and the heterogeneity of porous media. Shi et al. [31] investigated the basic physical mechanism of Newtonian fluid replacing non-Newtonian fluid in porous media, which revealed the displacement mechanism of Newtonian fluid to non-Newtonian fluid from a mesoscopic perspective. Lautenschlaeger et al. [32] proposed a homogenization method for simulating multiphase flow in heterogeneous porous media, which is based on the lattice Boltzmann method and combines the gray level with the multi-component Shan-Chen method. This method makes the fluid–fluid and solid–fluid interactions in pores less than numerical discretization, and has been successfully applied to solving various single-phase and two-phase flow problems.

Unstable flows will create adverse effects for oil production, carbon dioxide storage and other aspects. In order to reveal the law of the unstable flow of fluid in porous media, some scholars have obtained the relevant phase diagrams through a large number of experiments. Some scholars have tried to improve the unstable flow of fluid by designing the structure of porous media. However, how a wetting gradient impacts viscous fingering remains unknown. In this paper, we used the lattice Boltzmann method to simulate the displacement process of fluids in porous media with a wetting gradient. The influence of the wetting gradient was studied over a wide range of capillary numbers (7.5×10^{-6} to 1.8×10^{-4}), viscosity ratios (0.0025 to 0.04) and porosity (0.48 to 0.68). The flow results were quantitatively analyzed by means of quantitative parameters such as fractal dimension, displacement efficiency, saturation and interface length. Moreover, a phase diagram of flow stability results related to wetting gradient and capillary number was drawn.

2. Numerical Method

2.1. Mathematical Model

The lattice Boltzmann method is used for two-dimensional numerical simulation to simulate the displacement process of fluid in porous media. The lattice Boltzmann method is a mesoscopic simulation method between micro-molecular dynamics and macro-fluid dynamics. This method uses the Boltzmann transport equation to calculate two processes of collision and migration between micro-particles, replacing macro-particles with micro-particles. The macroscopic parameters of the system can be obtained by statistical averaging of a large number of particles without concern for the motion state of each particle. The motion behavior of the whole fluid is simulated and the corresponding macroscopic phenomena are analyzed. The Boltzmann transport equation is:

$$f_i(\mathbf{r} + \mathbf{e_i}, t + dt) = f_i(\mathbf{r}, t) + \Omega_i(f_0, \ldots, f_b), i = 0, \ldots, b. \tag{1}$$

$f_i(\mathbf{r}, t)$ is the distribution equation of particles in the direction i of position $\mathbf{r}$ at time t. $\Omega_i(f_0, \ldots, f_b)$ is collision operator.

Macro density ρ and velocity v on the node can be obtained by integration:

$$\rho = \sum_{i=0}^{b} f_i \tag{2}$$

$$\rho v = \sum_{i=0}^{b} f_i \mathbf{e}_i. \tag{3}$$

The discrete velocities model adopted was the D2Q9 model, and its discrete velocities are shown in Figure 1.

The D2Q9 model illustrated in Figure 1 involves nine velocity vectors defined by:

$$\vec{e}_i = \begin{cases} (0,0) & i = 0 \\ (1,0), (0,1), (-1,0), (0,-1) & i = 1,2,3,4 \\ (1,1), (-1,1), (-1,-1), (1,-1) & i = 5,6,7,8 \end{cases} \tag{4}$$

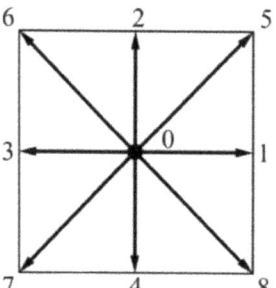

Figure 1. D2Q9 lattice model of velocities discretization.

The evolution equation of the particle distribution function of the multiple-relaxation time model is:

$$f_i(\mathbf{r}+\mathbf{e_i},t+dt) = f_i(\mathbf{r},t) - M_{ij}^{-1} S_{ij}\left(\mu_i^{eq} - \mu_i\right), \quad (5)$$

where $\mu_i = M_{ij} f_j$, f_j is the distribution equation of the j direction and $\mu_i^{eq} = M_{ij} f_j^{eq}$, f_j^{eq} is the equilibrium distribution function in the j direction.

$$\mu = (\mu_0, \mu_1, \ldots, \mu_8)^T = \left(\rho, e, \epsilon, j_x, q_x, j_y, q_y, p_{xx}, p_{yy}\right)^T, \quad (6)$$

where ρ is the density, e is the kinetic energy, ϵ is related to the kinetic energy square, j_x and j_y correspond to components of momentum, q_x and q_y correspond to the energy components, and p_{xx} and p_{yy} correspond to the symmetric traceless viscous stress tensors.

$$\begin{aligned}
\rho|_i &= |e_i|^0 = 1 \\
e|_i &= -4|e_i|^0 + 3\left(e_{i,x}^2 + e_{i,y}^2\right) \\
\epsilon|_i &= 4|e_i|^0 - 21/2\left(e_{i,x}^2 + e_{i,y}^2\right) + 9/2\left(e_{i,x}^2 + e_{i,y}^2\right)^2 \\
j_x|_i &= e_{i,x} \\
q_x|_i &= e_{i,x}\left[-5|e_i|^0 + 3\left(e_{i,x}^2 + e_{i,y}^2\right)\right] \\
j_y|_i &= e_{i,y} \\
q_y|_i &= e_{i,y}\left[-5|e_i|^0 + 3\left(e_{i,x}^2 + e_{i,y}^2\right)\right] \\
p_{xx}|_i &= \left(e_{i,x}^2 - e_{i,y}^2\right)\rho \\
p_{xy}|_i &= e_{i,x} e_{i,y}/\rho
\end{aligned} \quad (7)$$

Hence, the transformation matrix M is

$$M = \begin{pmatrix} \rho|_i \\ e|_i \\ \epsilon|_i \\ j_x|_i \\ q_x|_i \\ j_y|_i \\ q_y|_i \\ p_{xx}|_i \\ p_{xy}|_i \end{pmatrix} = \begin{pmatrix} 1 & 1 & 1 & 1 & 1 & 1 & 1 & 1 & 1 \\ -4 & -1 & -1 & -1 & -1 & 2 & 2 & 2 & 2 \\ 4 & -2 & -2 & -2 & -2 & 1 & 1 & 1 & 1 \\ 0 & 1 & 0 & -1 & 0 & 1 & -1 & -1 & 1 \\ 0 & -2 & 0 & 2 & 0 & 1 & -1 & -1 & 1 \\ 0 & 0 & 1 & 0 & -1 & 1 & 1 & -1 & -1 \\ 0 & 0 & -2 & 0 & 2 & 1 & 1 & -1 & -1 \\ 0 & 1 & -1 & 1 & -1 & 0 & 0 & 0 & 0 \\ 0 & 0 & 0 & 0 & 0 & 1 & -1 & 1 & -1 \end{pmatrix}. \quad (8)$$

$S_{ij} = \text{diag}(s_0, s_1, s_2, s_3, s_4, s_5, s_6, s_7, s_8)$ is the diagonal relaxation matrix. In this approach, the kinematic viscosity v and the bulk viscosity μ are related to the following relaxation parameters:

$$v = c_s^2 \left(\frac{1}{s_7} - \frac{1}{2} \right) = c_s^2 \left(\frac{1}{s_8} - \frac{1}{2} \right) \tag{9}$$

$$\mu = \frac{5 - 9c_s^2}{9} \left(\frac{1}{s_1} - \frac{1}{2} \right). \tag{10}$$

$c_s = c/\sqrt{3}$, c_s is the sound velocity of the lattice; $c = \delta x / \delta t$, c is the lattice velocity; δx is the length step; and δt is the time step.

Applying the transformation matrix to the equilibrium probability distribution function, the raw moments at the equilibrium are

$$\begin{aligned} e^{eq} &= -2\rho + 3\left(j_x^2 + j_y^2\right) \\ \epsilon^{eq} &= \rho - 3\left(j_x^2 + j_y^2\right) \\ q_x^{eq} &= -j_x \\ q_y^{eq} &= -j_y \\ p_{xx}^{eq} &= j_x^2 - j_y^2 \\ p_{xy}^{eq} &= j_x j_y \end{aligned} \tag{11}$$

The moments ρ^{eq}, j_x^{eq} and j_y^{eq} are not required as they will be multiplied by s_0, s_3, and s_5, which are zero.

The bounce-back boundary condition is used to implement the no-slip condition on a geometry wall. The incoming probability distribution functions at the wall node are reflected back to the initial fluid node, giving value to the unknown probability distribution functions. In the illustration in Figure 2, at a time t, the distribution function f_7 is reflected back to f_5, f_3 to f_1, and f_6 provides f_8 value.

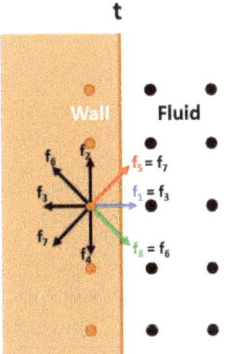

Figure 2. Bounce-back boundary condition.

The inlet and outlet boundary conditions used in the current simulations are velocity and pressure boundary conditions, respectively. The approach consists of the formulation of a linear system with mass and momentum conservation. This linear system will provide the value of the unknown probability distribution functions created after the streaming step as well as the value of ρ if a velocity boundary condition is imposed, or u if there is a pressure condition at the boundary. According to the Figure 2 example, after rearranging the moments in Equations (2) and (3) (density, x-momentum, y-momentum), the linear system can be written as:

$$\begin{aligned} f_1 + f_5 + f_8 &= \rho - (f_0 + f_2 + f_4 + f_3 + f_6 + f_7) \\ f_1 + f_5 + f_8 &= \rho u + f_3 + f_6 + f_7 \\ f_5 - f_8 &= \rho v + f_2 + f_4 - f_6 + f_7 \end{aligned} \quad (12)$$

A fourth equation is required to close the system and compute the value of velocity or pressure. A valid assumption is to apply the bounce-back rule for the nonequilibrium part of the particle distribution normal to the boundary. Therefore, considering that an inlet boundary is imposed on the left nodes of Figure 2, the fourth equation is

$$f_1 - f_1^{eq} = f_3 - f_3^{eq}. \quad (13)$$

Solving the linear system, the unknown probability distribution functions are

$$\begin{aligned} f_1 &= \tfrac{2}{3}\rho u \\ f_5 &= f_7 - \tfrac{1}{2}(f_2 - f_4) + \tfrac{1}{6}\rho u + \tfrac{1}{2}\rho v \\ f_8 &= f_6 - \tfrac{1}{2}(f_2 - f_4) + \tfrac{1}{6}\rho u - \tfrac{1}{2}\rho v \end{aligned} \quad (14)$$

For a velocity boundary condition, the linear system gives:

$$\rho = \frac{1}{1-v}[f_0 + f_2 + f_4 + 2(f_3 + f_6 + f_7)]. \quad (15)$$

For a pressure boundary condition, $p = \rho c_s^2$; taking $v = 0$, the u is defined by:

$$u = 1 - \frac{[f_0 + f_2 + f_4 + 2(f_3 + f_6 + f_7)]}{\rho}. \quad (16)$$

2.2. Simulation Setup

We used the computational fluid dynamics software XFlow to carry out the relevant numerical simulations. The phase field algorithm was employed for calculating the multi-phase flow. The temperature type was set as isothermal. The computational domain and boundary conditions are shown in Figure 3. The size of the computational domain was 16 mm × 10 mm. The cylinders represent an impermeable solid material, while the area formed between the cylinders are the pore channels. The diameter of the cylinder and the spacing between cylinders are described in Figure 3A. The displaced fluid initially filled in the pores of the porous media, and the displaced fluid flowed in at a certain speed from the bottom, forcing the displaced fluid to flow out at the top. The initial gauge pressure field was 0 Pa (the corresponding actual pressure was 1.013 × 10^5 Pa) and the initial velocity field was 0 m/s. The bounce-back boundary condition was used to implement the no-slip condition on the solid walls. The inlet and outlet boundary conditions used in the current numerical simulations were velocity and pressure boundary conditions. The displaced fluid was silicon oil (the wetting phase) and the displacing fluid was water (the non-wetting phase). The parameters of these two fluids were as follows: the density and viscosity of displacing fluid were set as ρ_1 = 998.3 kg/m^3, μ_1 = 1 mPa s. The density and viscosity of the displaced fluid were set to ρ_2 = 960 kg/m^3, μ_2 = 200 mPa s, and the viscosity ratio was M = 0.005. The surface tension σ = 28.2 mN/m. The inlet velocity was 0.001 m/s unless otherwise specified. The wetting gradient was set as follows: Take the wetting gradient $\Delta\theta$ = 4° as an example, as shown in Figure 3B. The wetting angle of the lowest row of cylinders was set to 90°, and the same row of cylinders had the same wetting angle. The wetting angle was increased by 4° for each row up to the last row, thus creating the porous media of wetting gradient $\Delta\theta$ = 4°. The wetting gradients $\Delta\theta$ = 1°, $\Delta\theta$ = 2° and $\Delta\theta$ = 3° were arranged in the same way. The wetting gradient $\Delta\theta$ = 0° meant that the overall wettability of the porous media was uniform and the overall wetting angle was 150°.

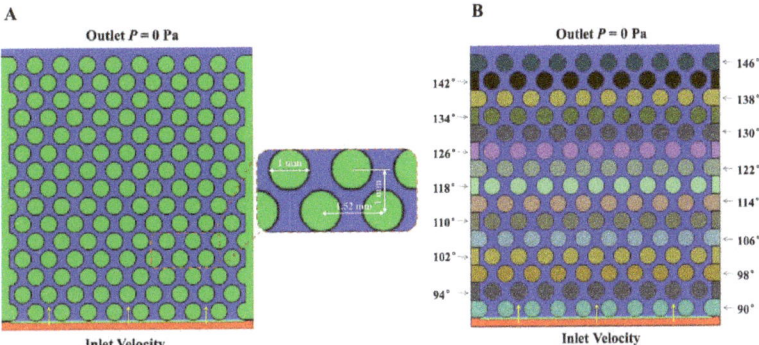

Figure 3. (**A**) Initial boundary conditions without wetting gradient. (**B**) Initial boundary conditions at the wetting gradient $\Delta\theta = 4°$. The red area is the displacing fluid, which is injected at a certain flow rate from below. The blue area is the displaced fluid, flowing out from the upper outlet. The outlet gauge pressure is 0 Pa. The horizontal line between the red area and the blue area represents the interface between the displacing fluid and the displaced fluid. The yellow arrow represents the flow direction. The green cylinder represents an impermeable solid material, while the area formed between the cylinder and the cylinder is a pore channel, and the fluid flows in the pore channel between the cylinders.

2.3. Model Verification

We first performed mesh independence verification and time step independence verification. The time steps selected for grid independence verification were all 1×10^{-6} s; the grid size selected for time step independence verification was 6×10^{-5} m. The total calculation time is the time required for the displacement fluid to flow to the porous media outlet. Figure 4 exhibits the time evolution of the saturation of displacing fluid and the interface length between the displacing fluid and the displaced fluid under different mesh sizes. The results show that the saturation and interface length tend to stabilize when the grid size is 6×10^{-5} m. Figure 5 presents the time evolution of the saturation of the displacing fluid and the interface length between the displacing fluid and the displaced fluid under different time steps. The results show that the saturation and interface length tend to be stable when the time step is 1×10^{-6} s. Thus, a grid size of 6×10^{-5} m and a time step of 1×10^{-6} s were employed in the subsequent numerical simulations.

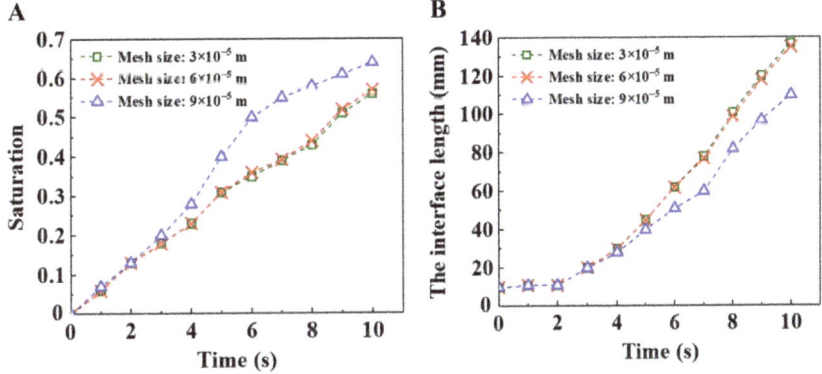

Figure 4. (**A**) Time evolution of the saturation of displacing fluid under different mesh sizes. (**B**) Time evolution of the interface length between the displacing fluid and the displaced fluid under different mesh sizes.

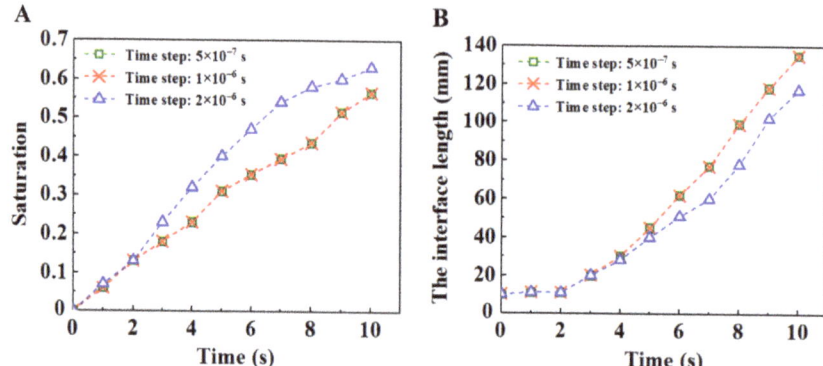

Figure 5. (**A**) Time evolution of the saturation of displacing fluid under different time steps. (**B**) Time evolution of the interface length between the displacing fluid and the displaced fluid under different time steps.

We selected the experimental results of Rabbani et al. [11] for experimental verification. Their experiment involves filling silicone oil into a porous medium arranged by cylinders. Then, water is injected into the porous medium to see how water displaces silicone oil in the porous medium. Among them, two different porous structures are used. One is the porous medium with uniform pores. The other is porous media with a pore gradient whose size increases from inlet to outlet. We selected four groups of experimental results to verify the current numerical model, including the fluid displacement processes in the porous mediums with and without pore gradient under the capillary number of 7.5×10^{-6} and 1.4×10^{-5}. In order to save computing time and resources, we simulated a part of the experimental model. The fluid parameters and the amplitude of the pore gradient decline are consistent with the parameters in the experiment. The comparisons between simulation results and experimental results are shown in Figures 6 and 7. The uniform medium shown in Figure 6 was composed of cylinders with the same diameter of 1 mm. The nonuniform medium in Figure 6 consisted of cylinders with the diameter decreasing row by row from the top to the bottom. The diameter difference between the adjacent row of cylinders was 0.0135 mm and the diameters of the cylinder in the top and bottom rows were 1 mm and 0.81 mm, respectively.

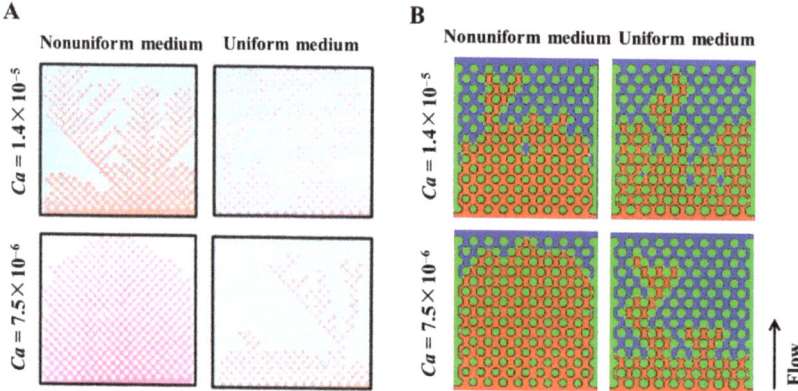

Figure 6. Results of the displacing fluid front morphologies at the time of the displacing fluid reaching the outlet under the condition of nonuniform medium and uniform medium with the capillary numbers of 7.5×10^{-6} and 1.4×10^{-5}. (**A**) Experimental results of Rabbani et al. [11]. (**B**) Simulation results of the current investigation.

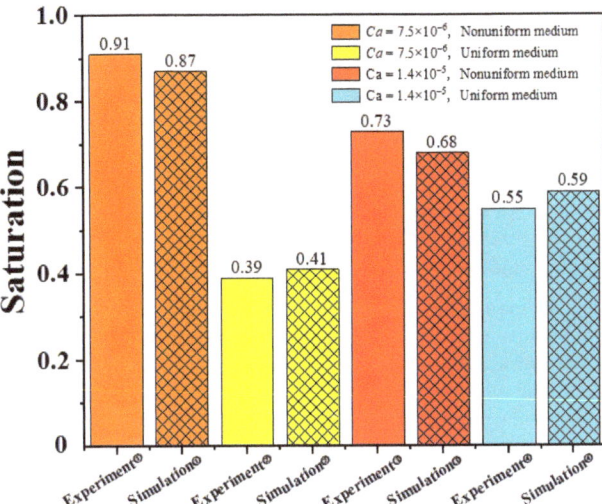

Figure 7. Comparison of the saturation between the simulation results and the experimental results under the condition of nonuniform medium and uniform medium with the capillary numbers of 7.5×10^{-6} and 1.4×10^{-5}.

It can be seen from the comparison between the simulation results and the experimental results [8] that, when the capillary number is 7.5×10^{-6} in the nonuniform medium (the left figure of A and B in Figure 6), it can be simulated that the flow of displacing fluid in porous media is stable. In the other three cases, it can also be simulated that the displacing fluid has an unstable flow. In order to compare the simulation results with the experimental results more clearly, we also drew a diagram comparing the saturation under various conditions, as shown in Figure 6. It can be seen from the figure that the simulation results are similar to the experimental results, which shows that the lattice Boltzmann method can simulate fluid flow in porous media.

3. Results and Discussion

3.1. Effect of Wettability

When the non-wetting fluid displaces the wetting fluid in a porous medium, it will flow forward only when the driving pressure of the displacement flow is greater than the capillary pressure threshold $\sim \sigma \cos \theta / r$, where r is the pore radius. While the capillary pressure threshold can be adjusted by regulating the wettability to further regulate the fluid displacement process. It can be seen from the above formula that, with the increase of wetting angle, the capillary pressure threshold increases and then the flow resistance increases. Based on this, a porous medium with the wettability gradients will hopefully be designed to improve the flow stability, as shown in Figure 8. When the fluid flows from a row of cylinders with a wetting angle of θ to the next row of cylinders with a larger wetting angle of θ', the flow resistance increases and the fluid will preferentially flow transversely and then forward.

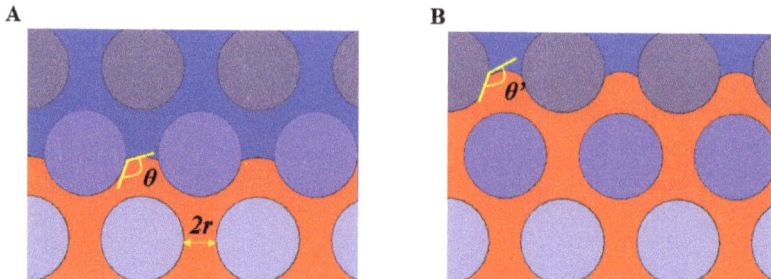

Figure 8. Schematic diagram of wetting angle change of a fluid in a porous medium with a wetting gradient. (**A**) The displacing fluid flows through a row of cylinders with a wetting angle of θ. (**B**) The displacing fluid flows through a row of cylinders with a wetting angle of θ' ($\theta' > \theta$). The displacing fluid is injected from the bottom upwards. The red and blue areas represent the displacing fluid and the displaced fluid, respectively. Cylinders with different colors represent the media with different wetting angles.

In order to verify the influence of wettability on the fluid displacement process in the porous medium, the variation curves of average flow rate and outflow time at different wetting angles were investigated, as shown in Figure 9. With the increase in the wetting angle, the average flow rate of the displacement fluid decreases and the time required for the displacement fluid to flow out of the porous media increases. Thus, the flow resistance of the displacement fluid in porous media increases with the increment in wetting angle, which is consistent with the above theoretical analysis results.

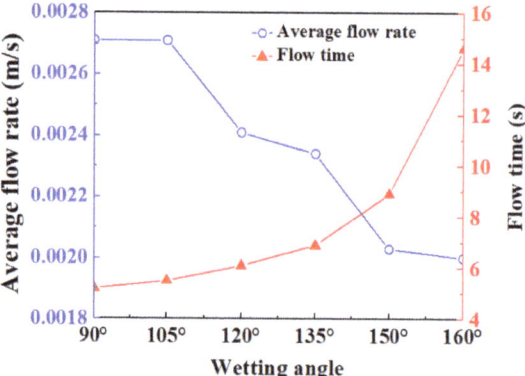

Figure 9. Variation curves of average flow rate and outflow time at different wetting angles.

3.2. Comparison between the Porous Media with and without Wetting Gradient

3.2.1. Effect of Capillary Number

Figure 10 shows the simulation results for different capillary numbers and wetting gradients. It can be seen that the flow pattern of the displacing fluid is unstable with or without wetting gradient when the capillary number is large. In the case of wetting gradient, the stability of the displacing fluid flow front is improved when the capillary number decreases. Without wetting gradient, the fluid still has more bifurcation. When the capillary number decreases to 7.5×10^{-6}, the result with the wetting gradient is the best. The resistance gradient due to the wetting gradient can be fully displayed. In the absence of a wetting gradient, the flow of the displacing fluid is unstable even when the capillary number is reduced. It can be clearly seen from the simulation results in Figure 9 that the wetting gradient plays a significant role in improving the stability of the flow front.

Furthermore, the smaller the capillary number (the lower the injection speed), the better the fluid filling, and the more stable the fluid front is.

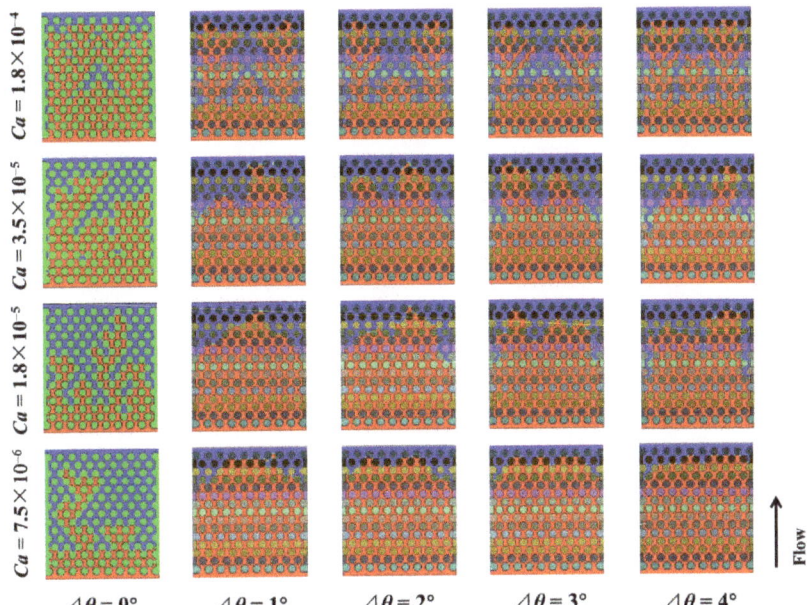

Figure 10. Simulation results of the displacing fluid front morphologies under different capillary numbers and wetting gradients at the time of the displacing fluid reaching the outlet. The viscosity ratio is 0.005.

To quantitatively analyze the flow results shown in Figure 11, three parameters were selected for the quantitative analysis. The three quantitative parameters were the fractal dimension (Figure 11A), the displacement efficiency (Figure 11B), and the interface length (Figure 11C). It can be seen from the results of Figure 11A that, on the whole, the fractal dimension with the wetting gradient is larger than that without the wetting gradient. This indicates the effect of the wetting gradient on improving the flow stability of displacing fluids; with the wetting gradient, the capillary number is smaller and the fractal dimension is larger. The smaller the capillary number, the better the stability of the fluid front. This also indicates that both the wetting gradient and the capillary number affect the stability of the flow front. From the comparison of the displacement efficiency results of Figure 11B, it can be seen that, on the whole, the displacement efficiency with the wetting gradient is higher than that without the wetting gradient, which indicates that the existence of the wetting gradient improves the filling degree of the fluid. As the capillary number decreases, the displacement efficiency increases, and the filling degree improves. From the comparison of the results in Figure 11C, it can be seen that the interface length of porous media with the wetting gradient is much shorter than that without the wetting gradient, which also indicates that the bending degree of the fluid front interface is reduced. In addition, the smaller the capillary number, the shorter the interfacial length of the fluid front, the smaller the bending degree of the fluid front interface and the better the stability of the fluid front.

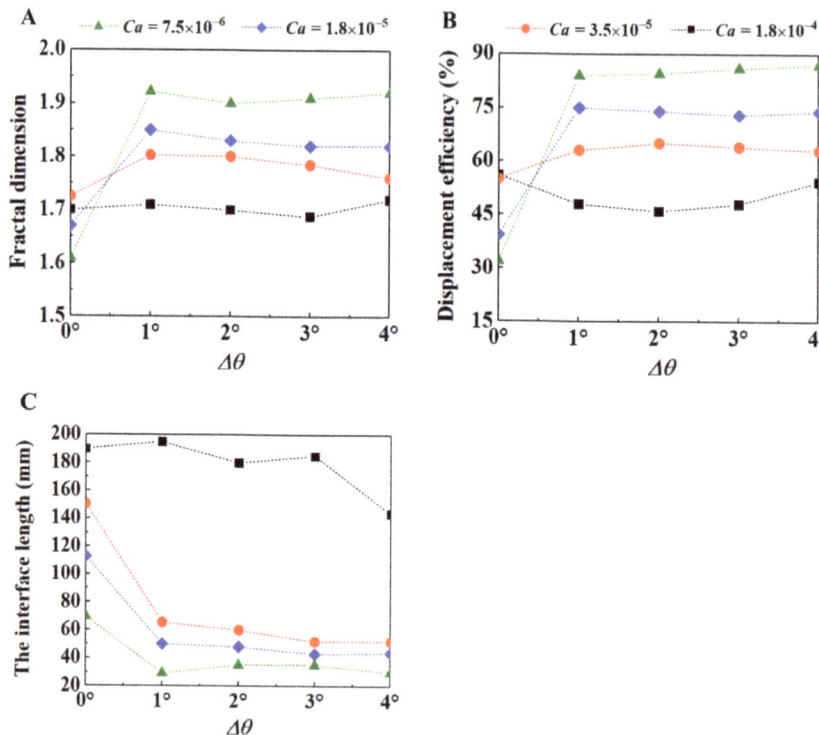

Figure 11. Effect of wetting gradient on the fractal dimension (**A**), the displacement efficiency (**B**), and the interface length (**C**) under different capillary numbers. The viscosity ratio is 0.005.

From the results of these quantitative analyses, when the capillary number increases to 1.8×10^{-4} under the condition of the wetting gradient, the overall flow stability is reduced and the filling effect is poor. This is because when the capillary number is 1.8×10^{-4}, the corresponding inlet flow rate is relatively large, and the displacing fluid is subjected to a large force in the vertical direction, which exceeds the threshold range of the resistance gradient. The flow front of the displacing fluid is prone to local preferential flow and, when there is a wetting gradient, due to the effect of the resistance gradient, part of the displacement fluid flows laterally, making the bifurcation phenomenon of the displacement fluid front more serious. However, in the case of the wetting gradient, when the capillary number is less than 3.5×10^{-5}, the overall filling effect is better and the stability of the flow is improved with the decrease of the capillary number. This is because the force of the displacement fluid in the vertical direction is less than the threshold range of the resistance gradient, which means the resistance gradient can inhibit the occurrence of local preferential flow.

We also plotted the phase diagram of the wetting gradient and capillary number, as shown in Figure 12. The identification of flow patterns in porous media can be considered comprehensively by way of parameters such as saturation, fractal dimension and finger width [33]. Therefore, we used this method to distinguish whether the flow of fluid in porous media is stable or unstable. It can be seen from the phase diagram that, in the case of the wetting gradient, the displacement process of fluid in porous media is mostly stable except for in the case of a large capillary number. The results of the fluid displacement process are unstable without the wetting gradient, which indicates that the wetting gradient plays a significant role in improving the flow stability of fluids in porous media.

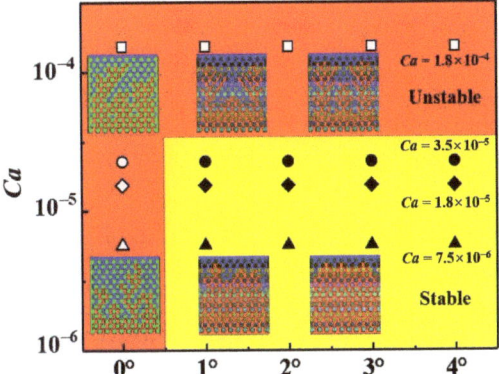

Figure 12. The phase diagram of the flow stability results related to wetting gradient and capillary number. The red area with solid symbols represents the unstable flow, and the yellow area with hollow symbols represents the steady flow. The viscosity ratio is 0.005.

In order to better show the effect of wettability gradient on fluid flow in porous media, we selected a group of typical simulation results for comparison at different times. Figure 13 shows the comparison of saturation and interface length at different times with and without wetting gradient when the capillary number was 1.8×10^{-5}. The time is normalized by the time required for the displacement fluid to flow to the porous media outlet, to characterize the different stages of the entire displacement process. The slope of the curve in Figure 13A shows that the filling speed of the fluid is faster when there is a wetting gradient, and the saturation result at any normalization time shows that the filling effect is better when there is a wetting gradient than when there is no wetting gradient. Figure 13B shows that the slope increases significantly in the case of no wetting gradient after the normalization time is 0.2, which indicates that there is unstable flow at the front of the fluid at this moment, especially the bifurcated finger-like flow, which makes the interface length of the front of the fluid significantly longer. In the case of the wetting gradient, the slope of the interface length is relatively flat, which means that the flow is relatively stable and there is no bifurcation phenomenon. So, the change of the interface length at the front of the fluid is not very large, which also means that porous media with a wetting gradient can better improve the stability of the flow front.

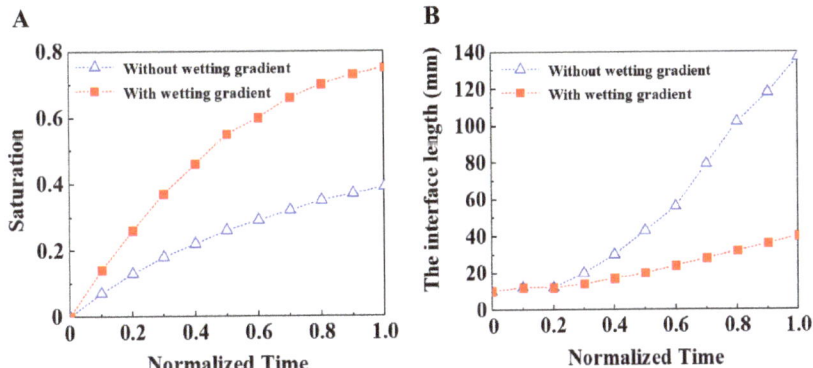

Figure 13. (**A**) Time evolution of the saturation of displacing fluid. (**B**) Time evolution of the interface length between the displacing fluid and the displaced fluid. In both cases, the capillary number is 1.8×10^{-5} and the viscosity ratio is 0.005. The solid and hollow symbols indicate the porous media with and without wetting gradient, respectively.

3.2.2. Effect of Viscosity Ratio

In order to clarify the influence of wetting gradient on the fluid displacement process with a wider range of parameters, we compared and studied the fluid displacement process of different fluid viscosities with and without a wetting gradient. We chose the case with a capillary number of 3.5×10^{-5} for our research. This is because under this capillary number, although there is a wetting gradient that plays a certain role in the stable flow of the fluid, there are still bifurcation phenomena at the front of the fluid, which affect the filling effect. Therefore, we tried to study how to improve the flow stability of the front of the fluid by changing the viscosity ratio. Figure 14A shows the comparison results of saturation with or without wetting gradient when the viscosity ratio is 0.0025, 0.005, 0.01, 0.02 and 0.04, respectively. We kept the viscosity of the displaced fluid unchanged at 1 mPa s, and changed the viscosity of the displaced fluid. The viscosities of the displaced fluid were set at 400 mPa s, 200 mPa s, 100 mPa s, 50 mPa s and 25 mPa s. From the comparison results in the figure, it can be seen that the change of viscosity ratio does have a great impact on the flow filling effect. On the whole, the saturation value under the condition of the wetting gradient is higher than that under the condition of no wetting gradient, which indicates that the filling effect under the condition of the wetting gradient is better. In addition, under the condition of the wetting gradient, the greater the viscosity ratio, the better the stability of the fluid front. Figure 14B shows the comparison results of interface length under different viscosity ratios. On the whole, the interface length under the condition of the wetting gradient is smaller than that under the condition of no wetting gradient, indicating that the flow stability under the condition of the wetting gradient is higher. The reason for this result is that when the viscosity ratio increases, the viscosity of the displaced fluid decreases, which makes the flow resistance of the injected fluid relatively lower; for the second half of the fluid flowing into the porous medium especially, the flow resistance decreases to a reasonable size, so that the effect of the resistance gradient can be reflected—the effect of the resistance gradient is to prevent the partial preferential breakthrough when the fluid flows in the porous medium. Therefore, due to the effect of the resistance gradient, the bifurcation phenomenon is reduced. In the case of the wetting gradient, the effect of the viscosity ratio on the wetting gradient is more obvious, so the flow of fluid in the porous media is more stable. In the case of no wetting gradient, the flow resistance of the corresponding injected fluid decreases due to the increase of viscosity ratio. At this time, there is no effect of resistance gradient, and the flow front is more prone to local breakthrough, leading to local preferential flow and affecting the stability of the displacement process.

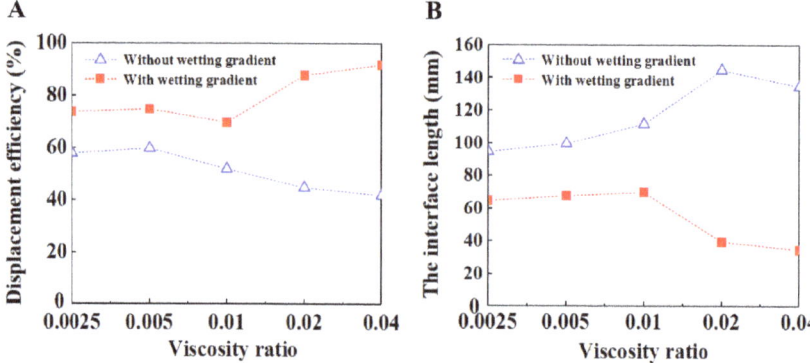

Figure 14. The effect of viscosity ratio on the displacement efficiency (**A**) and the interface length (**B**). In both cases, the capillary number is 3.5×10^{-5} and the porosity is 0.48. The solid and hollow symbols indicate the porous media with and without wetting gradient, respectively.

We selected two groups with a viscosity ratio M of 0.01 and 0.02 for comparative analysis at different times, as shown in Figure 15. On the whole, the saturation with a wetting gradient is higher than that without a wetting gradient, indicating that the filling effect is better with a wetting gradient. In terms of interface length, the interface length with the wetting gradient is smaller than that without the wetting gradient, which indicates that the stability of the fluid front is better with a wetting gradient; after the normalization time is 0.2, the interface length without the wetting gradient increases faster, which indicates that the unstable flow phenomenon begins at this time, and the finger-like flow becomes more and more obvious in the later stage, making the interface length increase faster. At the same time, the saturation of the fluid increases slowly. On the whole, the filling condition with the wetting gradient is better than that without a wetting gradient, and the filling effect with a high viscosity ratio is better than that with a low viscosity ratio.

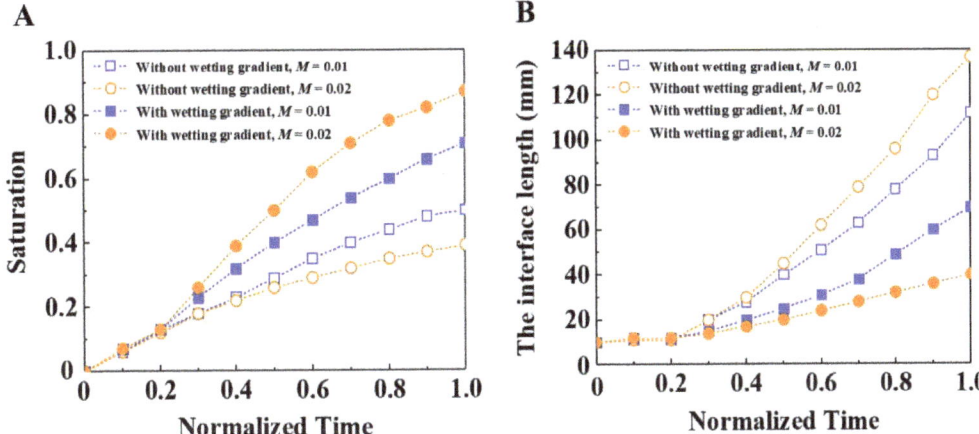

Figure 15. (A) Time evolution comparison of the saturation of displacing fluid under different viscosity ratios. (B) Time evolution comparison of the interface length between the displacing fluid and the displaced fluid under different viscosity ratios. In both cases, the capillary number is 3.5×10^{-5} and the porosity is 0.48. The solid and hollow symbols indicate the porous media with and without wetting gradient, respectively.

3.2.3. Effect of Porosity

The size of the porosity also has a great impact on the flow of fluid in porous media., The size of porosity especially affects the function of the wetting gradient. Figure 16 shows the saturation and interface length of different porosities at the normalized time. It can be seen that, in the case of a wetting gradient, when the porosity is large, the slope of the fluid saturation is large, indicating that the flow front is relatively stable. This is because in the case of the wetting gradient, with the increase of porosity, the upper and lower spacings between cylinders become wider, and the contact area between fluid and porous media decreases. This can reduce the resistance gradient to an appropriate size, and then make the resistance gradient play a maximum role, which inhibits the occurrence of local breakthrough. On the whole, the filling condition with the wetting gradient is better than that without a wetting gradient, and the larger the porosity, the better the filling effect.

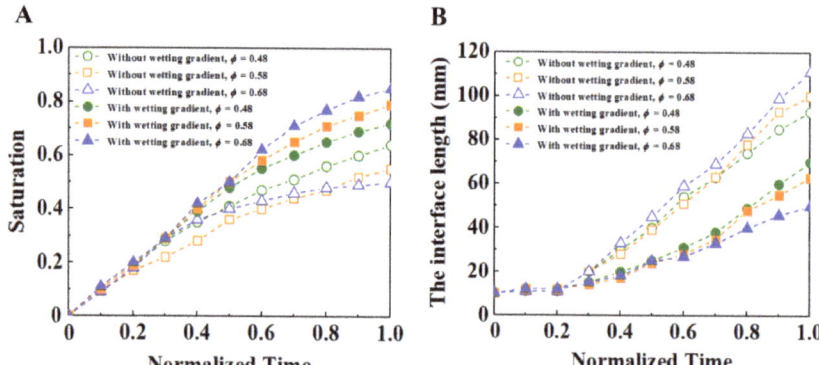

Figure 16. (**A**) Time evolution comparison of the saturation of displacing fluid under different porosities. (**B**) Time evolution comparison of the interface length between the displacing fluid and the displaced fluid under different porosities. In both cases, the capillary number is 3.5×10^{-5} and the viscosity ratio is 0.02. The solid and hollow symbols indicate the porous media with and without wetting gradient, respectively.

3.2.4. Effect of Nonuniformity

We randomly adjusted the diameter of some cylinders to obtain the nonuniform medium. In order to maintain the same porosity as the uniform medium, we adopted the method of expanding and shrinking the same number of cylinder diameters, respectively, to make the porosity of the nonuniform medium consistent with that of the uniform medium, as shown in Figure 17. In addition, we chose to compare the uniform media with a capillary number of 3.5×10^{-5}, a viscosity ratio of 0.02 and a porosity of 0.48. From the comparison of saturation and interface length in Figure 18, it can be seen that under the condition of the wetting gradient, the flow stability of the displacement fluid in the nonuniform medium is indeed not as good as that in the uniform medium. This is because the nonuniformity of pores in the inhomogeneous medium causes the displacement fluid to flow preferentially in local areas, which makes the flow stability of the displacement fluid worse. In addition, under the condition of no wetting gradient, because there is no resistance gradient, this local preferential flow phenomenon is more likely to occur, resulting in the flow stability of the displacement fluid in the nonuniform medium being worse. However, from the comparison diagram, even in the case of nonuniform media, the filling effect of the displacement fluid with the wetting gradient is still better than that without a wetting gradient, which also shows the role of the wetting gradient in improving the stability of the displacement process.

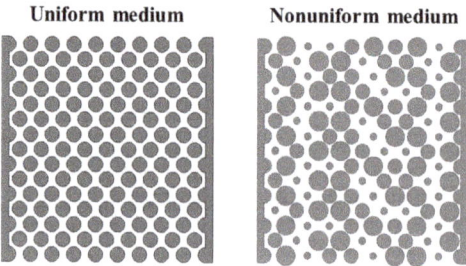

Figure 17. Schematic diagram of uniform and nonuniform medium, with the same porosity of 0.48. In order to maintain the same porosity as the uniform medium, the same number of cylinder diameters have been expanded and shrunk.

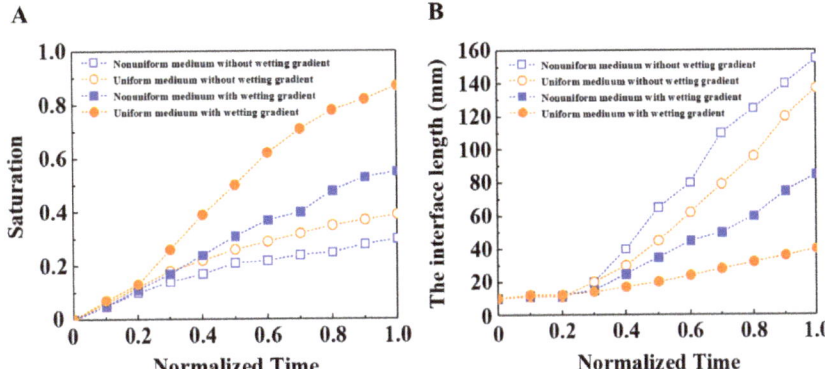

Figure 18. (**A**) Time evolution comparison of the saturation of displacing fluid in uniform medium and nonuniform medium. (**B**) Time evolution comparison of the interface length between the displacing fluid and the displaced fluid in uniform medium and nonuniform medium. In both cases, the capillary number is 3.5×10^{-5}, the viscosity ratio is 0.02 and the porosity is 0.48. The solid and hollow symbols indicate the porous media with and without wetting gradient, respectively.

4. Conclusions

In the present study, the displacement of immiscible fluid in porous media was numerically investigated using the lattice Boltzmann method. Our results demonstrate that the displacement pattern and efficiency can be controlled by the wetting gradient. The flow front stability can be improved by setting a wetting gradient on the porous media, which is confirmed over a wide parameter range of different wetting gradients, capillary numbers, viscosity ratios and porosities. When the capillary number is less than 3.5×10^{-5}, the viscous fingering can be successfully suppressed and the transition from unstable to stable displacement can be achieved by the wetting gradient. While increasing the wetting gradient has little effect on the fluid displacement process, under the conditions of high viscosity ratio ($M > 0.01$) and large porosity ($\Phi > 0.58$), the improvement of flow front stability is more significant when adding a wetting gradient. In addition, for the porous media with a wetting gradient, decreasing the capillary number, increasing the viscosity ratio or increasing porosity can improve flow front stability and filling efficiency, which is consistent with that of homogeneous wetting media. The present findings could be helpful for the design of porous products to suppress viscous fingering, which will be of great significance for industrial applications such as composite material preparation and oil exploration.

Author Contributions: Conceptualization, C.Y., X.W. and J.W.; methodology, C.Y. and X.W.; software, X.W. and C.Y.; validation, X.W. and C.Y.; formal analysis, X.W. and C.Y.; investigation, X.W. and C.Y.; resources, J.W., K.Z. and C.Y.; data curation, J.W., K.Z. and C.Y.; writing—original draft preparation, X.W., C.Y., Z.T. and Y.X.; writing—review and editing, X.W., C.Y., Z.T. and Y.X.; visualization, X.W. and C.Y.; supervision, X.W. and C.Y.; project administration, C.Y., K.Z., Z.Z. and J.W.; funding acquisition, C.Y., K.Z., Z.Z. and J.W. All authors have read and agreed to the published version of the manuscript.

Funding: This research was funded by the National Natural Science Foundation of China (No. 52006041), the Guangdong Academy of Sciences (Nos. 2021GDASYL-20210103101, 2020GDASYL-20200101001, 2022GDASZH-2022010109), the National Key Research and Development Program of China (No. 2021YFB3701204), and the Guangdong Major Project of Basic and Applied Basic Re-search (No. 2020B0301030006).

Institutional Review Board Statement: Not applicable.

Informed Consent Statement: Not applicable.

Data Availability Statement: Not applicable.

Conflicts of Interest: The authors declare no competing financial interest.

References

1. Simjoo, M.; Dong, Y.; Andrianov, A.; Talanana, M.; Zitha, P.L.J. CT Scan Study of Immiscible Foam Flow in Porous Media for Enhancing Oil Recovery. *Ind. Eng. Chem. Res.* **2013**, *52*, 6221–6233. [CrossRef]
2. Kim, Y.; Wan, J.; Kneafsey, T.J.; Tokunaga, T.K. Dewetting of Silica Surfaces upon Reactions with Supercritical CO2 and Brine: Pore-Scale Studies in Micromodels. *Environ. Sci. Technol.* **2012**, *46*, 4228–4235. [CrossRef] [PubMed]
3. Chen, L.; Kang, Q.; Tao, W. Pore-scale numerical study of multiphase reactive transport processes in cathode catalyst layers of proton exchange membrane fuel cells. *Int. J. Hydrogen Energy* **2021**, *46*, 13283–13297. [CrossRef]
4. Zhang, D.; Cai, Q.; Gu, S. Three-dimensional lattice-Boltzmann model for liquid water transport and oxygen diffusion in cathode of polymer electrolyte membrane fuel cell with electrochemical reaction. *Electrochim. Acta* **2018**, *262*, 282–296. [CrossRef]
5. Sauter, C.; Zahn, R.; Wood, V. Understanding Electrolyte Infilling of Lithium Ion Batteries. *J. Electrochem. Soc.* **2020**, *167*, 10. [CrossRef]
6. Kota, A.K.; Kwon, G.; Choi, W.; Mabry, J.M. Hygro-responsive membranes for effective oil-water separation. *Nat. Commun.* **2012**, *3*, 1025. [CrossRef] [PubMed]
7. Paliwal, S.; Panda, D.; Bhaskaran, S.; Vorhauer-Huget, N.; Tsotsas, E.; Surasani, V.K. Lattice Boltzmann method to study the water-oxygen distributions in porous transport layer (ptl) of polymer electrolyte membrane (pem) electrolyser. *Int. J. Hydrogen Energy* **2021**, *46*, 22747–22762. [CrossRef]
8. Lautenschlaeger, M.P.; Prifling, B.; Kellers, B.; Weinmiller, J.; Danner, T.; Schmidt, V. Understanding electrolyte filling of lithium-ion battery electrodes on the pore scale using the lattice Boltzmann method. *Batter. Supercaps* **2022**, *5*, e202200090. [CrossRef]
9. Lake, L.W. *Enhanced Oil Recovery*; Prentice Hall: Hoboken, NJ, USA, 2010.
10. Berg, S.; Ott, H. Stability of CO_2-brine immiscible displacement. *Int. J. Greenh. Gas Control* **2013**, *11*, 188–203. [CrossRef]
11. Rabbani, H.S.; Or, D.; Liu, Y.; Lai, C.Y. Suppressing viscous fingering in structured porous media. *Proc. Natl. Acad. Sci. USA* **2018**, *115*, 4833–4838. [CrossRef]
12. Tsuji, T.; Jiang, F.; Christensen, K.T. Characterization of immiscible fluid displacement processes with various capillary numbers and viscosity ratios in 3D natural sandstone. *Adv. Water Resour.* **2016**, *95*, 3–15. [CrossRef]
13. Sun, B.L. Fractal dimension and its measurement method. *J. Northeast For. Univ.* **2004**, *032*, 116–119.
14. Lenormand, R.; Touboul, E.; Zarcone, C. Numerical models and experiments on immiscible displacements in porous media. *J. Fluid Mech.* **1988**, *189*, 165–187. [CrossRef]
15. Zhang, C.Y.; Mart, O.; Wietsma, T.M.; Grate, J.W.; Warner, M.G. Influence of Viscous and Capillary Forces on Immiscible Fluid Displacement: Pore-Scale Experimental Study in a Water-Wet Micromodel Demonstrating Viscous and Capillary Fingering. *Energy Fuels* **2011**, *25*, 3493–3505. [CrossRef]
16. Zheng, X.L.; Mahabadi, N.; Yun, T.S.; Jang, J. Effect of capillary and viscous force on CO_2 saturation and invasion pattern in the microfluidic chip. *J. Geophys. Res. Solid Earth* **2017**, *122*, 1634–1647. [CrossRef]
17. Singh, K.; Jung, M.; Brinkmann, M.; Seemann, R. Capillary-Dominated Fluid Displacement in Porous Media. *Annu. Rev. Fluid Mech.* **2019**, *51*, 429–449. [CrossRef]
18. Zhao, B.; MacMinn, C.W.; Juanes, R. Wettability control on multiphase flow in patterned microfluidics. *Proc. Natl. Acad. Sci. USA* **2016**, *113*, 10251–10256. [CrossRef]
19. Jung, M.; Brinkmann, M.; Seemann, R.; Hiller, T.; Herminghaus, S. Wettability controls slow immiscible displacement through local interfacial instabilities. *Phys. Rev. Fluids* **2016**, *1*, 7. [CrossRef]
20. Lei, W.H.; Lu, X.K.; Liu, F.L.; Wang, M. Non-monotonic wettability effects on displacement in heterogeneous porous media. *J. Fluid Mech.* **2022**, *942*, R5. [CrossRef]
21. Golmohammadi, S.; Ding, Y.; Schlueter, S.; Kuechler, M.; Reuter, D. Impact of Wettability and Gravity on Fluid Displacement and Trapping in Representative 2D Micromodels of Porous Media (2D Sand Analogs). *Water Resour. Res.* **2021**, *57*, 10. [CrossRef]
22. Hu, R.; Zhou, C.X.; Wu, D.S.; Yang, Z.B. Roughness Control on Multiphase Flow in Rock Fractures. *Geophys. Res. Lett.* **2019**, *46*, 12002–12011. [CrossRef]
23. Lu, N.B.; Browne, C.A.; Amchin, D.B. Controlling capillary fingering using pore size gradients in disordered media. *Phys. Rev. Fluids* **2019**, *4*, 78. [CrossRef]
24. Aidun, C.K.; Clausen, J.R. Lattice-Boltzmann Method for Complex Flows. *Annu. Rev. Fluid Mech.* **2010**, *42*, 439–472. [CrossRef]
25. Yin, C.C.; Wang, T.Y.; Che, Z.Z.; Jia, M.; Sun, K. Critical and Optimal Wall Conditions for Coalescence-Induced Droplet Jumping on Textured Superhydrophobic Surfaces. *Langmuir* **2019**, *35*, 16201–16209. [CrossRef] [PubMed]
26. Raabe, D. Overview of the lattice Boltzmann method for nano-and microscale fluid dynamics in materials science and engineering. *Model. Simul. Mater. Sci. Eng.* **2004**, *12*, 11–15. [CrossRef]
27. Chen, H.D.; Kandasamy, S.; Orszag, S. Extended Boltzmann Kinetic Equation for Turbulent Flows. *Science* **2003**, *307*, 633–636. [CrossRef]
28. Afra, B.; Delouei, A.A.; Mostafavi, M.; Tarokh, A. Fluid-structure interaction for the flexible filament's propulsion hanging in the free stream. *J. Mol. Liq.* **2021**, *323*, 114941. [CrossRef]

29. Jalali, A.; Amiri Delouei, A.; Khorashadizadeh, M.; Golmohammadi, A.M.; Karimnejad, S. Mesoscopic Simulation of Forced Convective Heat Transfer of Carreau-Yasuda Fluid Flow over an Inclined Square: Temperature-dependent Viscosity. *J. Appl. Comput. Mech.* **2020**, *6*, 307–319. [CrossRef]
30. Liu, H.; Zhang, Y.; Valocchi, A.J. Lattice Boltzmann simulation of immiscible fluid displacement in porous media: Homogeneous versus heterogeneous pore network. *Phys. Fluids* **2015**, *27*, 121776–121840. [CrossRef]
31. Shi, Y.; Tang, G.H. Non-Newtonian rheology property for two-phase flow on fingering phenomenon in porous media using the lattice Boltzmann method. *J. Non-Newton. Fluid Mech.* **2016**, *229*, 86–95. [CrossRef]
32. Lautenschlaeger, M.P.; Weinmiller, J.; Kellers, B.; Danner, T.; Latz, A. Homogenized lattice Boltzmann model for simulating multi-phase flows in heterogeneous porous media. *Adv. Water Resour.* **2022**, *170*, 104320. [CrossRef]
33. Hu, Y.X.; Patmonoaji, A.; Zhang, C.W. Experimental study on the displacement patterns and the phase diagram of immiscible fluid displacement in three-dimensional porous media. *Adv. Water Resour.* **2020**, *140*, 103584. [CrossRef]

Disclaimer/Publisher's Note: The statements, opinions and data contained in all publications are solely those of the individual author(s) and contributor(s) and not of MDPI and/or the editor(s). MDPI and/or the editor(s) disclaim responsibility for any injury to people or property resulting from any ideas, methods, instructions or products referred to in the content.

Article

Turbulent CFD Simulation of Two Rotor-Stator Agitators for High Homogeneity and Liquid Level Stability in Stirred Tank

Cuicui Yin [1], Kaihong Zheng [1], Jiazhen He [1,*], Yongnan Xiong [1], Zhuo Tian [1], Yingfei Lin [1] and Danfeng Long [2]

1 Guangdong Provincial Key Laboratory of Metal Toughening Technology and Application, Institute of New Materials, Guangdong Academy of Sciences, National Engineering Research Center of Powder Metallurgy of Titanium and Rare Metals, Guangzhou 510651, China
2 Institute of Intelligent Manufacturing, Guangdong Academy of Sciences, Guangzhou 510651, China
* Correspondence: hejiazhen@gdinm.com; Tel.: +86-020-61086608

Citation: Yin, C.; Zheng, K.; He, J.; Xiong, Y.; Tian, Z.; Lin, Y.; Long, D. Turbulent CFD Simulation of Two Rotor-Stator Agitators for High Homogeneity and Liquid Level Stability in Stirred Tank. *Materials* 2022, 15, 8563. https://doi.org/10.3390/ma15238563

Academic Editor: Gee-Soo Lee

Received: 24 October 2022
Accepted: 25 November 2022
Published: 1 December 2022

Publisher's Note: MDPI stays neutral with regard to jurisdictional claims in published maps and institutional affiliations.

Copyright: © 2022 by the authors. Licensee MDPI, Basel, Switzerland. This article is an open access article distributed under the terms and conditions of the Creative Commons Attribution (CC BY) license (https://creativecommons.org/licenses/by/4.0/).

Abstract: Good solid-liquid mixing homogeneity and liquid level stability are necessary conditions for the preparation of high-quality composite materials. In this study, two rotor-stator agitators were utilized, including the cross-structure rotor-stator (CSRS) agitator and the half-cross structure rotor-stator (HCSRS) agitator. The performances of the two types of rotor-stator agitators and the conventional A200 (an axial-flow agitator) and Rushton (a radial-flow agitator) in the solid-liquid mixing operations were compared through CFD modeling, including the homogeneity, power consumption and liquid level stability. The Eulerian–Eulerian multi-fluid model coupling with the RNG k–ε turbulence model were used to simulate the granular flow and the turbulence effects. When the optimum solid-liquid mixing homogeneity was achieved in both conventional agitators, further increasing stirring speed would worsen the homogeneity significantly, while the two rotor-stator agitators still achieving good mixing homogeneity at the stirring speed of 600 rpm. The CSRS agitator attained the minimum standard deviation of particle concentration σ of 0.15, which was 42% smaller than that achieved by the A200 agitators. Moreover, the average liquid level velocity corresponding to the minimum σ obtained by the CSRS agitator was 0.31 m/s, which was less than half of those of the other three mixers.

Keywords: solid-liquid mixing; rotor-stator agitator; suspension quality; liquid level stability; power consumption; CFD

1. Introduction

The solid-liquid mixing operation is an important process in the research of particle reinforced metal matrix composites prepared by stirring casting [1]. Due to the obvious density difference between the reinforcing phase and the matrix melt, the problems of particle sinking and agglomeration are serious. In addition, the stability of the liquid surface will also affect the quality of the composite. When the vortex of the liquid surface is large, gas and inclusions will be introduced to pollute the melt. Therefore, good solid-liquid mixing homogeneity and liquid level stability are necessary conditions for the preparation of high-quality composite materials. The design of the agitator needs to meet the following requirements: (1) To provide an intensive melt shearing for the dispersion of agglomerated particles; (2) To generate a relatively homogeneous macro flow in the stirred tank for the homogeneous distribution of particles; (3) To avoid a large liquid level velocity to ensure that gas and other contaminants do not enter the melt from the liquid surface.

The particle suspension is the result of the balance between the driving forces generated by agitator rotation and the particle gravity. In particular, the driving forces of particle suspension are the drag force imposed by the moving fluid and the lifting force generated by the turbulent eddies bursting [2]. The suspension quality of solid particles in the stirred tank is controlled by the process parameters, including particle size/density/loading capacity, melt density/viscosity/liquid level height, stirring speed, size and structure of the agitator, etc. Among them, the optimal design of the agitator determines the flow pattern and turbulence intensity in the stirred tank, which is a significant way to improve the suspension quality of solid particles and to reduce the power consumption [3]. Good uniformity of particle concentration distribution is able to improve product quality and to enhance heat transfer [4,5].

According to the flow patterns, the agitators are mainly classified as radial-flow and axial-flow agitators. Most previous studies have shown that axial-flow agitators are more suitable for solid-liquid mixing than radial-flow agitators [6–8]. Among four axial-flow impellers of Lightnin A100, A200, A310 and A320, the A320 was the most effective impeller [9]. The four-bladed 45° PBT impeller was the most energy-efficient comparing with the Lightnin A310 and PF3 impeller [10]. Downward-flow PBT agitators were more efficient than upward-flow PBT agitators [6]. Zhao et al. showed that the blade shape had a great effect on the trailing vortex characteristics and the large curvature led to the longer residence time of the vortex at the impeller tip [11]. In addition to the studies on the performances of agitators with traditional structures, it is believed that many scholars have recently proposed agitators with new structures. The power consumption for the turbine agitator with V cuts has been found to be less than that of the conventional turbine agitator [12]. Mishra et al. demonstrated that the Maxblend impeller attained a higher maximum homogeneity compared to the A200 and the Rushton impellers [13,14]. The punched rigid-flexible impeller was more efficient in suspending solid particles compared with the rigid impeller and rigid-flexible impeller at the same power consumption [15]. The fractal impeller also has reduced power consumption compared to the regular impeller due to the breaking up of the trailing vortices [16,17]. The impeller with zigzag punched blades was developed to enhance the mixing of non-Newtonian fluids by producing impinging jet streams from face-to-face holes [18]. The punched-bionic impellers were proposed to improve the energy efficiency and homogeneity in solid-liquid mixing processes [19].

Rotor-stator agitators are widely used in dispersion, emulsification and homogenization processes because of the high shear rate created by the small rotor-stator gap. The break-up and dispersion of nanoparticle clusters [20–23], droplet break-up mechanisms [24], droplets size distribution [25], the scale up of the equilibrium drop size [26], the emulsification of a high viscosity oil in water [27] and the dispersion of water into oil [28,29] in rotor–stator mixers were investigated thoroughly by scholars. In contrast, the research on the application of rotor–stator agitators in solid-liquid mixing process is rare. Moreover, most of the current optimum design studies on agitators are aimed at improving the efficiency and homogeneity of solid-liquid mixing, but little attention is paid to the stability of liquid level.

The computational fluid dynamics (CFD) has been widely employed to investigate the solid-liquid mixing process [30–38]. There are two main methods for solving the solid-liquid multiphase flow in stirred tank based on Navier-Stokes equations, i.e., Eulerian–Lagrangian (E–L) and Eulerian–Eulerian (E–E) methods [39]. Among them, the E–L method tracks the motion of each particle, which has a large computational resource consumption [40]. The E–E method treats the solid particle mathematically as a continuous phase considering the interpenetration and interaction of the solid and liquid phases [41]. This method has been widely employed by scholars due to the relatively low requirements of computing resources and it was validated by comparing the computational results to the experimental results [10,30,42,43]. Turbulence effects are important for solid-liquid mixing processes and need to be considered in the creating of mathematical models. The Reynolds–averaged Navier–Stokes (RANS) approach has been widely used in large industrial processes due to the good economy and calculation accuracy. The Reynolds stress model establishes six equations for Reynolds stress tensor and one equation for dissipation rate, which can predict all Reynolds stresses correctly, while the equations are difficult to converge. The standard k–ε turbulence model estimates the turbulent viscosity by solving the turbulent kinetic energy (k) and the turbulent kinetic energy dissipation rate (ε). However, there exists a large error in the calculation of non-uniform turbulence problems. The RNG

k–ε turbulence model adopts the framework of two equations, which are derived from the original governing equations of momentum transfer by using the renormalization group method. It is more accurate in predicting the rapidly strained flows by adding an additional term in its ε equation comparing with the standard k–ε turbulence model [30]. Siddiqui et al. [44] also demonstrated the RNG k–ε model can fairly predict the velocity contours and streamlines. Based on the E–E model along with the RNG k–ε turbulence model, scholars have conducted multiple verification work by comparing the simulation data with the experimental results and a large amount of work on multiphase mixing operations ([30,43]).

Most of the current optimum design studies on agitators are aimed at improving the efficiency and homogeneity of solid-liquid mixing, but little attention is paid to the stability of liquid level. In this paper, two types of rotor-stator agitators were utilized to improve the homogeneity and liquid level stability at the same time for the preparation of high-quality composite materials. The flow field, turbulent energy dissipation, pressure field and particle concentration distribution were predicted using the 3D E–E multiphase fluid model along with RNG k–ε turbulence model. And the solid-liquid mixing homogeneity, liquid level stability and power consumption of the two types of rotor-stator agitators were evaluated by comparing with those of A200 and Rushton turbine agitators.

2. Numerical Modeling

In this investigation, all the numerical simulations are carried out in a stirred tank employing four types of agitators, which including the cross structure rotor-stator (CSRS) agitator, the half cross structure rotor-stator (HCSRS) agitator, the A200 and Rushton turbine, the details of the geometries and dimensions of the stirred tank and agitators are depicted in Figure 1 and Table 1. The specific operating conditions and physical parameters are presented in Table 2. The liquid density and viscosity corresponds to that of AZ91 magnesium alloy at 650 °C, in addition, the particle density corresponds to that of pure titanium.

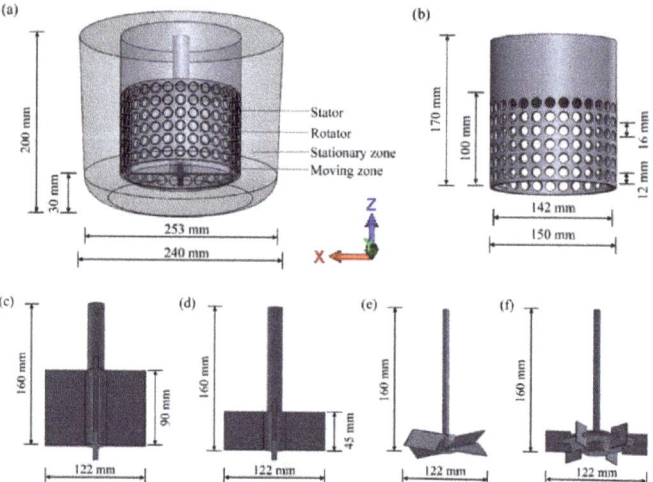

Figure 1. Geometries and dimensions of the stirred tank and agitators; (**a**) computational domain of the stirred tank; (**b**) the stator of the rotor-stator agitators; (**c**) the rotor of the CSRS agitator; (**d**) the rotor of the HCSRS agitator; (**e**) A200 agitator; (**f**) Rushton turbine agitator.

Table 1. Geometric parameters of the stirred tank and agitators.

Geometric Parameter (mm)	Values
Diameter of top of stirring tank	253
Diameter below stirring tank	240
Stirring tank height	200
Height of hole area	100
Total height of stator	170
Hole center spacing of stator	16
Hole diameter of stator	12
Inner diameter of stator	142
Outer diameter of stator	150

Table 2. Operating conditions and physical parameters.

Variable	Values
Agitator type	CSRS, HCSRS, A200, Rushton
Stirring speed (rpm)	200, 300, 400, 500, 600
Particle size (μm)	100
Solids weight fraction (wt.%)	10
Grain density (kg/m^3)	4506
Liquid density (kg/m^3)	1650
Liquid viscosity (Pa*s)	0.00139

2.1. The Governing Equations

The 3D E–E multiphase model is used for solving the continuity and momentum equations for the liquid and solid phases [45].

The continuity equation is:

$$\frac{\partial(\alpha_q\rho_q)}{\partial t} + \nabla \cdot (\alpha_q\rho_q \vec{u}_q) = 0 \tag{1}$$

where α_q is the volume fraction of phase q, ρ_q is the density, t is the time and u_q is the velocity of phase q.

The momentum equation for phase q is:

$$\frac{\partial(\alpha_q\rho_q \vec{u}_q)}{\partial t} + \nabla \cdot (\alpha_q\rho_q \vec{u}_q \vec{u}_q) = -\alpha_q\nabla p + \nabla\overline{\overline{\tau}}_q + \alpha_q\rho_q \vec{g} + \sum_{p=1}^{n}(\vec{R}_{pq} + \dot{m}_{pq}\vec{u}_{pq} - \dot{m}_{qp}\vec{u}_{qp}) \tag{2}$$

where p is the pressure, g is the gravitational acceleration and R_{pq} is the interaction force between phase p and phase q, m_{pq} represents the mass transfer from phase p to phase q, m_{qp} represents the mass transfer from phase q to phase p, $\overline{\overline{\tau}}_q$ is the stress-strain tensor defined as:

$$\overline{\overline{\tau}}_q = \alpha_q\mu_q(\nabla \vec{u}_q + \nabla \vec{u}_q^T) + \alpha_q(\lambda_q - \frac{2}{3}\mu_q)\nabla \vec{u}_q I \tag{3}$$

here, μ_q and λ_q are the shear and bulk viscosity of phase q, respectively.

The interaction force between phase p and phase q is computed as follows:

$$\sum_{p=1}^{n}\vec{R}_{pq} = \sum_{p=1}^{n}k_{pq}(\vec{u}_p - \vec{u}_q) \tag{4}$$

where k_{pq} is the momentum exchange coefficient between phase p and phase q. The fluid-solid momentum exchange is mainly controlled by drag, lift, virtual mass and Basset forces. Studies have shown that the lift, virtual mass and Basset forces are irrespective for modeling the solid holdup profiles [46]. Hence, only the drag force is considered in the current calculation. The k_{pq} is obtained by the Gidaspow model [41]:

$$k_{pq} = \begin{cases} \dfrac{3}{4} C_D \dfrac{\alpha_p \alpha_q \rho_p \left|\vec{u}_p - \vec{u}_q\right|}{d_p} \alpha_q^{-2.65} & (\alpha_q > 0.8) \\ \dfrac{150 \alpha_p (1-\alpha_q) \mu_q}{\alpha_q d_p^2} + \dfrac{1.75 \alpha_p \rho_q \left|\vec{u}_p - \vec{u}_q\right|}{d_p} & (\alpha_q \leq 0.8) \end{cases} \quad (5)$$

where d_p is the particle diameter and C_D is the drag coefficient, which can be written as:

$$C_D = \dfrac{24\left[1 + 0.15(\alpha_q Re)^{0.687}\right]}{\alpha_q Re} \quad (6)$$

where Re is the relative Reynolds number which can be computed via:

$$Re = \dfrac{\rho_q d_p \left|\vec{u}_p - \vec{u}_q\right|}{\mu_q} \quad (7)$$

The RNG k–ε turbulence model [47] is employed for simulating the turbulence effect. The turbulent kinetic energy (k) and the turbulent kinetic energy dissipation rate (ε) are expressed as follows:

$$\dfrac{\partial}{\partial t}(\rho k) + \dfrac{\partial}{\partial x_i}(\rho k u_i) = \dfrac{\partial}{\partial x_j}\left(\alpha_k \mu_t \dfrac{\partial k}{\partial x_j}\right) - \overline{\rho u'_i u'_j} \dfrac{\partial u_j}{\partial x_i} - \rho \varepsilon - 2\rho \varepsilon \dfrac{k}{a^2}; \quad (8)$$

$$\dfrac{\partial}{\partial t}(\rho \varepsilon) + \dfrac{\partial}{\partial x_i}(\rho \varepsilon u_i) = \dfrac{\partial}{\partial x_j}\left(\alpha_\varepsilon \mu_t \dfrac{\partial \varepsilon}{\partial x_j}\right) - C_{1\varepsilon} \dfrac{\varepsilon}{k} \overline{\rho u'_i u'_j} \dfrac{\partial u_j}{\partial x_i} - C_{2\varepsilon} \rho \dfrac{\varepsilon^2}{k} - R_\varepsilon. \quad (9)$$

where μ_t is the turbulent viscosity and R_ε is the additional term in the ε equation that takes:

$$\mu_t = \rho C_\mu \dfrac{k^2}{\varepsilon} f(\alpha_s, \Omega, \dfrac{k}{\varepsilon}); \quad (10)$$

$$R_\varepsilon = \dfrac{C_\mu \rho \eta^3 (1 - \eta/\eta_0)}{1 + \beta \eta^3} \dfrac{\varepsilon^2}{k}. \quad (11)$$

where C_μ is 0.0845 derived by RNG theory, α_s is a swirl constant depending on the strength of swirling flows and is set to 0.07 for mildly swirling flows, Ω is a characteristic swirl number evaluated within Fluent, $\beta = 0.012$, $\eta_0 = 4.38$, $\eta = S_k/\varepsilon$, S is the total entropy. The constants $C_{1\varepsilon} = 1.42$ and $C_{2\varepsilon} = 1.68$ are used by default.

2.2. Simulation Setup

The simulations of solid-liquid mixing in the stirred tank are conducted using the CFD method. The 3D Eulerian–Eulerian multiphase model along with the RNG k–ε turbulence model are used for simulating the granular flow and the turbulence effects. Moreover, the moving reference frame (MRF) technique is employed in this work [48]. The computational domain can be divided into stationary and moving zones, and the moving zone is specified frame motion at a rotating speed. The near wall zones are modeled by employing the enhanced wall function. The unstructured tetrahedral grids are used throughout the computational domain, and the grid distribution in the stirred tank is shown in Figure 2. The grid independence test is carried out by using three sets of grids: 107,732 elements, 263,709 elements, 439,085 elements, 599,290 elements, 827,636 elements and 1,179,019 elements, as shown in Figure 3. The deviation of the relative standard deviation of particle concentration σ and the mean velocity in the computational domain between 599,290 and 1,179,019 cells are less than 1%. Thus, the grid of 599,290 elements is used in the current simulations. The solid particles were initially patched in a defined region at the bottom of tank. The convergence criterion of all transport equations is specified as the residual values below 0.0001. The calculated time step is set to 0.001 s and the maximum number of iterations to be performed per time step is 20. A typical simulation with such grid resolution takes about 72 h real time with paralleling 32 Intel Xeon CPU cores (2.3 GHz).

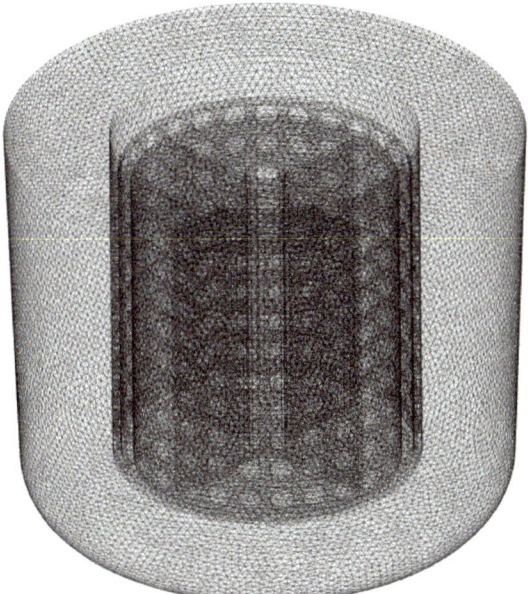

Figure 2. Grid distribution in the stirred tank.

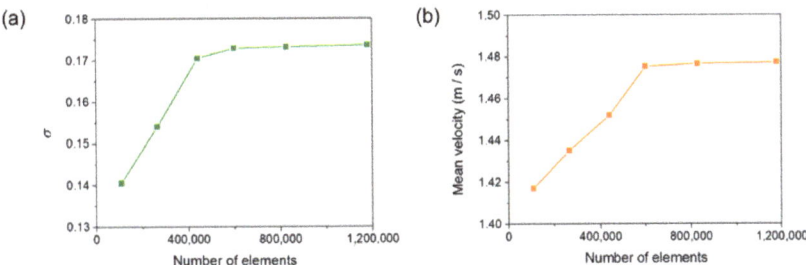

Figure 3. Grid independence test; (**a**) Effect of the number of elements on the relative standard deviation of particle concentration; (**b**) Effect of the number of elements on the mean velocity of the computational domain.

To evaluate the homogeneity of solid particles distribution in the stirred tank, the suspension quality is defined by the relative standard deviation of particle concentration σ that can be expressed as [49]:

$$\sigma = \sqrt{\frac{1}{n}\sum_{i=1}^{n}\left(\frac{\alpha_i - \alpha_{avg}}{\alpha_{avg}}\right)^2} \qquad (12)$$

where α_{avg} is the averaged volume fraction of solid particles in the whole computational domain, i denotes the discrete volume element, n is the total number of computational cells.

Homogeneous suspension conditions can be achieved when $\sigma < 0.2$, while the incomplete suspension is resulted at $\sigma > 0.8$. The mixing energy level (*MEL*) denotes the power consumption per unit volume, which is calculated as:

$$MEL = P/v \qquad (13)$$

where v is the volume of working fluids, and P is the power consumption of agitator which can be calculated by:

$$P = 2\pi NT/60 \qquad (14)$$

where N is the stirring speed, T is the torque of agitator.

2.3. Model Validation

To validate the accuracy of the current numerical model, we simulate the dispersions and holdups of sand in the glycerite–sand system from the experiments of Wang et al. [30]. Figure 4 shows the comparison of numerical and experimental results for axial profiles of the normalized solid volume fraction. The numerical results have showed a good agreement with the experimental data. The slight deviations could root in the experimental error of solid holdup which is basically generated from the fluctuation of the turbulent flow and the sampling [30]. Therefore, the present liquid-solid multiphase model can predict well the distribution of particles in the stirred tank.

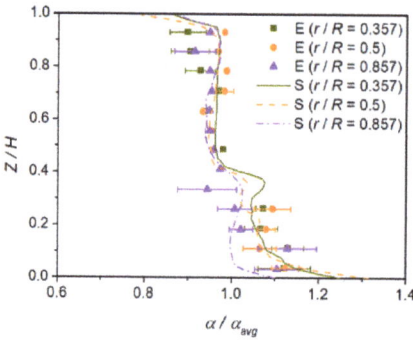

Figure 4. Comparison of numerical and experimental [30] results for axial profiles of the normalized solid volume fraction.

3. Results and Discussion

3.1. Flow Fields

First, the flow fields of the two new rotor-stator agitators of CSRS and HCSRS with that of the conventional agitators of the A200 and Rushton are compared. Figure 5 shows the velocity vectors of solid particles for the four types of agitators, the color and arrows in the figures indicate the magnitude and direction of the velocity vector, respectively. For the CSRS agitator, as shown in Figure 5a, after the fluid is discharged through the blade tip, part of the fluid forms a squeeze flow in the gap between the stator and rotor and circulates inside the stator. Part of the fluid passes through the circular hole of the stator structure after being discharged by the paddle (see Figure 6a), then impacts the side wall of the stirred tank and is divided into two upward and downward streams. The upper stream forms a small annular flow and merges with the lower fluid and finally enters the stator under the agitator to form a complete circulation. It can be seen that the velocity distribution in the whole stirred tank is relatively homogeneous, and there is almost no flow dead zone. For the HCSRS agitator, as described in Figure 5b, the fluid is discharged obliquely downward after being discharged from the blade tip. After being blocked by the wall surface, it forms a large circulation upward along the side wall of the stirred tank, and then enters the stator through the circular holes on the side wall of the stator structure (see

Figure 6b). A small part of the fluid forms an internal circulation above the rotor, and most of the fluid flows to the rotor, forming a complete flow cycle. The velocity distribution in the whole stirred tank is relatively homogeneous, but the flow at the bottom of the agitator is weak. For the A200 agitator, as exhibited in Figure 5c, after the fluid is discharged from the tip of the paddle, it will obliquely impact the side wall of the stirred tank and flow upward along the side wall to form a circulation. It can be seen that the flow under the agitator, near the stirring rod and above the stirred tank is very weak, and there are small vortices near the wall surface above the stirred tank, which are not conducive to solid-liquid mixing. For the Rushton agitator, as descripted in Figure 5d, the fluid is discharged from the tip of the blade and impinges on the side wall of the stirred tank. It is divided into two streams, one of which forms a large circulation above the agitator, and the other forms a small circulation below the agitator along the wall surface. Similar to the A200 agitator, the flow under the agitator, near the agitator bar and above the agitator tank is very weak, and there are small vortices on the wall above the agitator tank. In addition, the low-speed zone near the agitator shaft is wider, which is not conducive to solid-liquid mixing. In conclusion, the HCSR agitator has a more homogeneous velocity distribution, which is conducive to the homogeneous dispersion of particles in the melt.

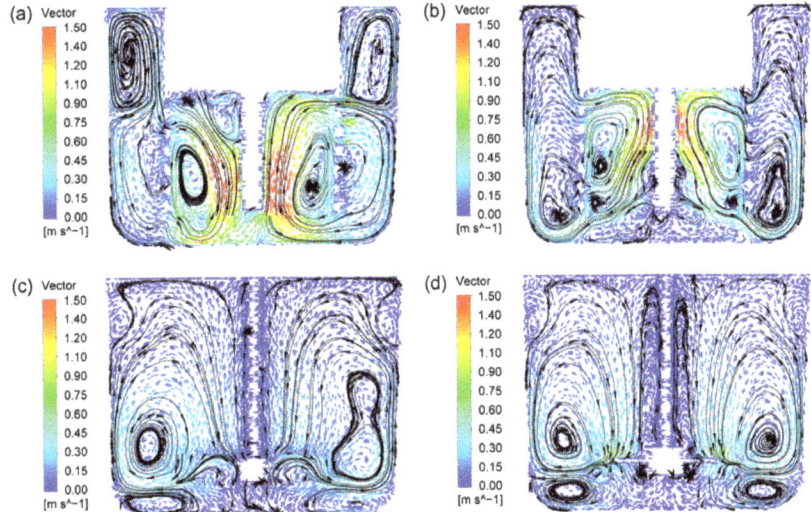

Figure 5. Velocity vectors of solid particles for the four types of agitators at the stirring speed of 400 rpm; (**a**) CSRS agitator; (**b**) HCSRS agitator; (**c**) A200 agitator; (**d**) Rushton agitator.

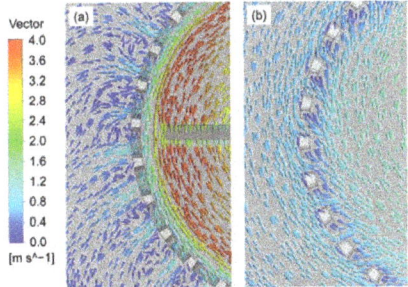

Figure 6. Velocity vectors near the stator sidewall holes for the two types of rotor-stator agitators at the horizontal plane of $Z = 0.105$ m under the stirring speed of 400 rpm; (**a**) CSRS agitator; (**b**) HCSRS agitator.

The stability of liquid surface is very important to the quality of composite materials in the process of stirring preparation. Stable liquid surface can ensure that gas and other contaminants cannot enter the melt from the liquid surface. Thus, the liquid level velocity for the four types of agitators are investigated in this study, as shown in Figure 7. It can be attained that the liquid level velocities obtained by the two new types of rotor-stator agitators are far lower than that obtained by the two conventional agitators. Additionally, the liquid level velocity obtained by the CSRS agitator is the minimum, while the liquid level velocity obtained by the Rushton agitator is the maximum. Since the pressure distributions outside the stator of the rotor-stator agitators are more uniform, and the specific pressure distribution will be discussed later.

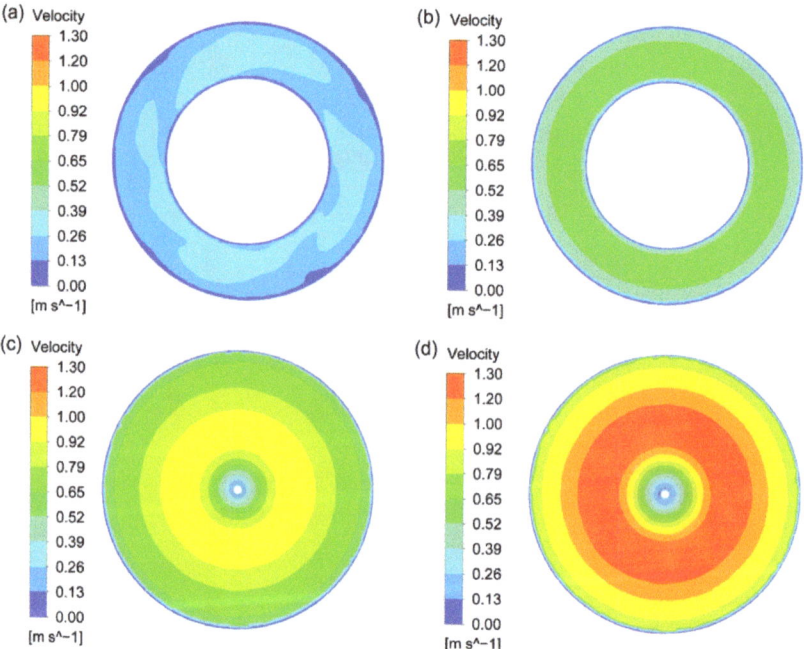

Figure 7. Liquid level velocity for the four types of agitators at the stirring speed of 400 rpm; (**a**) CSRS agitator; (**b**) HCSRS agitator; (**c**) A200 agitator; (**d**) Rushton agitator.

3.2. Turbulent Energy Dissipation

Turbulent energy dissipation rate is an important parameter in particle dispersion system, which determines the size distribution of bubbles, droplets and agglomerated particles. Therefore, the turbulent energy dissipation distributions on the vertival center plane and horizontal plane across the rotor at the stirring speed of 400 rpm for the four types of agitators were explored, as shown in Figures 8 and 9. It can be found that, for the two new types of rotor-stator agitators, the high turbulent energy dissipation areas are mainly distributed in the gaps between the stator and rotor and the circular hole channels on the stator. Moreover, the turbulent energy dissipation near the inner wall of the solid connection part between the circular holes on the stator is the highest due to the strong shear effect on the fluid between the high-speed rotating rotor and the stationary wall. For the CSRS agitator, the turbulent energy dissipation in the area between the upper surface of the rotor blade and the inner wall at the top of the stator is also very high. For the two conventional agitators, the high turbulent energy dissipation area is located at the place where the blade tips pass through, and the turbulent energy dissipation generated by A200 agitator is significantly higher than that generated by Rushton agitator. In a comprehensive comparison, the high turbulent energy dissipation areas generated by the two new types of rotor-stator agitators are significantly more than those generated by the two conventional agitators and the high turbulent energy dissipation areas generated by the CSRS agitator distribute more widely.

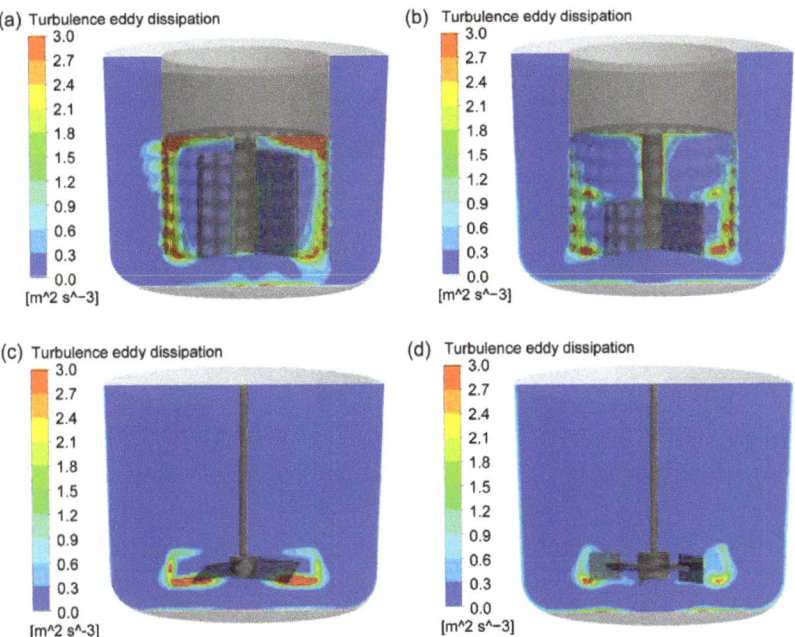

Figure 8. The distribution of turbulent energy dissipation rate on the vertival plane at the stirring speed of 400 rpm for the four types of agitators; (**a**) CSRS agitator; (**b**) HCSRS agitator; (**c**) A200 agitator; (**d**) Rushton agitator.

3.3. Pressure Fields

In order to better understand the reasons why different agitators produce different flow patterns, the pressure fields of the four agitators are investigated, as shown in Figure 10. For the CSRS agitator (see Figure 10a), the fluid follows the rotation of the rotor and passes through the circular holes on the side wall of the stator under the action of centrifugal force. Since the external fluid cannot enter the stator structure in time, a large negative pressure is formed inside the stator structure. The fluid outside the stator enters the stator from the opening below the stator under the pressure difference, which is also the reason for the large upward speed near the middle and lower parts of the rotor. For the HCSRS agitator (see Figure 10b), the fluid flows out from the opening below the stator as the blade rotates. Since the external fluid cannot enter the stator structure in time, an obvious negative pressure field appears in the area above the rotor near the mixing shaft. Therefore, the fluid around the stator enters the stator through the circular holes on the stator side wall under the pressure difference. For the A200 agitator and the Rushton agitator, as exemplified in Figure 10c,d, the fluid is pushed to the side wall of the stirred tank under the action of the rotor. The pressure distribution in the stirred tank is gradually increased from the center of the stirred tank to the outside, which is detrimental to the liquid level stability. The Rushton agitator has a stronger radial liquid discharge capacity, resulting in greater pressure in the stirred tank. For CSRS agitator, there is an obvious pressure gradient inside the stator, but the pressure distribution outside the stator is very uniform, which is very conducive to obtaining a stable liquid level.

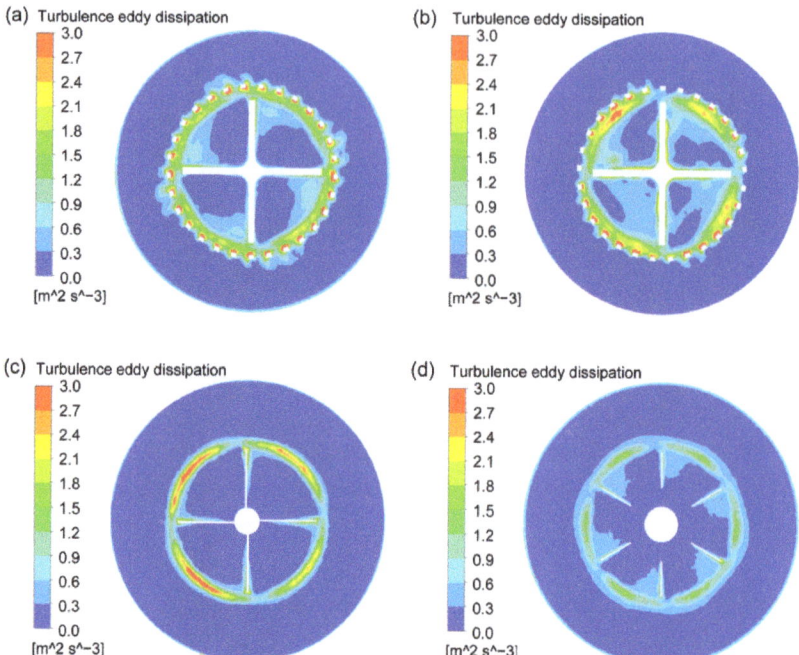

Figure 9. The distribution of turbulence energy dissipation rate on the horizontal plane ($Z = 0.038$ m) at the stirring speed of 400 rpm for the four types of agitators; (**a**) CSRS agitator; (**b**) HCSRS agitator; (**c**) A200 agitator; (**d**) Rushton agitator.

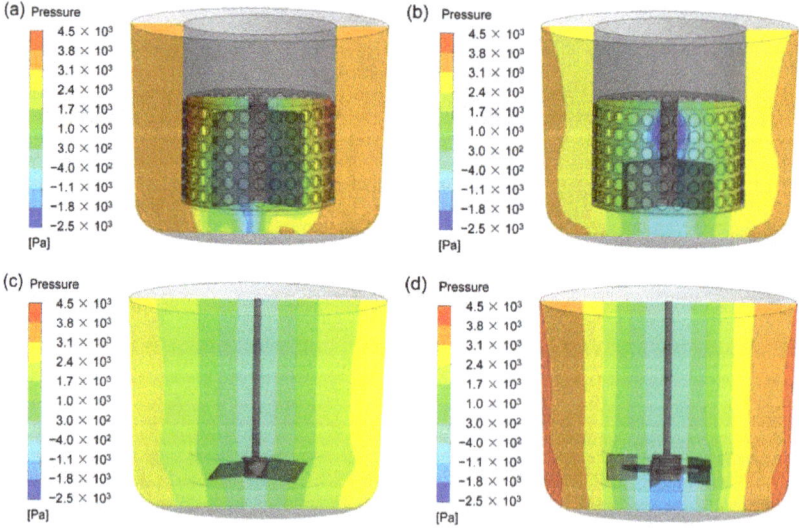

Figure 10. Pressure fields at the vertical plane for the four types of agitators at the stirring speed of 400 rpm; (**a**) CSRS agitator; (**b**) HCSRS agitator; (**c**) A200 agitator; (**d**) Rushton agitator.

3.4. Solid Particle Distribution

To compare the performances for the four types of agitators, the distribution of solid volume fraction at the vertical and horizontal planes are exhibited in Figures 11 and 12.

Additionally, the quantitative study on the axial distribution of solid volume fraction is shown in Figure 8. The averaged solid volume fraction of the horizontal plane ($\alpha_{ave\text{-}p}$) is normalized by that of the whole stirred tank (α_{ave}). For the CSRS agitator, it can be seen that the solid particles are evenly distributed throughout the stirred tank due to the relatively homogeneous velocity distribution (as shown in Figure 11a), but the concentration of particles increases slightly at the bottom of the stirred tank and the problem of particle accumulation is more serious in the bottom central area (as shown in Figure 12a). For the HCSRS agitator, the distribution of solid particles in the whole stirred tank is also basically homogeneous (as shown in Figure 11b), except that the particles in the bottom central area of the stirred tank are relatively sparse and the concentration of particles increases slightly above the stirred tank. The closer the particles are to the wall, the higher the concentration can be seen from the radial distribution (as shown in Figure 12b). For the A200 agitator, the distribution of particles in the stirred tank is not homogeneous, the particles are heavily deposited at the bottom of the stirred tank due to the flow dead zone, and the particles near the stirring shaft are extremely sparse due to the weak flow (as shown in Figures 11c and 12c). For the Rushton agitator, the distribution of particles in the stirred tank is very uneven, the particles accumulate slightly under the impeller blades, and the particles near the stirring shaft and below the stirring shaft are extremely sparse (as shown in Figure 11d). It can be seen from the radial distribution that particles can also accumulate near the wall above the stirred tank (as shown in Figure 12d). In conclusion, the use of two rotor-stator agitators can significantly improve the homogeneity of solid particle distribution in the stirred tank compared with the conventional agitators of A200 and Rushton agitators. Similar conclusions can also be obtained by quantitative analysis of the axial distribution of solid particles, as exhibited in Figure 13.

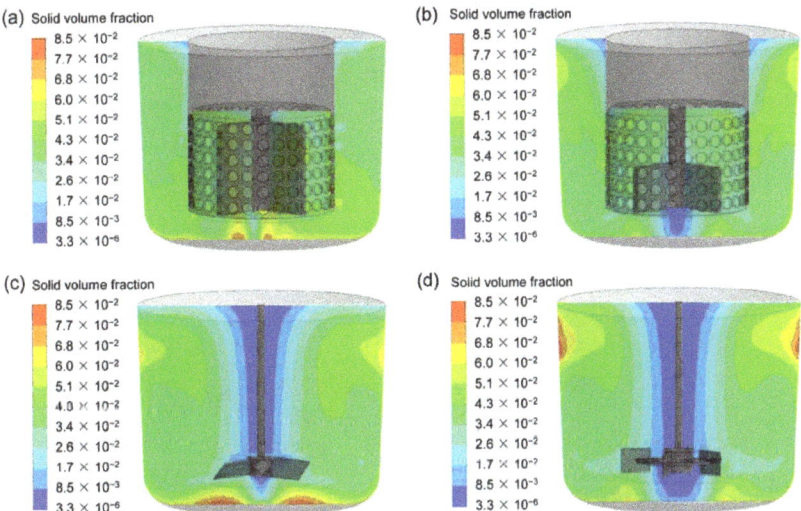

Figure 11. The distribution of solid volume fraction at the vertical plane for the four types of agitators at the stirring speed of 400 rpm; (**a**) CSRS agitator; (**b**) HCSRS agitator; (**c**) A200 agitator; (**d**) Rushton agitator.

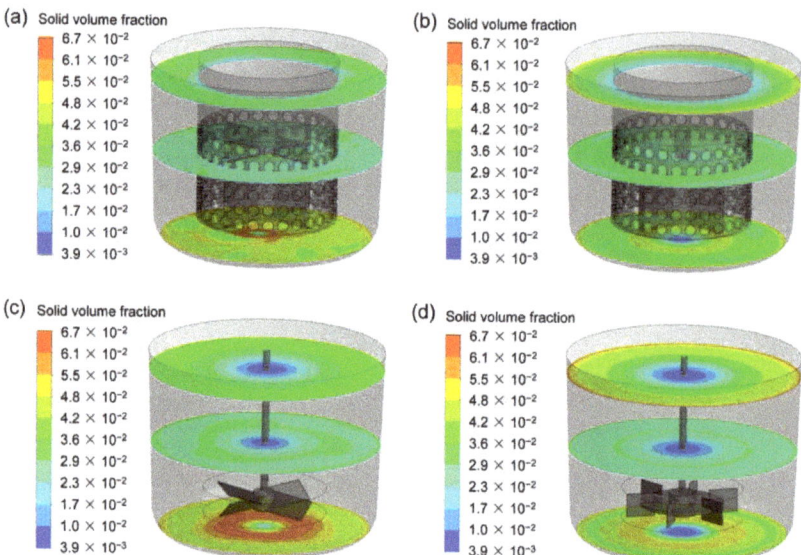

Figure 12. The distribution of solid volume fraction at different horizontal planes for the four types of agitators at the stirring speed of 400 rpm; (**a**) CSRS agitator; (**b**) HCSRS agitator; (**c**) A200 agitator; (**d**) Rushton agitator.

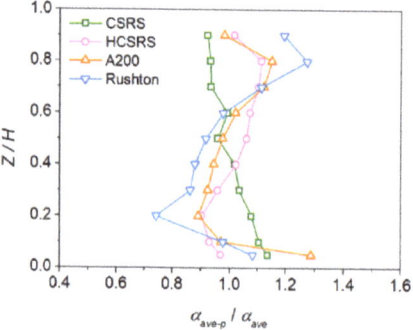

Figure 13. The axial distribution of the normalized averaged solid volume fraction on the horizontal planes (α_{ave-p}) for the four types of agitators at the stirring speed of 400 rpm.

3.5. Effect of Stirring Speed

It is known that stirring speed has a great influence on the homogeneity of solid-liquid mixing. Figure 14 compares the relative standard deviation of particle concentration of four agitators at different stirring speeds. It can be found that the minimum σ achieved by the two rotor-stator agitators of CSRS and HCSRS are much lower than that of the two conventional agitators of A200 and Rushton. The CSRS agitator yielded the best solid-liquid mixing homogeneity, while the Rushton agitator yielded the worst. For all agitators, the influence of stirring speed on the homogeneity of solid-liquid mixing shows a rule of first promoting and then suppressing. However, for the two conventional agitators, after achieving the best mixing homogeneity, increasing the stirring speed will worsen the mixing homogeneity significantly. For the two rotor-stator agitators, after achieving the best mixing homogeneity, increasing the stirring speed has little effect on the mixing homogeneity, especially for the CSRS agitator. It is very important for the dispersion of agglomerated particles and the grain refinement to enhance the shearing effect by increasing the stirring speed, while ensuring a good homogeneity of solid-liquid mixing in the stirring casting process.

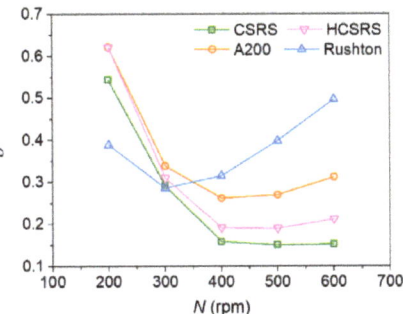

Figure 14. Effect of stirring speed on the relative standard deviation of particle concentration for the four types of agitators.

3.6. Evaluation of Power Consumption and Surface Stability

In order to evaluate the performances of the two rotor-stator agitators in solid-liquid mixing operation, the solid-liquid mixing homogeneity verse the power consumption for the four agitators was investigated, as shown in Figure 15. It can be found that the power consumption required by the conventional agitators to achieve the minimum σ are much lower than that of the two rotor-stator agitators, because the addition of the stator structure increases the flow resistance, which in turn significantly increases the power consumption required to achieve the same speed. At low power consumption, the two new agitators could not perform very well and the resulting homogeneity of solid-liquid mixing is inferior to that of the two conventional agitators. When the optimum solid-liquid mixing homogeneity is achieved in both conventional agitators, increasing power consumption will worsen the homogeneity significantly, while the two new agitators will continue to optimize the mixing homogeneity, even when the optimum mixing homogeneity is reached, the influence of increasing power consumption on the mixing homogeneity is very small. In addition, at the same power consumption, the homogeneity of solid-liquid mixing obtained by A200 is always lower than that of Rushton.

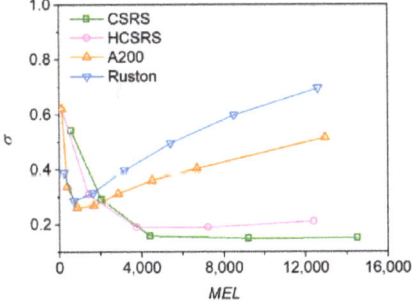

Figure 15. Effect of power consumption per unit volume on the relative standard deviation of particle concentration for the four types of agitators.

The liquid level velocities corresponding to the four types of agitators when reaching the minimum standard deviation of particle concentration are evaluated in Figure 16. The two rotor-stator agitators can not only obtain better mixing homogeneity due to the more homogeneous velocity distributions in the whole stirring tanks, but also gain lower liquid level velocities due to the more uniform pressure distributions outside the stators. Among them, the liquid level stability of the CSRS agitator is much better than that of the other three agitators.

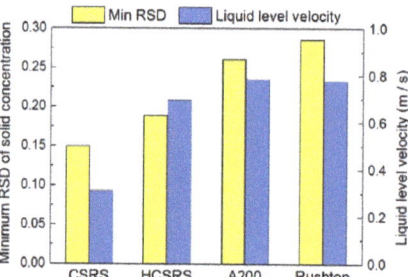

Figure 16. The minimum relative standard deviation of particle concentration and the corresponding average liquid level velocity obtained by the four agitators.

4. Conclusions

The performances of the two rotor-stator agitators (CSRS and HCSRS) for solid-liquid mixing operations in the stirred tank comparing with that of the conventional agitators of A200 and Rushton were investigated using the CFD modeling to improve the mixing homogeneity and liquid level stability.

The minimum σ achieved by the two rotor-stator agitators were much lower than that achieved by the two conventional agitators. The CSRS agitator yielded the best solid-liquid mixing homogeneity due to the relatively homogeneous velocity distribution in the whole stirred tank. For all agitators, the influence of stirring speed on the homogeneity of solid-liquid mixing showed a rule of first promoting and then suppressing. When the optimum solid-liquid mixing homogeneity was achieved in both conventional agitators, increasing stirring speed would worsen the homogeneity significantly, while the two rotor-stator agitators still achieving good mixing homogeneity under high shear conditions. In addition, the high turbulent energy dissipation areas generated by the CSRS agitator were the most widely distributed, which was very advantageous to particle dispersion and grain refinement during the preparation of composite materials.

The liquid level velocities corresponding to the minimum standard deviation of particle concentration obtained by the two rotor-stator agitators were also far less than that obtained by the two conventional agitators. The specific numerical order is: CSRS < HCSRS < A200 < Rushton. The above findings are of great significance for promoting the solid-liquid mixing homogeneity, enhancing the shear effect on the agglomerated particles and melt and stabilizing the liquid surface, thereby improving the quality of the composite materials.

Author Contributions: C.Y.: Conceptualization, methodology, validation, investigation, data curation, formal analysis, visualization, writing—original draft preparation, funding acquisition, project administration. K.Z.: Conceptualization, writing–review and editing, funding acquisition, project administration. J.H.: Methodology, formal analysis, visualization, supervision. Y.X.: Methodology, software, investigation. Z.T.: Resources, project administration. Y.L.: Formal analysis, supervision. D.L.: Software, validation. All authors have read and agreed to the published version of the manuscript.

Funding: This research was funded by the National Natural Science Foundation of China (No. 52006041), the Guangdong Academy of Sciences (Nos. 2021GDASYL-20210103101, 2020GDASYL-20200101001, 2022GDASZH-2022010109) and the Guangdong Major Project of Basic and Applied Basic Research (No. 2020B0301030006).

Institutional Review Board Statement: Not applicable.

Informed Consent Statement: Not applicable.

Data Availability Statement: Not applicable.

Conflicts of Interest: The authors declare no competing financial interest.

References

1. Yang, H.; Huang, Y.; Song, B.; Kainer, K.U.; Dieringa, H. Enhancing the creep resistance of AlN/Al nanoparticles reinforced Mg-2.85Nd-0.92Gd-0.41Zr-0.29Zn alloy by a high shear dispersion technique. *Mater. Sci. Eng. A* **2019**, *755*, 18–27. [CrossRef]
2. Paul, E.L.; Atiemo-obeng, V.A.; Kresta, S.M. *Handbook of Industrial Mixing: Science and Practice*; John Wiley & Sons, Inc.: New York, NY, USA, 2004.
3. Kumaresan, T.; Joshi, J.B. Effect of impeller design on the flow pattern and mixing in stirred tanks. *Chem. Eng. J.* **2006**, *115*, 173–193. [CrossRef]
4. Selimefendigil, F.; Öztop, H.F. Numerical study of MHD mixed convection in a nanofluid filled lid driven square enclosure with a rotating cylinder. *Int. J. Heat Mass Transf.* **2014**, *78*, 741–754. [CrossRef]
5. Mak, A.T.-C. *Solid-Liquid Mixing in Mechanically Agitated Vessels*; University of London: London, UK, 1992.
6. Frijlink, J.J.; Bakker, A.; Smith, J.M. Suspension of solid particles with gassed impellers. *Chem. Eng. Sci.* **1990**, *45*, 1703–1718. [CrossRef]
7. Ayranci, I.; Kresta, S.M. Design rules for suspending concentrated mixtures of solids in stirred tanks. *Chem. Eng. Res. Des.* **2011**, *89*, 1961–1971. [CrossRef]
8. Devarajulu, C.; Loganathan, M. Effect of impeller clearance and liquid level on critical impeller speed in an agitated vessel using different axial and radial impellers. *J. Appl. Fluid Mech.* **2016**, *9*, 2753–2761. [CrossRef]
9. Hosseini, S.; Patel, D.; Ein-Mozaffari, F.; Mehrvar, M. Study of solid-liquid mixing in agitated tanks through electrical resistance tomography. *Chem. Eng. Sci.* **2010**, *65*, 1374–1384. [CrossRef]
10. Kazemzadeh, A.; Ein-Mozaffari, F.; Lohi, A. Effect of impeller type on mixing of highly concentrated slurries of large particles. *Particuology* **2020**, *50*, 88–99. [CrossRef]
11. Zhao, J.; Gao, Z.; Bao, Y. Effects of the blade shape on the trailing vortices in liquid flow generated by disc turbines. *Chin. J. Chem. Eng.* **2011**, *19*, 232–242. [CrossRef]
12. Ankamma Rao, D.; Sivashanmugam, P. Experimental and CFD simulation studies on power consumption in mixing using energy saving turbine agitator. *J. Ind. Eng. Chem.* **2010**, *16*, 157–161. [CrossRef]
13. Mishra, P.; Ein-Mozaffari, F. Using tomograms to assess the local solid concentrations in a slurry reactor equipped with a Maxblend impeller. *Powder Technol.* **2016**, *301*, 701–712. [CrossRef]
14. Mishra, P.; Ein-Mozaffari, F. Using computational fluid dynamics to analyze the performance of the Maxblend impeller in solid-liquid mixing operations. *Int. J. Multiph. Flow* **2017**, *91*, 194–207. [CrossRef]
15. Gu, D.; Liu, Z.; Qiu, F.; Li, J.; Tao, C.; Wang, Y. Design of impeller blades for efficient homogeneity of solid-liquid suspension in a stirred tank reactor. *Adv. Powder Technol.* **2017**, *28*, 2514–2523. [CrossRef]
16. Başbuğ, S.; Papadakis, G.; Vassilicos, J.C. Reduced power consumption in stirred vessels by means of fractal impellers. *AIChE J.* **2018**, *64*, 1485–1499. [CrossRef]
17. Gu, D.; Cheng, C.; Liu, Z.; Wang, Y. Numerical simulation of solid-liquid mixing characteristics in a stirred tank with fractal impellers. *Adv. Powder Technol.* **2019**, *30*, 2126–2138. [CrossRef]
18. Yang, J.; Zhang, Q.; Mao, Z.S.; Yang, C. Enhanced micromixing of non-Newtonian fluids by a novel zigzag punched impeller. *Ind. Eng. Chem. Res.* **2019**, *58*, 6822–6829. [CrossRef]
19. Zhang, W.; Gao, Z.; Yang, Q.; Zhou, S.; Xia, D. Study of novel punched-bionic impellers for high efficiency and homogeneity in pcm mixing and other solid-liquid stirs. *Appl. Sci.* **2021**, *11*, 9883. [CrossRef]
20. Padron, G.; Eagles, W.; Özcan-Taskin, G.; McLeod, G.; Xie, L. Effect of particle properties on the break up of nanoparticle clusters using an in-line rotor-stator. *J. Dispers. Sci. Technol.* **2008**, *29*, 580–586. [CrossRef]
21. Özcan-Taşkin, N.G.; Padron, G.; Voelkel, A. Effect of particle type on the mechanisms of break up of nanoscale particle clusters. *Chem. Eng. Res. Des.* **2009**, *87*, 468–473. [CrossRef]
22. Bałdyga, J.; Orciuch, W.; Makowski, Ł.; Malski-Brodzicki, M.; Malik, K. Break up of nano-particle clusters in high-shear devices. *Chem. Eng. Process. Process Intensif.* **2007**, *46*, 851–861. [CrossRef]
23. Bałdyga, J.; Orciuch, W.; Makowski, Ł.; Malik, K.; Özcan-Taşkin, G.; Eagles, W.; Padron, G. Dispersion of nanoparticle clusters in a rotor-stator mixer. *Ind. Eng. Chem. Res.* **2008**, *47*, 3652–3663. [CrossRef]
24. Hall, S.; Cooke, M.; El-Hamouz, A.; Kowalski, A.J. Droplet break-up by in-line Silverson rotor-stator mixer. *Chem. Eng. Sci.* **2011**, *66*, 2068–2079. [CrossRef]
25. Qin, C.; Chen, C.; Xiao, Q.; Yang, N.; Yuan, C.; Kunkelmann, C.; Cetinkaya, M.; Mülheims, K. CFD-PBM simulation of droplets size distribution in rotor-stator mixing devices. *Chem. Eng. Sci.* **2016**, *155*, 16–26. [CrossRef]
26. James, J.; Cooke, M.; Kowalski, A.; Rodgers, T.L. Scale-up of batch rotor-stator mixers. Part 2—Mixing and emulsification. *Chem. Eng. Res. Des.* **2017**, *124*, 321–329. [CrossRef]
27. Gallassi, M.; Gonçalves, G.F.N.; Botti, T.C.; Moura, M.J.B.; Carneiro, J.N.E.; Carvalho, M.S. Numerical and experimental evaluation of droplet breakage of O/W emulsions in rotor-stator mixers. *Chem. Eng. Sci.* **2019**, *204*, 270–286. [CrossRef]

28. Rueger, P.E.; Calabrese, R.V. Dispersion of water into oil in a rotor-stator mixer. Part 1: Drop breakup in dilute systems. *Chem. Eng. Res. Des.* **2013**, *91*, 2122–2133. [CrossRef]
29. Rueger, P.E.; Calabrese, R.V. Dispersion of water into oil in a rotor-stator mixer. Part 2: Effect of phase fraction. *Chem. Eng. Res. Des.* **2013**, *91*, 2134–2141. [CrossRef]
30. Wang, L.; Zhang, Y.; Li, X.; Zhang, Y. Experimental investigation and CFD simulation of liquid-solid-solid dispersion in a stirred reactor. *Chem. Eng. Sci.* **2010**, *65*, 5559–5572. [CrossRef]
31. Fathi Roudsari, S.; Dhib, R.; Ein-Mozaffari, F. Using a novel CFD model to assess the effect of mixing parameters on emulsion polymerization. *Macromol. React. Eng.* **2016**, *10*, 108–122. [CrossRef]
32. Tamburini, A.; Cipollina, A.; Micale, G.; Brucato, A.; Ciofalo, M. Influence of drag and turbulence modelling on CFD predictions of solid liquid suspensions in stirred vessels. *Chem. Eng. Res. Des.* **2014**, *92*, 1045–1063. [CrossRef]
33. Joshi, J.B.; Nere, N.K.; Rane, C.V.; Murthy, B.N.; Mathpati, C.S.; Patwardhan, A.W.; Ranade, V.V. CFD simulation of stirred tanks: Comparison of turbulence models. Part I: Radial flow impellers. *Can. J. Chem. Eng.* **2011**, *89*, 23–82. [CrossRef]
34. Wadnerkar, D.; Utikar, R.P.; Tade, M.O.; Pareek, V.K. CFD simulation of solid-liquid stirred tanks. *Adv. Powder Technol.* **2012**, *23*, 445–453. [CrossRef]
35. Hosseini, S.; Patel, D.; Ein-Mozaffari, F.; Mehrvar, M. Study of solid-liquid mixing in agitated tanks through computational fluid dynamics modeling. *Ind. Eng. Chem. Res.* **2010**, *49*, 4426–4435. [CrossRef]
36. Derksen, J.J. Numerical simulation of solids suspension in a stirred tank. *AIChE J.* **2003**, *49*, 2700–2714. [CrossRef]
37. Wadnerkar, D.; Tade, M.O.; Pareek, V.K.; Utikar, R.P. CFD simulation of solid–liquid stirred tanks for low to dense solid loading systems. *Particuology* **2016**, *29*, 16–33. [CrossRef]
38. Qi, N.; Zhang, H.; Zhang, K.; Xu, G.; Yang, Y. CFD simulation of particle suspension in a stirred tank. *Particuology* **2013**, *11*, 317–326. [CrossRef]
39. Tamburini, A.; Cipollina, A.; Micale, G.; Brucato, A.; Ciofalo, M. CFD simulations of dense solid-liquid suspensions in baffled stirred tanks: Prediction of solid particle distribution. *Chem. Eng. J.* **2013**, *223*, 875–890. [CrossRef]
40. Sommerfeld, M.; Decker, S. State of the art and future trends in CFD simulation of stirred vessel hydrodynamics. *Chem. Eng. Technol.* **2004**, *27*, 215–224. [CrossRef]
41. Gidaspow, D. *Multiple Flow and Fluidization: Continuum and Kinetic Theory Descriptions*; Academic Press: Cambridge, MA, USA, 1994. [CrossRef]
42. Khopkar, A.R.; Kasat, G.R.; Pandit, A.B.; Ranade, V.V. Computational fluid dynamics simulation of the solid suspension in a stirred slurry reactor. *Ind. Eng. Chem. Res.* **2006**, *45*, 4416–4428. [CrossRef]
43. Zhang, Y.; Yu, G.; Siddhu, M.A.H.; Masroor, A.; Ali, M.F.; Abdeltawab, A.A.; Chen, X. Effect of impeller on sinking and floating behavior of suspending particle materials in stirred tank: A computational fluid dynamics and factorial design study. *Adv. Powder Technol.* **2017**, *28*, 1159–1169. [CrossRef]
44. Siddqui, M.I.H.; Jha, P.K. Assessment of turbulence models for prediction of intermixed amount with free surface variation using coupled level set volume of fluid method. *ISIJ Int.* **2014**, *54*, 2578–2587. [CrossRef]
45. Murthy, B.N.; Ghadge, R.S.; Joshi, J.B. CFD simulations of gas-liquid-solid stirred reactor: Prediction of critical impeller speed for solid suspension. *Chem. Eng. Sci.* **2007**, *62*, 7184–7195. [CrossRef]
46. Vojir, D.J.; Michaelides, E.E. Effect of the history term on the motion of rigid spheres in a viscous fluid. *Int. J. Multiph. Flow* **1994**, *20*, 547–556. [CrossRef]
47. Yakhot, V.; Orszag, S.A. Renormalization group analysis of turbulence. I. Basic theory. *J. Sci. Comput.* **1986**, *1*, 3–51. [CrossRef]
48. Ochieng, A.; Lewis, A.E. CFD simulation of solids off-bottom suspension and cloud height. *Hydrometallurgy* **2006**, *82*, 1–12. [CrossRef]
49. Bohnet, M.; Niesmak, G. Distribution of solids in stirred suspension. *Ger. Chem. Eng.* **1980**, *3*, 57–65. [CrossRef]

Article

Controlling Charge Transport in Molecular Wires through Transannular π–π Interaction

Jianjian Song [1,2,3], Jianglin Zhu [2,*], Zhaoyong Wang [4,*] and Gang Liu [4]

1. School of Petroleum Engineering, Yangtze University, Wuhan 430100, China
2. Southern Marine Science and Engineering Guangdong Laboratory (Zhanjiang), Zhanjiang 524000, China
3. Key Laboratory of Drilling and Production Engineering for Oil and Gas, Hubei Province, Wuhan 430100, China
4. China Oilfield Services Ltd. (Blue Ocean BD Hi-Tech Co., Ltd.), Quanzhou 362800, China
* Correspondence: zhujl@zjblab.com (J.Z.); wangzhy16@cosl.com.cn (Z.W.)

Abstract: This paper describes the influence of the transannular π–π interaction in controlling the carrier transport in molecular wires by employing the STM break junction technique. Five pentaphenylene-based molecular wires that contained [2.2]paracyclophane-1,9-dienes (PCD) as the building block were prepared as model compounds. Functional substituents with different electronic properties, ranging from strong acceptors to strong donors, were attached to the top parallel aromatic ring and used as a gate. It was found that the carrier transport features of these molecular wires, such as single-molecule conductance and a charge-tunneling barrier, can be systematically controlled through the transannular π–π interaction.

Keywords: charge-transfer; molecular electronics; transannular π–π interaction; [2.2]paracyclophane-1,9-dienes (PCD)

Citation: Song, J.; Zhu, J.; Wang, Z.; Liu, G. Controlling Charge Transport in Molecular Wires through Transannular π–π Interaction. *Materials* **2022**, *15*, 7801. https://doi.org/10.3390/ma15217801

Academic Editor: Leonid Gurevich

Received: 12 October 2022
Accepted: 2 November 2022
Published: 4 November 2022

Publisher's Note: MDPI stays neutral with regard to jurisdictional claims in published maps and institutional affiliations.

Copyright: © 2022 by the authors. Licensee MDPI, Basel, Switzerland. This article is an open access article distributed under the terms and conditions of the Creative Commons Attribution (CC BY) license (https://creativecommons.org/licenses/by/4.0/).

1. Introduction

Single-molecule electronics (SME) is the study of electrical procedures measured or controlled on a molecular scale. Charge-transport property regulation is one of the main challenges in the field of molecular electronics, and molecular-scale tuning is the most effective way to address this. Therefore, controlling the charge transport through single-molecule wires is very important for both the understanding of mechanisms and for practical single-molecule applications [1–4]. It is well-known that both electronic and geometric changes could result in a considerable effect on electric conductance of symmetric single-molecule wire, and the molecule rectification effect of asymmetric single-molecule wire could be rendered by dipole [5–7]. Although there are some publications concerning both experimental measurements [8–10] and theoretical calculations [11–15] on charge-transport properties, regulating the transport of the charge in a controllable manner (e.g., dc/protonation, photon absorption, isomerization, oxidation/reduction) with single molecules is still a challenging subject [16–20].

Cyclophane chemistry has attracted much interest over the last 5 decades due to its unique geometric structure and electronic properties. It has been widely used in various aspects, such as molecular self-assembly [21,22], molecular recognition [23], and optoelectronic polymers [24]. One of the most widely investigated molecules is [2.2]paracyclophane-1,9-dienes (PCD), which was first reported by Cram in 1958 [25], where the two double-bond bridges between the two aromatic rings of the PCD molecule greatly shorten the transannular π–π distance and strengthen the intramolecular π–π interaction of the two aromatic rings, in contrast with the saturated bridges of [2.2]paracyclophanes [26,27]. The strong transannular π–π interaction also leads to a distorted π-electron system and a highly strained molecular structure, which is of importance as PCD molecules can be ring-opened by ring-opening metathesis polymerization to produce poly(1,4-phenylenevinylenes)-based conjugated polymers for the fabrication of functional devices [24,28,29]. Recently, the through-space charge-transfer property of dioctyloxy diperfluorohexyl-substituted PCD molecules was investigated by Yu et al., which also implies the importance of the transannular π–π interaction in regulating the electronic property of PCD molecules [30].

However, aromatic π–π interaction, which is a noncovalent interaction, has been rarely used as a tuning factor to control the charge-transport process in molecular electronics [31–33]. In this paper, we have designed and prepared five single-molecule wires bearing a PCD building motif (see our detailed synthesis of ESI†), and we investigated their molecular electronic properties using the STM-BJ technique. Figure 1 shows the designed structure of these molecular wires, where the top parallel aromatic ring is attached to the conjugated pentaphenylene part by two vinyl groups. Both the π-orbital of the top phenyl ring and the vinyl groups are orthogonal to the charge-transport direction along the molecular wire (Figure 1). This structural design allowed us to adjust the electronic interaction between the parallel aromatic ring and the charge-transport channel (pentaphenyl backbone) by adjusting the electronic properties of different functional substituents (R groups) on the top aromatic ring. The top parallel aromatic system can act as a simulated chemical gate to control the conductance of the single-molecule wire, the energy level of the orbit, and the tunneling barrier of the charge from the molecular wires.

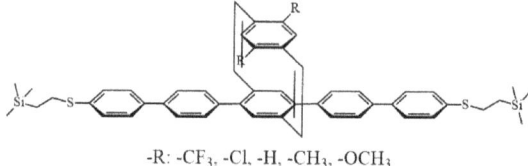

-R: -CF$_3$, -Cl, -H, -CH$_3$, -OCH$_3$

Figure 1. Chemical structure of the PCD molecular wires.

2. Materials and Methods

2.1. Materials

The commercially available chemicals were bought from TCI, except for the tetrakis (triphenylphosphine) palladium, which was purchased from Aldrich. Toluene and tetrahydrofuran (THF) were dried over sodium/benzophenone and freshly distilled prior to use. Other reagents were used without further purification. Compounds 6a, 6b, 6c, 6d, 6e, and 7 (Figure S1, ESI†) were prepared according to the standard methods in the literature [34].

2.2. Characterizations

A KSV (Helsinki, Finland) CAM 200 contact angle device was selected to detect the contact angle. An Escalab-250Xi (Thermo-Fisher, Waltham, MA, USA) containing monochromatic Al Kα as a radiation source was used to study the X-ray photoelectron spectroscopy (XPS). ^{1}H nuclear magnetic resonance (^{1}H NMR) spectra were recorded on a Bruker AV400 spectrometer (Billerica, MA, USA). Matrix-assisted laser desorption/ionization with time-of-flight (MALDI-TOF) mass spectrometry was carried out on a Bruker Ultraflextreme MALDI-TOF/TOF spectrometer using a matrix of dithranol.

2.3. Fabrication of Self-Assembled Monolayers (SAMs)

First, a propane flame was used to anneal the chosen substrate, and then the SAMs of the isolated molecules were fabricated according to a previous publication. Briefly, tetrabutylammonium fluoride (TBAF) was used as the deprotecting agent with a tetrahydrofuran (THF) solution containing ca. 1×10^{-4} M of the target molecule. After 4 h of immersion in PCD solution, the gold substrate was then washed with THF and ethanol, dried with N$_2$, and used immediately.

2.4. Single-Molecule Junctions by Break Junction (STM-BJ)

Briefly, a 0.25 mm gold wire was prepared as the gold tip by mechanical cutting. All the experiments were conducted in degassed mesitylene in order to decrease the chance of surface contamination. After that, the STM Teflon solvent holder was sonicated in acetone and dried under N_2. Then, the solvent holder was placed over the gold–SAM surface, and the designated amount of toluene was added during the measurement. It is estimated that about ~2000 current-distance traces are required in a typical BJ experiment.

2.5. I-V Recording of Single-Molecule Junctions

The *I-V* recording experiments were conducted at 100 mV bias voltage in toluene. There are three steps, include tapping, conductance step detection, and *I-V* recording. For the tapping step, when the current increases to a predefined value, the tip will move to the substrate, and then it will retract until reaching a lower, pre-set current. During each retraction, the conductance step detection was applied, and we selected the measured conductance range according to a previously measured conductance histogram. The *I-V* recording step was performed immediately upon the application of the conductance step, and the whole procure included three sub-steps. The first was the immediate holding in position of the tip. The second was the quick (10 Hz) current–voltage curve recording (from +1.5 V to −1.5 V (or +1.8 V to −1.8 V)). The last sub-step occurred when the *I-V* curve detection was completed; the tip would be removed from the substrate once the current decreased to a pre-set value, and the test was started again. For the data selection, *I-V* curves that were incomplete or contained large switching noises were detected and removed from the statistical analysis. This procedure allowed us to obtain complete *I-V* curves for statistical analysis.

Energy offsets φ_0 and the voltage division factor γ of the molecular junctions were calculated, following the work by Baldea, with the Equation [35]:

$$\varphi_0 = eV_{t,a} = \frac{2e|V_{t,n}V_{t,p}|}{\sqrt{V_{t,p}^2 + 10V_{t,p}V_{t,n}/3 + V_{t,n}^2}} \quad (1)$$

$$\gamma = -(\varphi_0/V_{t,p} + \varphi_0/V_{t,n})/4 \quad (2)$$

3. Results

3.1. Synthesis and Characterization of PCD-Based Molecular Wires

The synthetic routes for the PCD-based molecular wires with different substituents are shown in the Supporting Materials in detail (Figure S1, ESI†). First, [3.3]dithiacyclophanes intermediates (compounds 8a–e) were prepared through the cyclization reaction of benzylic dithiols (compounds 6a-e) and benzylic dithiol (compound 7) in a dilute solution. Then, Stevens rearrangement, oxidation, and pyrolysis treatments led to dibromo-[2.2]paracyclophane-1,9-diene (compounds 9a–e) in a yield of ~15% after flash column chromatography purification. A Suzuki coupling reaction between B pinacol (Bpin) functionalized biphenyl compounds (compound 5) and dibromo-[2.2]paracyclophane-1,9-diene (compounds 9a–e) furnished the target PCD-based molecular wires (compounds 10a–e). The top parallel phenyl rings were substituted with several groups possessing different electronic demands, which may play a significant role in controlling the transannular π–π interaction between the two parallel aromatic systems to adjust the charge transport of the molecular junction. More specifically, substituents -CF_3, -Cl, -H, -CH_3, and -OCH_3 were immobilized on the top phenyl ring, ranging from a strong electron acceptor to a strong electron donor. For the convenience of discussion, the molecular wires have been named "PCD-X", where X represents the substituent group. Detailed characterizations of the intermediate products and the final PCD-X molecular wires were carried out by combining ^{1}H-NMR (Figures S2–S4 ESI†), MALDI-TOF-MS (Figures S5–S9 ESI†), and UV–Vis spectroscopy (Figure S10 ESI†).

In principle, two isomers of dibromo-[3.3]dithiacyclophanes (compounds 8a–e) intermediates are expected as products during the cyclization process, namely, pseudo-gemini (eclipsed conformation of substituents on the aromatic rings) and pseudo-ortho isomers. However, only one of the possible isomers was recognized on the TLC plate and successfully purified as the main product after column chromatography, with an overall yield of around 40–50%. The clear ^{1}H-NMR spectra for all five of the dibromo-[3.3]dithiacyclophanes (Figure S2 ESI†) and the corresponding dibromo-[2.2]para cyclophane-1,9-dienes (Figure S3 ESI†) also supported the finding that only one isomer was obtained.

To unambiguously verify the atom configuration of the prepared isomer, the crystal structures of dichloro-dibromo-[2.2]paracyclophane-1,9-diene (compound 9b) were investigated, and the crystallographic data can be found in ESI†. Single crystals were grown through the slow diffusion of MeOH into the concentrated CH_2Cl_2 solutions of both compounds. The X-ray crystal structure (Figure S11 ESI†) of compound 9b clearly confirmed the pseudo-ortho isomer configuration. The selectivity for this specific isomer during the cyclization process appears to be determined by the steric influence between the bulky bromine atoms and the functional groups (CF_3, Cl, CH_3, and OCH_3), which excludes the influence of isomerization on the electronic property of the prepared molecular wires. In addition, two phenyl groups in cyclophane are closely stacked with a short distance of 2.97 Å; therefore, the π electron clouds of the phenyl rings can strongly overlap, and electronic tuning can be performed more effectively. UV–Vis measurement showed that the absorption edges for the PCD-Cl, PCD-H, PCD-CH_3, and PCD-OCH_3 molecular wires were around ~350 nm, while a 20 nm red shift was observed for PCD-CF_3, which is strong evidence that the R group on the top phenyl ring can successfully affect the pentaphenylene backbone via the effective π–π overlap. The trend agrees well with the DFT calculation results (Table S1 ESI†).

3.2. Self-Assembled Monolayer Films on Gold

Self-assembled monolayer (SAM) films were fabricated on a gold surface, with the reaction of the thiol–gold used to test the conductance of the single molecules, and the molecular junction was based on the reaction of two molecules' terminal thiol groups with two gold electrodes. Successful immobilization of the molecular wires onto the gold substrates was confirmed by combining contact angle and XPS measurements of the SAMs (Figure 2). Namely, the contact angles (Figure 2a) for the SAMs were found to be 10~15° smaller than that of bare gold, and the similar contact angles observed for all the SAMs indicated that the outermost thiol group dominated the wetting property of the SAM. XPS characterization (Figure 2b) of PCD-H SAM showed that the C/S ratio was 43/2, consistent with the theoretical ratio of 40/2, quantitatively verifying the successful immobilization of PCD molecular wires on the gold surfaces.

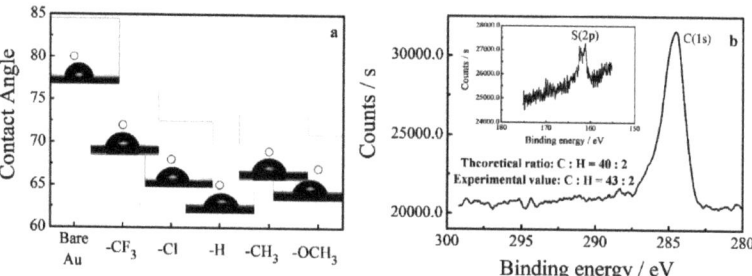

Figure 2. (a) Contact angles of the PCD-based self-assembled monolayers, where the statistically average values, based on 3 measurements, were used; (b) XPS characterization of the elements C and S of the PCD-H self-assembled monolayer.

The STM-based break junction technique was used to measure the single-molecule conductance [36,37]. With the help of repeatedly forming and breaking the gold point contacts, single-molecule junctions were established at a small bias voltage of 100 mV. The LabVIEW program was utilized to record the conductance traces of the substituted molecules, and Figure 3a exhibits the typical conductance–distance curves of the five PCD molecular junctions. The constant conductance plateaus were between $10^{-6} \sim 10^{-4}$ G_0 in these stretching traces, which is consistent with the first signature in a single-molecule junction. (G_0 is the quantum of conductance $2 e^2/h$, e is the electron charge, and h is the Planck constant). The plateau length is commonly around 0.5–1.0 nm, and the conductance fluctuation below 10^{-6} G_0 may be caused by the noise floor. Figure 3b describes the corresponding conductance histograms calculated from ~400 effective conductance decay curves, which show unambiguous peaks for all of the molecules measured. In addition, the conductance values for the junction were determined by fitting the peak of the Gaussian curve. More specifically, the single-molecule conductance values of the molecular wires PCD-CF$_3$, PCD-Cl, PCD-H, PCD-CH$_3$, and PCD-OCH$_3$ were 5.8×10^{-6} G_0, 8.2×10^{-6} G_0, 2.5×10^{-5} G_0, 3.2×10^{-5} G_0, and 8.1×10^{-5} G_0, respectively. The conductance systematically changed as the substituent varied from strong acceptor (CF$_3$) to strong donor (OCH$_3$).

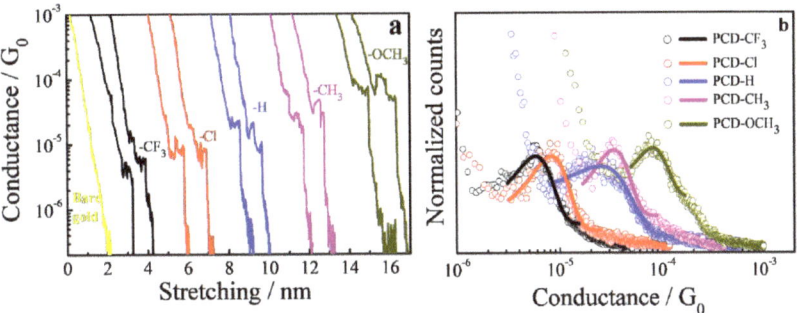

Figure 3. (a) Typical conductance–distance traces, and (b) conductance histograms of the five PCD molecular wires, where the solid lines represent the Gaussian fitting curves.

Keeping in mind that bias voltage range plays an important role in the charge transport of the single-molecule junction, thousands of current–voltage (I-V) curves were recorded for each sample. Thus, the corresponding transition voltage spectroscopy (TVS) results can be obtained. The incomplete and switching I-V curves resulting from the breakdown and instability of the molecular junctions were automatically detected but did not appear in the I-V histogram. The I-V and G-V histograms of the five substituted molecules are presented in the supporting information (Figure S12 ESI†). The so-called Fowler–Nordheim (FN) single-molecule TVS plot was obtained by transforming each I-V curve into an $\ln(I/V^2)$ versus $1/V$ curve [38–41]. The transition voltage Vt minimum value was compatible with that of the FN plot, and the transition voltage of 1D TVS histograms could be created from these minimum values.

Figure 4a–e shows the 1D transition voltage histograms of the five PCD molecules, which present two distinguishing characteristics. One is that all of the histograms are asymmetric, which was previously reported by Kushmerick and Tao [36,37]. The other one is that the absolute values of the experimental positive and negative transition voltages ($V_{t,p}$ and $V_{t,n}$) decreased from 1.4 V to 0.9 V and from 1.6 V to 1.3 V when the substituent is altered from PCD-CF$_3$ to PCD-OCH$_3$, and this could be explained as the contact of symmetry molecule and metal-molecule. In addition, the asymmetry could also be caused by the solvent polarity [42].

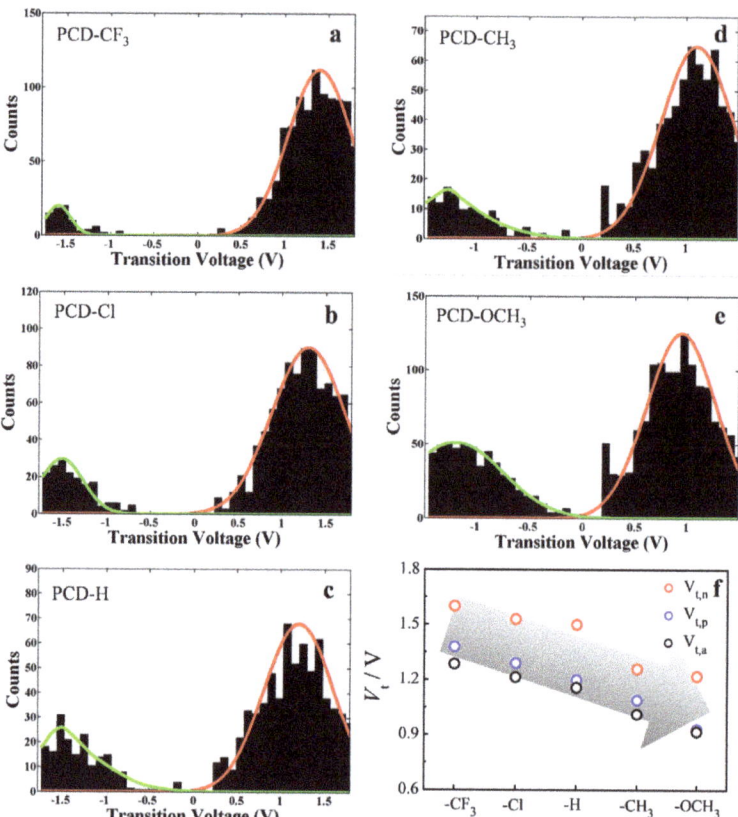

Figure 4. One-dimensional transition voltage histograms of (**a**) PCD-CF$_3$, (**b**) PCD-Cl, (**c**) PCD-H, (**d**) PCD-CH$_3$, and (**e**) PCD-OCH$_3$; (**f**) The negative, positive, and calculated ambipolar transition voltages of the five PCD molecular wires.

Considering that the offset energy (φ_0) between the electrodes' Fermi level and molecular orbitals are seriously affected by the applied bias V, $\varphi_0 \equiv \varphi_0(V)|V = 0 \to \varphi_0(V)$, voltage division factor γ was used to offset the effect, where $\varphi_0(V) = \varphi_0 + \gamma eV$ [13,30], In particular, the experimentally measured transition voltages $V_{t,p}$ and $V_{t,n}$ for bias polarities could be utilized to evaluate the corresponding ambipolar transition voltage ($V_{t,a} = \varphi_0/e$); the correct energy offset φ_0 and the voltage division factor γ based on the equations were established by Baldea (Equations (S1) and (S2) ESI†) [41,42]. The calculated $V_{t,a}$ and γ are summarized in Figure 4f and Figure S13 and in Table S1. As shown in Figure 4f, the $V_{t,a}$ value decreases from 1.3 V to 0.9 V as the substituents change from electron-withdrawing group -CF$_3$ to electron-donating group -OCH$_3$. In addition, the tunneling barrier heights of the PCD molecules with diverse groups (-CF$_3$, -Cl, -H, -CH$_3$, and -OCH$_3$) were deduced as being around 1.28 eV, 1.21 eV, 1.15 eV, 1.01 eV, and 0.91 eV, respectively. We can thus conclude that the transition voltage and the corresponding charge-tunneling barrier increase with substituent varying from electron donor to electron acceptor. Finally, we calculated the voltage division factor γ, and the values ranged from 0.03~0.06 (Figure S12 ESI†), which confirms a slightly larger potential drop at the soft contact (e.g., the STM tip).

All of the results discussed above indicate that the top parallel phenyl ring can modulate the molecular conductance and the charge-tunneling barrier of the pentaphenylene conducting channel through the transannular π–π interaction. DFT calculations based on the RB3LYP method, with basis set of 6-31G*, were carried out to obtain insight into the energy-level alignment of these PCD molecular wires. Figure 5a shows that the HOMO starts to shift from the pentaphenylene backbone to the top parallel benzene ring as the substituent varies from electron acceptor -CF$_3$ to electron donor -OCH$_3$, while the LUMO simultaneously extends to the pentaphenylene backbone. Figure 5b shows that both the HOMO and LUMO energy levels of the molecular wire decrease when the substituent on the top parallel phenyl ring varies from strong donor to strong acceptor. Since the backbone of the molecule is a p-type semiconductor, the hole transport is much easier with the HOMO molecule, which thus benefits the single-molecule conductance of electron-donating groups on the top parallel phenyl [43]. Figure 5c proves how the tunneling barrier φ_0 changes with the HOMO energy level of the PCD-based molecular wires. It is clear that the φ_0 decreases as the HOMO energy level increases, indicating that the charge-transport mechanism in these PCD molecular wires is indeed through hole transport and can be regulated by the intramolecular π–π interaction.

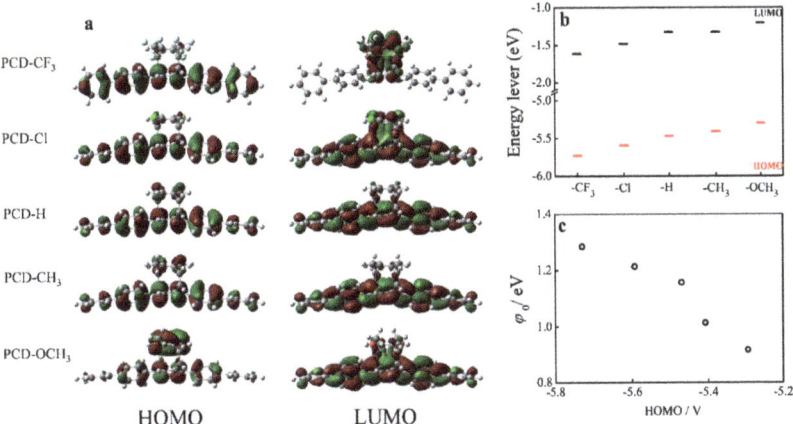

Figure 5. (a) Molecular orbital diagrams of PCD molecules, (b) Calculated HOMO and LUMO energy levels of these PCD molecules, (c) Plot of tunneling barriers ($V_{t,a}$) vs. HOMO energy level, where $\varphi_0 = eV_{t,a}$.

4. Conclusions

In summary, we have successfully prepared a series of molecular wires that demonstrated how the intramolecular transannular π–π interaction affects the charge-transport property in molecular wires. Both the single-molecule conductance and current–voltage characteristics can be systematically tuned through the transannular π–π interaction. Transition voltage spectroscopy measurement and DFT calculation suggest that the effect is manifested via tuning the HOMO energy level and orbital spatial distribution of these PCD molecules.

Supplementary Materials: The following supporting information can be downloaded at: https://www.mdpi.com/article/10.3390/ma15217801/s1: Figure S1: Synthetic route of the molecular wires; Figure S2: 1H-NMR spectra of compounds **8a–8e**; Figure S3: 1H-NMR spectra of compounds **9a–9e**; Figure S4: 1H-NMR spectra of compounds **10a–10e**; Figure S5: MALDI-TOF-MS spectrum of compound **10a**; Figure S6: MALDI-TOF-MS spectrum of compound **10b**; Figure S7: MALDI-TOF-MS spectrum of compound **10c**; Figure S8: MALDI-TOF-MS spectrum of compound **10d**; Figure S9: MALDI-TOF-MS spectrum of compound **10e**; Figure S10: UV–Vis spectra of the intermediate and final molecular wires in deuterated chloroform; Figure S11: Single-crystal structure of paracyclophane-1,9-dienes; Figure S12: (a) Current–voltage 2D histograms, (b) conductance–voltage 2D histograms, (c) transition voltage 2D histograms, and (d) transition voltage 1D histograms of the five molecular wires; Figure S13: Calculated voltage division factor (γ) for the five PCD molecular wires; Table S1: Experimental and DFT calculation results for PCD molecular wires and calculated values of φ_0 and γ, based on Equations (1) and (2) as developed by Baldea.

Author Contributions: Writing—original draft preparation and funding acquisition, J.Z.; investigation and data curation, Z.W. and G.L., formal analysis and methodology, J.Z. and J.S. All authors have read and agreed to the published version of the manuscript.

Funding: The financial support of the Funding of Southern Marine Science and Engineering Guangdong Laboratory (012S22004-006) is acknowledged.

Institutional Review Board Statement: Not applicable.

Informed Consent Statement: Not applicable.

Data Availability Statement: The data can be supported when required.

Conflicts of Interest: The authors declare no conflict of interest.

References

1. Ioffe, Z.; Shamai, T.; Ophir, A.; Noy, G.; Yutsis, I.K.K.; Cheshnovsky, O.; Selzer, Y. Detection of heating in current-carrying molecular junctions by Raman scattering. *Nat. Nanotechnol.* **2008**, *3*, 727. [CrossRef] [PubMed]
2. Díez-Pérez, I.; Hihath, J.; Lee, Y.; Yu, L.; Adamska, L.; Kozhushner, M.A.; Oleynik, I.I.; Tao, N. Rectification and stability of a single molecular diode with controlled orientation. *Nat. Chem.* **2009**, *1*, 635. [CrossRef] [PubMed]
3. McCold, C.E.; Domulevicz, L.; Cai, Z.; Lo, W.-Y.; Hihath, S.; March, K.; Mohammad, H.M.; Anantram, M.P.; Yu, L. Molecular Control of Charge Carrier and Seebeck Coefficient in Hybrid Two-Dimensional Nanoparticle Superlattices. *J. Hihath. J. Phys. Chem. C* **2020**, *124*, 17. [CrossRef]
4. Tanaka, Y.; Kiguchi, M.; Akita, M. Inorganic and Organometallic Molecular Wires for Single-Molecule Devices. *Chem. Eur. J.* **2017**, *23*, 4741. [CrossRef] [PubMed]
5. Jiang, P.; Morales, G.M.; You, W.; Yu, L. Synthesis of diode molecules and their sequential assembly to control electron transport. *Angew. Chem. Int. Edit.* **2004**, *43*, 4471. [CrossRef]
6. Morales, G.M.; Jiang, P.; Yuan, S.W.; Lee, Y.G.; Sanchez, A.; You, W.; Yu, L. Inversion of the rectifying effect in diblock molecular diodes by protonation. *J. Am. Chem. Soc.* **2005**, *127*, 10456. [CrossRef]
7. Lo, W.Y.; Zhang, N.; Cai, Z.; Li, L.; Yu, L. Beyond molecular wires: Design molecular electronic functions based on dipolar effect. *Acc. Chem. Res.* **2016**, *49*, 1852. [CrossRef]
8. Duan, P.; Liu, J.; Wang, J.-Y.; Qu, K.; Cai, S.; Wang, F.; Chen, L.; Huang, X.; Li, R.; Shi, J.; et al. Enhancing single-molecule conductance of platinum (II) complexes through synergistic aromaticity-assisted structural asymmetry. *Sci. China Chem.* **2020**, *63*, 467. [CrossRef]
9. Guo, S.; Zhou, G.; Tao, N. Single molecule conductance, thermopower, and transition voltage. *Nano Lett.* **2013**, *13*, 4326. [CrossRef]
10. Kovalchuk, A.; Abu-Husein, T.; Fracasso, D.; Egger, D.A.; Zojer, E.; Zharnikov, M.; Terfort, A.; Chiechi, R.C. The fabrication of a supra-amphiphile for dissipative self-assembly. *Chem. Sci.* **2016**, *7*, 781. [CrossRef]
11. Bâldea, I.J. Interpretation of stochastic events in single-molecule measurements of conductance and transition voltage spectroscopy. *Am. Chem. Soc.* **2012**, *134*, 7958. [CrossRef]
12. Bâldea, I. Effects of stochastic fluctuations at molecule–electrode contacts in transition voltage spectroscopy. *Chem. Phys.* **2012**, *400*, 65. [CrossRef]
13. Bâldea, I. Protocol for disentangling the thermally activated contribution to the tunneling-assisted charge transport. Analytical results and experimental relevance. *Phys. Chem. Chem. Phys.* **2005**, *17*, 20217. [CrossRef]
14. Bléger, D.; Kreher, D.; Mathevet, F.; Attias, A.J.; Arfaoui, I.; Metgé, G.; Douillard, L.; Fiorini-Debuisschert, C.; Charra, F. Periodic Positioning of Multilayered [2.2]Paracyclophane-Based Nanopillars. *Angew. Chem. Int. Edit.* **2008**, *47*, 8412. [CrossRef]
15. Guo, S.; Hihath, J.; Díez-Pérez, I.; Tao, N. Measurement and statistical analysis of single-molecule current–voltage characteristics, transition voltage spectroscopy, and tunneling barrier height. *J. Am. Chem. Soc.* **2011**, *133*, 19189. [CrossRef]
16. Liu, Y.; Qiu, X.; Soni, S.; Chiechi, R.C. The chemical landscape of Chemical Physics Reviews. *Chem. Phys. Rev.* **2021**, *2*, 21303. [CrossRef]
17. Feng, A.; Zhou, Y.; Al-Shebami, M.A.Y.; Chen, L.; Pan, Z.; Xu, W.; Zhao, S.; Zeng, B.; Xiao, Z.; Yang, Y.; et al. Sigma-sigma Stacked supramolecular junctions. *Nat. Chem.* **2022**, *14*, 1158. [CrossRef]
18. Majumdar, P.; Tharammal, F.; Gierschner, J.; Varghese, S. Tuning Solid-State Luminescence in Conjugated Organic Materials: Control of Excitonic and Excimeric Contributions through π Stacking and Halogen Bond Driven Self-Assembly. *Chem. Phys. Chem.* **2020**, *21*, 616. [CrossRef]
19. Che, Y.; Perepichka, D.F. Quantifying Planarity in the Design of Organic Electronic Materials. *Angew. Chem. Int. Ed. Engl.* **2021**, *60*, 1364. [CrossRef]
20. Bennett, T.L.R.; Alshammari, M.; Au-Yong, S.; Almutlg, A.; Wang, X.; Wilkinson, L.A.; Albrecht, T.; Jarvis, S.P.; Cohen, L.F.; Ismael, A.; et al. Multi-component self-assembled molecular-electronic films: Towards new high-performance thermoelectric systems. *Chem. Sci.* **2022**, *13*, 5176. [CrossRef]
21. Asakawa, M.; Ashton, P.R.; Hayes, W.; Janssen, H.M.; Meijer, E.W.; Menzer, S.; Pasini, D.; Stoddart, J.F.; White, A.J.P.; Williams, D.J. Constitutionally Asymmetric and Chiral [2]Pseudorotaxanes[1]. *J. Am. Chem. Soc.* **1998**, *120*, 920. [CrossRef]

22. Agrawal, Y.K.; Sharma, C.R. Supramolecular assemblies and their applications. *Rev. Anal. Chem.* **2005**, *24*, 35. [CrossRef]
23. Neelakandan, P.P.; Hariharan, M.; Ramaiah, D. A supramolecular ON–OFF–ON fluorescence assay for selective recognition of GTP. *J. Am. Chem. Soc.* **2006**, *128*, 11334. [CrossRef] [PubMed]
24. Porz, M.; Mäker, D.; Brödner, K.; Bunz, U.H.F. Poly(para-phenylene vinylene) and Polynorbornadiene Containing Rod-Coil Block Copoylmers via Combination of Acyclic Diene Metathesis and Ring-Opening Metathesis Polymerization. *Macro Rapid Commun.* **2013**, *34*, 873. [CrossRef]
25. Cram, D.J.; Steinberg, H. Macro rings. I. Preparation and spectra of the paracyclophanes. *J. Am. Chem. Soc.* **1951**, *73*, 5691. [CrossRef]
26. De Meijere, A.; Kozhushkov, S.I.; Rauch, K.; Schill, H.; Verevkin, S.P.; Kümmerlin, M.; Beckhaus, H.D.; Rüchardt, C.; Yufit, D.S. Heats of formation of [2.2]paracyclophane-1-ene and [2.2]paracyclophane-1,9-diene—An experimental study. *J. Am. Chem. Soc.* **2003**, *125*, 15110. [CrossRef]
27. Dodziuk, H.; Demissie, T.B.; Ruud, K.; Szymański, S.; Jaźwiński, J.; Hopf, H. Structure and NMR spectra of cyclophanes with unsaturated bridges (cyclophenes). *Magn. Reson. Chem.* **2012**, *50*, 449. [CrossRef]
28. Yu, C.Y.; Horie, M.; Spring, A.M.; Tremel, K.; Turner, M.L. Homopolymers and Block Copolymers of p-Phenylenevinylene-2,5-diethylhexyloxy-p-phenylenevinylene and m-Phenylenevinylene-2,5-diethylhexyloxy-p-phenylenevinylene by Ring-Opening Metathesis Polymerization. *Macromolecules* **2009**, *43*, 222. [CrossRef]
29. Menk, F.; Mondeshki, M.; Dudenko, D.; Shin, S.; Schollmeyer, D.; Ceyhun, O.; Choi, T.L.; Zentel, R. Reactivity Studies of Alkoxy-Substituted [2.2]Paracyclophane-1,9-dienes and Specific Coordination of the Monomer Repeating Unit during ROMP. *Macromolecules* **2015**, *48*, 7435. [CrossRef]
30. Yu, C.Y.; Sie, C.H.; Yang, C.Y. Synthesis and through-space charge transfer of dioctyloxy diperfluorohexyl substituted [2.2]paracyclophane-1,9-diene. *New J. Chem.* **2014**, *38*, 5003. [CrossRef]
31. Schneebeli, S.T.; Kamenetska, M.; Cheng, Z.; Skouta, R.; Friesner, R.A.; Venkataraman, L.; Breslow, R. Single-molecule conductance through multiple π–π-stacked benzene rings determined with direct electrode-to-benzene ring connections. *J. Am. Chem. Soc.* **2011**, *133*, 2136. [CrossRef] [PubMed]
32. Lo, W.Y.; Bi, W.; Li, L.; Jung, I.H.; Yu, L. Edge-on gating effect in molecular wires. *Nano Lett.* **2015**, *15*, 958. [CrossRef] [PubMed]
33. Zhang, N.; Lo, W.Y.; Cai, Z.; Li, L.; Yu, L. Molecular Rectification Tuned by Through-Space Gating Effect. *Nano Lett.* **2017**, *17*, 308. [CrossRef]
34. Lidster, B.J.; Kumar, D.R.; Spring, A.M.; Yu, C.Y.; Helliwell, M.; Raftery, J.; Turner, M.L. Alkyl substituted [2.2]paracyclophane-1,9-dienes. *Org. Biomol. Chem.* **2016**, *14*, 6079–6087. [CrossRef] [PubMed]
35. Bâldea, I. Ambipolar transition voltage spectroscopy: Analytical results and experimental agreement. *Phys. Rev. B* **2012**, *85*, 35442. [CrossRef]
36. Xu, B.; Tao, N.J. Measurement of single-molecule resistance by repeated formation of molecular junctions. *Science* **2003**, *301*, 1221. [CrossRef]
37. Xu, B.; Xiao, X.; Tao, N.J. Measurements of single-molecule electromechanical properties. *J. Am. Chem. Soc.* **2003**, *125*, 16164. [CrossRef]
38. Wielopolski, M.; Molina-Ontoria, A.; Schubert, C.; Margraf, J.T.; Krokos, E.; Kirschner, J.; Gouloumis, A.; Clark, T.; Guldi, D.M.; Martin, N. Blending through-space and through-bond pi-pi-coupling in [2,2']-paracyclophane-oligophenylenevinylene molecular wires. *J. Am. Chem. Soc.* **2013**, *135*, 10372. [CrossRef]
39. Beebe, J.M.; Kim, B.; Gadzuk, J.W.; Frisbie, C.D.; Kushmerick, J.G. Transition from direct tunneling to field emission in metal-molecule-metal junctions. *Phys. Rev. Lett.* **2006**, *97*, 26801. [CrossRef]
40. Huisman, E.H.; Guédon, C.M.; van Wees, B.J.; van der Molen, S.J. Interpretation of transition voltage spectroscopy. *Nano Lett.* **2009**, *9*, 3909. [CrossRef]
41. Stern, T.E.; Gossling, B.S.; Fowler, R.H. Further studies in the emission of electrons from cold metals. *Proc. R. Soc. Lon. Ser.-A* **1929**, *124*, 699.
42. Capozzi, B.; Xia, J.; Adak, O.; Dell, E.J.; Liu, Z.F.; Taylor, J.C.; Neaton, J.B.; Campos, L.M.; Venkataraman, L. Single-molecule diodes with high rectification ratios through environmental control. *Nat. Nanotechnol.* **2015**, *10*, 522. [CrossRef]
43. Reddy, P.; Jang, S.; Segalman, R.A.; Majumdar, A. Thermoelectricity in molecular junctions. *Science* **2007**, *315*, 1568. [CrossRef]

Editorial

Special Issue: Advanced Science and Technology of Polymer Matrix Nanomaterials

Liguo Xu [1], Jintang Zhou [2,*], Zibao Jiao [2,*] and Peijiang Liu [3,*]

[1] College of Light Chemical Industry and Materials Engineering, Shunde Polytechnic, Foshan 528333, China; 21099@sdpt.edu.cn
[2] College of Materials Science and Technology, Nanjing University of Aeronautics and Astronautics, Nanjing 211100, China
[3] Reliability Research and Analysis Centre, China Electronic Product Reliability and Environmental Testing Research Institute, Guangzhou 510610, China
* Correspondence: imzjt@nuaa.edu.cn (J.Z.); jiaozibao@126.com (Z.J.); cz2343222@163.com (P.L.)

Citation: Xu, L.; Zhou, J.; Jiao, Z.; Liu, P. Special Issue: Advanced Science and Technology of Polymer Matrix Nanomaterials. *Materials* **2022**, *15*, 4735. https://doi.org/10.3390/ma15144735

Received: 30 June 2022
Accepted: 5 July 2022
Published: 6 July 2022

Publisher's Note: MDPI stays neutral with regard to jurisdictional claims in published maps and institutional affiliations.

Copyright: © 2022 by the authors. Licensee MDPI, Basel, Switzerland. This article is an open access article distributed under the terms and conditions of the Creative Commons Attribution (CC BY) license (https://creativecommons.org/licenses/by/4.0/).

Nanotechnology has witnessed an incredible resonance and a substantial number of new applications in various areas during the past three decades [1]. The resulting basic paradigm shifts have opened up new possibilities towards materials science, and have caused dramatic developments. Basically, nanotechnology necessarily relies on the presence or supply of novel nanomaterials that form the prerequisite for any ongoing progress in this interdisciplinary area of technology and science [2]. Among other nanomaterials, in the quest for eliminating the inherent shortcomings of pristine polymers, polymer matrix nanomaterials are fabricated through the introduction of nanomaterials with uniform distribution in pure polymer matrices [3,4].

Polymer matrix nanomaterial is, thereby, an active coupling of nanomaterials (other fillers may also be present) and polymers, where at least one phase is preserved in the nano-sized regime (within 100 nm) in the resultant materials. As the existence of nanomaterials in the polymer matrix features particular properties that are characteristic of this kind of material, and that are correlative with surface and quantum effects, it may intrinsically develop a fresh set of properties hinging on the nanomaterials utilized. Additionally, nanomaterials supply a significant number of interfacial areas within the matrix and accordingly at sufficiently low concentrations improve the material properties, which implies lowering the product gross weight further. Therefore, the union of nanotechnology and nanoscience with polymer technology and science has accelerated multifaceted application-oriented uses for polymer matrix nanomaterials. Numerous studies have revealed that the conductivity, mechanical, magnetic, optical, dielectric, electronic and biological characteristics of several inorganic nanomaterials dramatically change as their sizes reduce from the macroscale to the micro and nano scale. In the field of polymer matrix nanomaterials, researchers and experts have been focusing on the reinforced characteristics (mechanical strength, impact resistance, conductivity, biodegradability) and many diverse functionalities (self-healing, anti-fouling, electro-optical properties, flame resistance, controlled substance release, energy absorption applications and others) that afford them with certain properties, performance, or applications of considerable industrial interest. Therefore, polymer matrix nanomaterials have engraved an inimitable role in the niche of advanced materials and technologies. A genius multidisciplinary cooperation of material science with physics, biology, chemistry, nanotechnology, engineering, and medical science is inevitable for the actual exploration of such a new class of materials. In this regard, the development of modulated polymer systems will enable us to tackle technical and scientific challenges at the same time as satisfying the globally increasing demand.

The current Special Issue entitled "Advanced Science and Technology of Polymer Matrix Nanomaterials" is engaged in uniting researchers and scientists working at research institutes, laboratories, universities and industries to discuss cutting-edge developments and research on processing new polymer matrix nanomaterials in which nanoscale particle materials, including graphene, single-walled and multiwalled carbon nanotubes, inorganic layered clay, metal and metal oxide nanoparticles, MXene and others, have been introduced [5–8]. The subjects of this issue aim to uncover the potential improvements in the synthesis, properties and performance of polymer matrix nanomaterials regarding new preparation techniques, sensing, electromagnetic interference shielding, self-healing, microwave absorption, switching, structural modulation, mechanical reinforcement, drug delivery and other biomedical applications etc. [9–16].

As Guest Editors, it is our honor to invite contributions in the form of original research articles or reviews about this subject.

Funding: This work was financially supported by the National Natural Science Foundation of China (21905097) and Featured Innovation Projects of General Colleges and Universities in Guangdong Province (ZX2022002502).

Acknowledgments: The Guest Editors wish to acknowledge the authors for their upcoming vital contributions to this Special Issue, and the editorial staff of *Materials* for their extraordinary support.

Conflicts of Interest: The authors declare that they have no known competing financial interests or personal relationships that could have appeared to influence the work reported in this paper.

References

1. Hulla, J.E.; Sahu, S.C.; Hayes, A.W. Nanotechnology: History and future. *Hum. Exp. Toxicol.* **2015**, *34*, 1318–1321. [CrossRef] [PubMed]
2. Zhu, W.; Bartos, P.J.M.; Porro, A. Application of nanotechnology in construction. *Mater. Struct.* **2004**, *37*, 649–658. [CrossRef]
3. Chanda, S.; Bajwa, D.S. A review of current physical techniques for dispersion of cellulose nanomaterials in polymer matrices. *Rev. Adv. Mater. Sci.* **2021**, *60*, 325–341. [CrossRef]
4. Pourhashem, S.; Saba, F.; Duan, J.; Rashidi, A.; Guan, F.; Nezhad, E.G.; Hou, B. Polymer/Inorganic nanocomposite coatings with superior corrosion protection performance: A review. *J. Ind. Eng. Chem.* **2020**, *88*, 29–57. [CrossRef]
5. Liu, P.; Peng, J.; Chen, Y.; Liu, M.; Tang, W.; Guo, Z.-H.; Yue, K. A general and robust strategy for in-situ templated synthesis of patterned inorganic nanoparticle assemblies. *Giant* **2021**, *8*, 100076. [CrossRef]
6. Dulińska-Litewka, J.; Dykas, K.; Felkle, D.; Karnas, K.; Khachatryan, G.; Karewicz, A. Hyaluronic Acid-Silver Nanocomposites and Their Biomedical Applications: A Review. *Materials* **2022**, *15*, 234. [CrossRef] [PubMed]
7. Dowbysz, A.; Samsonowicz, M.; Kukfisz, B. Modification of Glass/Polyester Laminates with Flame Retardants. *Materials* **2021**, *14*, 7901. [CrossRef] [PubMed]
8. Pal, K.; Sarkar, P.; Anis, A.; Wiszumirska, K.; Jarzębski, M. Polysaccharide-Based Nanocomposites for Food Packaging Applications. *Materials* **2021**, *14*, 5549. [CrossRef] [PubMed]
9. Liu, P.; Yao, Z.; Zhou, J.; Yang, Z.; Kong, L.B. Small magnetic Co-doped NiZn ferrite/graphene nanocomposites and their dual-region microwave absorption performance. *J. Mater. Chem. C* **2016**, *4*, 9738–9749. [CrossRef]
10. Thakur, V.K.; Kessler, M.R. Self-healing polymer nanocomposite materials: A review. *Polymer* **2015**, *69*, 369–383. [CrossRef]
11. Nag, A.; Afsarimanesh, N.; Nuthalapati, S.; Altinsoy, M.E. Novel Surfactant-Induced MWCNTs/PDMS-Based Nanocomposites for Tactile Sensing Applications. *Materials* **2022**, *15*, 4504. [CrossRef]
12. Ranakoti, L.; Gangil, B.; Mishra, S.K.; Singh, T.; Sharma, S.; Ilyas, R.A.; El-Khatib, S. Critical Review on Polylactic Acid: Properties, Structure, Processing, Biocomposites, and Nanocomposites. *Materials* **2022**, *15*, 4312. [CrossRef] [PubMed]
13. Fodor, C.; Kali, G.; Thomann, R.; Thomann, Y.; Iván, B.; Mülhaupt, R. Nanophasic morphologies as a function of the composition and molecular weight of the macromolecular cross linker in poly(N vinylimidazole) l poly(tetrahydrofuran) amphiphilic conetworks: Bicontinuous domain structure in broad composition ranges. *RSC Adv.* **2017**, *7*, 6827–6834. [CrossRef]
14. Stumphauser, T.; Kasza, G.; Domján, A.; Wacha, A.; Varga, Z.; Thomann, Y.; Thomann, R.; Pásztói, B.; Trötschler, T.M.; Kerscher, B.; et al. Nanoconfined Crosslinked Poly(ionic liquid)s with Unprecedented Selective Swelling Properties Obtained by Alkylation in Nanophase-Separated Poly(1-vinylimidazole)-l-poly(tetrahydrofuran) Conetworks. *Polymers* **2020**, *12*, 2292. [CrossRef] [PubMed]
15. Mugemana, C.; Grysan, P.; Dieden, R.; Ruch, D.; Bruns, N.; Dubois, P. Self-Healing Metallo-Supramolecular Amphiphilic Polymer Conetworks. *Macromol. Chem. Phys.* **2020**, *221*, 1900432. [CrossRef]
16. Pásztor, S.; Becsei, B.; Szarka, G.; Thomann, Y.; Thomann, R.; Mülhaupt, R.; Iván, B. The Scissors Effect in Action: The Fox-Flory Relationship between the Glass Transition Temperature of Crosslinked Poly(Methyl Methacrylate) and Mc in Nanophase Separated Poly(Methyl Methacrylate)-l-Polyisobutylene Conetworks. *Materials* **2020**, *13*, 4822. [CrossRef] [PubMed]

MDPI
St. Alban-Anlage 66
4052 Basel
Switzerland
www.mdpi.com

Materials Editorial Office
E-mail: materials@mdpi.com
www.mdpi.com/journal/materials

Disclaimer/Publisher's Note: The statements, opinions and data contained in all publications are solely those of the individual author(s) and contributor(s) and not of MDPI and/or the editor(s). MDPI and/or the editor(s) disclaim responsibility for any injury to people or property resulting from any ideas, methods, instructions or products referred to in the content.